LEGUMES OF WEST ASIA

Legumes of West Asia

A CHECK-LIST

J.M. LOCK
K. SIMPSON

© Copyright Royal Botanic Gardens, Kew 1991

First published 1991

This check-list has been prepared at the Royal Botanic Gardens, Kew, as a contribution to the International Legume Database and Information Service (ILDIS). The aim of the first phase of ILDIS is to produce a computerised database containing basic nomenclatural, distributional and descriptive information on the legumes of the world. At the time of writing, the main data set has been produced by merging an African data set prepared at the Royal Botanic Gardens, Kew, a European data set from the ESFDS Database, and a data set for the Americas provided by the Missouri Botanical Garden. The merged data set is being edited and maintained in content and taxonomic consistency by an international team of experts. The ILDIS project is co-ordinated by Dr F.A.Bisby (University of Southampton), Dr R.M.Polhill (Royal Botanic Gardens, Kew) and Dr J.L.Zarucchi (Missouri Botanical Garden). It is sponsored by BIOSIS, the CEC, IBPGR, IUCN, Missouri Botanic Garden, Royal Botanic Gardens, Kew, the Science and Engineering Research Council (U.K.), and the University of Southampton. The ILDIS Co-ordinating Centre may be contacted at the Department of Biology, School of Biological Sciences, Bassett Crescent East, Southampton SO9 3TU, U.K.

Cover design by Media Resources, RBG, Kew.

Typeset at Royal Botanic Gardens, Kew by P.Arnold, B.Carey, C.Beard, M.Newman, H.O'Brien and P.Rosen.

ISBN 0 947643 29 X

Printed and Bound in Great Britain by Whitstable Litho Ltd., Whitstable, Kent.

CONTENTS

Introduction vii

Caesalpinioideae 1

Mimosoideae 9

Papilionoideae 17

Geographical Bibliography 220

Bibliography 221

Index 233

INTRODUCTION

This check-list of the legumes of West Asia has been compiled at the Royal Botanic Gardens, Kew, as a contribution to the International Legume Database and Information Service (ILDIS). The data that have been used to prepare this book are held on computer at the Royal Botanic Gardens, Kew, and at the ILDIS Coordinating Centre, whence they are accessible to accredited users.

The area covered is Afghanistan, Iran, Iraq and the Arabian Peninsula. Countries bordering the Mediterranean are covered by the Med-Checklist Project, based at Berlin and Geneva, and are not included here. Pakistan (including Baluchistan) will be covered later together with the rest of the Indian sub-continent. The U.S.S.R will also be dealt with as a separate project.

The check-list has been prepared in the following stages:
1. A search of current Floras of the region, starting with the most recent.
2. Checking of any monographic accounts revealed during (1).
3. A search of the collections in the Herbarium of the Royal Botanic Gardens, Kew, including cultivated material.
4. Checking of any names which appear in the collections but not in regional Floras or monographs.
5. Checking Index Kewensis, Kew Record of Taxonomic Literature, and Kew Current Awareness Lists.
6. Checking of local floristic lists for country records.

The data obtained from these sources were accumulated onto check-sheets, and then entered into an IBM-compatible micro-computer using the program 'ALICE', devised and written by Dr R.Allkin and Mr P.J.Winfield. The completed database contains some 2015 taxa (species and infraspecific taxa). It contains some extra material not included in this printed version, which was produced from the database using the ALICE Report Generator Program. The resulting text was then edited on a word processor and typeset at the Royal Botanic Gardens, Kew.

Any worker on this region has to make use of Parsa's Flore de l'Iran (1948, ref.1015). This remarkable work was prepared and printed in Iran, and suffers from poor proof-reading and a lack of contact with workers elsewhere. Species are often included without definite records for Iran because they occur close to its borders. Whenever possible, confirmation of all Parsa's records has been sought but for some groups there is no other source of information. This should be borne in mind when records are encountered for which Parsa is the sole authority.

Information on all the genera except *Astragalus* and its segregate *Astracantha* were compiled and entered at Kew by Karen Simpson during a sandwich year of her course at Coventry Polytechnic. The rest of the material was collected and entered by Mike Lock while working as a Research Fellow at the Royal Botanic Gardens, Kew, financed by Kew through the Bentham-Moxon Trust.

During the compilation of this check-list both authors received much help and guidance from Dr R.M.Polhill and members of the Legume Section at the Royal Botanic Gardens, Kew. Bob Allkin and Peter Winfield provided invaluable help and advice with computing matters.

Most tribes have been checked by ILDIS taxonomic co-ordinators. Our thanks go to the following: G.P.Lewis (Caesalpinieae), K.Larsen (Cercideae), J.Vassal (Acacieae), I.Nielsen (Ingeae), B.Verdcourt (Abreae), V.Rudd (Aeschynomeneae), L.J.G.van der Maesen (Cicereae, Phaseoleae), R.M.Polhill (Crotalarieae), D.Podlech (Galegeae), G.Cristofolini (Genisteae), B.D.Schrire (Indigofereae, Millettieae), R.Maréchal (Phaseoleae), C.H.Stirton (Psoraleeae, Sophoreae), C.Heyn (Trifolieae), D.Goyder (Vicieae) & N.Maxted (Vicieae).

THE DATA CATEGORIES

The categories of data in the database are those defined for ILDIS by Dr S. Hollis (1990): Type One data fields: detailed specification. Version 9. (ILDIS, University of Southampton). Not all of these data types are included in this printed checklist.

NOMENCLATURE

Accepted Names
These are species names accepted in one or more standard Floras within the region. Such names are only rejected if more recent relevant monographic accounts show this to be necessary.
Subspecies are included in this checklist and given a full entry. The data entered under a species should be the summation of the data entered under all of its subspecies. Varieties are listed in the 'Notes' section (see below).
There are standard floras for most of the countries in the region. Not all, however, are complete. Where there is no available standard flora, names have been obtained from all possible sources. In many cases such sources are old, and in some cases are regarded as provisional (see below). Every accepted name in the list is accompanied by the number of a reference (see Bibliography) in which the name is accepted.
Names for which there is no recent information, and about which there is doubt as to their distinctness, or validity of publication, are marked as 'Provisional'. In the African Check-list (Lock 1989) species not seen for many years, or known only from a few specimens collected long ago, were often marked as provisional. Revisional work in Africa is in many cases more advanced than in West Asia; in particular, only about one third of the sections of the vast genus *Astragalus* have been revised in their entirety in recent years, and the region is a major centre of diversity. It must be said that many taxa in those sections of *Astragalus* lacking a recent revision would probably have been marked as provisional in the African Check-list (Lock 1989).
In the vast majority of cases the generic names are those accepted by Gunn (1983–ref. 1090).

Synonyms
These are names which have been in use either in standard works or in herbaria since 1940, and which are treated as synonyms of an accepted name in the most recent standard works, or in recent generic or sectional revisions. They are italicized in the check-list. As a rule, this check-list does not purport to present an exhaustive synonymy, and names used in the literature only before 1940 are not included. However, in sections of *Astragalus* which have been revised recently, a fuller synonymy is given to help workers in this enormous and complex genus. Otherwise, reference to the standard works will be needed if a full synonymy is required.
In a few cases, standard Floras or revisions provisionally place a name in synonymy. This is usually because the type specimen has been lost or destroyed, and the original identification is inadequate. Such names are marked as 'suspected synonyms' ('doubtful synonyms' in the ILDIS system).

Misapplied Names
These are names which have been applied to a taxon which does not include the type of the name. In the database, misapplied names are labelled as such, but

in this list they are distinguishable by the word *sensu* before the authority, as in (for example) *sensu* Parsa, or *sensu auctt*. In the former case only one author has misapplied the name; in the latter, the name has been persistently misapplied. In the first case, two references are attached, one to the original place of misapplication and one to the reference in which the mistake was corrected. In the second case, only one reference is given, to the place where the confusion was cleared up.

CHARACTERISTICS

Four categories of information are included here. First, is the plant a herb, a shrub or a tree? Herbs are regarded here as plants producing stems which do not persist from one year to the next. Herbaceous climbers are included here, as are suffrutices, by which is meant plants with a perennial and sometimes woody underground base from which annual herbaceous shoots arise. Few plants of this region fall into this life-form category, which is more characteristic of regularly burned grasslands in regions with highly seasonal rainfall. Shrubs are woody plants which are branched at or near the base. Lianas (woody climbers) — which are virtually absent from this region — are treated as climbing shrubs. Many plants of this region are, however, intermediate between herbs and shrubs; many of the cushion-forming species of *Astragalus* and *Astracantha* are here marked as 'Herb/Shrub'. Some species of variable size and form are also marked as 'Shrub/Tree'.

Secondly, does the plant climb? Those that do so, however they do it, are so marked. Those that may or may not climb, depending on their situation, are marked as 'Climbing/Not climbing'.

Thirdly, what is the normal lifespan of the plant? Does it persist from one year to the next (Perennial), or does it complete its life cycle within a single season and persist to the next as seeds (Annual)? Once again, variable taxa are marked as intermediate (Annual/Perennial).

Finally, there is, in a few cases, a note of the conservation status of the plant in the wild, following the conventions applied by the International Union for the Conservation of Nature and Natural Resources (IUCN). The categories used are: Extinct, Endangered, Vulnerable, Indeterminate, Insufficiently known, and Not Threatened. In this check-list very few species have been coded, because there is generally far too little information available to make any reasonable judgement of status.

ECONOMIC IMPORTANCE TO MAN

There is no entry for this category unless there is a recorded use. The classes that have been used are:

Chemical Products includes: All chemicals used in industry, including gums.
Domestic includes: Water purifiers/clarifiers; soap & soap substitutes; cosmetics, chewing sticks and tooth cleaners, hunting gear, brushes, thatching.
Environmental includes: Hedging, ornamental, green manure, shelter, shade, cover crop.
Fibre includes: Cordage, textiles, cork, fibre filling.
Food or drink includes: Vegetables and fruits, beverages, cooking fats and oils, vitamins.
Forage includes: Grazing, browse, fodder, bee plants, host plants.
Medicine includes: Medical and veterinary products.

Miscellaneous includes: Any economic use not otherwise covered. (Such entries are usually accompanied by a note).
Toxins includes: Materials toxic to any type of living organism.
Weed includes: Agricultural, aquatic, forestry and garden weeds.
Wood includes: Fuel, sawn timber, pole timber, carpentry and construction timber.

Most of these categories are self-explanatory. They represent the top level of a hierarchical classification of plant uses developed at Kew by Mrs Frances Cook (Economic & Conservation Section). In the computer database, all such records are linked to a source reference.

HABITAT AND PHYTOGEOGRAPHY

In "Legumes of Africa — a Check-list" (Lock 1989, ref.1005) each taxon was given a code corresponding to the phytochorion and the major vegetation types in which it occurs. It has not been found possible to do this consistently or satisfactorily for West Asian legumes, partly because neither of the compilers has field experience in the area, but also because of a lack of agreement on the classification of vegetation types and plant geography in the region. A brief outline can, however, be given here

Following Takhtajian's (1986) classification, the area is split between two Kingdoms, the Holarctic and the Palaeotropical. The latter is confined to the extreme south of the area, with the Yemen mainly falling within the South Arabian Province of the Eritraeo-Arabian Subregion of the Sudano-Zambezian Region, and Oman and southern Iran forming two provinces within the Omano-Sindian Subregion of the same. Most of the area falls within the Tethyan Sub-Kingdom of the Holarctic. Within this Sub-Kingdom, the central and northern parts of the Arabian Peninsula form part of the Egyptian-Arabian Province of the Saharo-Arabian Region, while the rest of area lies within various provinces (Mesopotamian, Armeno-Iranian, Hyrcanian, Turanian and West Himalayan) of the West Asiatic Subregion of the Irano-Turanian Region.

Various descriptions of the vegetation of parts of the region are available, but none is universal. Using White's (1983) vegetation categories, true grasslands appear to cover a rather small proportion of the region. Shrublands are probably the most widespread vegetation type. Bushlands and thicket are frequent in the moister regions but, like all woody vegetation types in the area, are under intense pressure for fuel and timber. True forest is probably restricted to the north-facing slopes of the Elburz Mountains, with perhaps some patches in northeast Afghanistan. Open woodlands, often with an understorey of shrubs, are frequent in the northern parts; they are illustrated by Guest (1966).

GEOGRAPHY

The countries included under 'West Asia' are Iran, Iraq, Afghanistan, Kuwait, Saudi Arabia, Qatar, Bahrain, United Arab Emirates, Oman and North and South Yemen. The last two countries united in 1990 but this was after data entry had finished and it has not been possible to amalgamate the records for this publication; besides, information would be lost by joining them.

Records from adjacent areas are listed as follows. Only countries sharing a common border with 'West Asia' are listed as such; other records are listed under continent or 'sub-continent' only. 'Middle East' includes Turkey, Syria, Lebanon,

Jordan, Israel and Sinai. Only India and Pakistan are entered under 'South Asia'. The U.S.S.R. is listed as a 'sub-continent', with no attempt to list occurrence within the several republics which share a border with 'West Asia'. 'East Asia' here includes only Tibet, which is listed as an area (Xizang Zizhiqu) within the country China.

Status within countries is indicated as Native (N), Introduced by man (I), or of Uncertain status (U).

LITERATURE POINTERS

This section lists references in which may be found a good description, a good illustration, and a distribution map of the whole range of the taxon. Up to three references are listed in each category; the numbers refer to the list at the end of the book.

References to descriptions do not necessarily include the original description; this is normally only included where nothing else is available. Most are in English or German, but some in Latin and French have also been included.

In selecting illustrations, priority is given to those showing the whole plant. Line drawings have been preferred, but some references have also been given to colour photographs of whole plants. A number of illustrations from floras of the U.S.S.R. and its republics have been included, as these greatly increase the number of species for which an illustration can be found.

Only maps showing the complete distribution of a taxon have been included; few are available for the flora of the region.

NOTES

Where notes were entered in the computer database, these have been included. They explain items in other sections, such as the reasons for considering a name to be provisional, or uses entered as 'Miscellaneous'. They also include listings of any subspecies or varieties that may have been described. A few vernacular names have been included in the computer database. These have not been reproduced here. No attempt has been made to compile an exhaustive list.

COMPILERS' NOTE

This check-list claims to include all taxa recognized in works received at Kew by the end of 1990. There will, of course, be errors and omissions. We would be most grateful if J.M.Lock could be informed of these, since this material will form part of the ILDIS World Legume Database, which will be continually updated to take account of new taxa, new country records, and other material.

CAESALPINIOIDEAE

AMHERSTIEAE

TAMARINDUS L.

The single species of this genus has been very widely introduced in the Old World tropics for its refreshing sweet-sour fruit-pulp. It may possibly be native in southern Madagascar, but could also be African or Indian in origin.

T. indica L. [1009]
Tamarindus indicus sensu Miller
Not climbing; Tree; Perennial.
West Asia: Bahrain(I), Iran(I), Iraq(I), North Yemen(I), Oman(I), Saudi Arabia(I), South Yemen(I); Africa; South Asia: India(N).
Description: 1007, 1009, 1016; Illustration: 1001, 1016.
Domestic 1016; Environmental 1009, 1016; Food or drink 1009, 1016; Forage 1016;Medicine 1009, 1016; Wood 1009.
Indian Date; Tamarind 1009, 1016.
The Iraq record is somewhat doubtful [1009].

CAESALPINIEAE

CAESALPINIA L.

A pantropical genus of about 100 species. Many have been widely planted for ornament; only one species is native in West Asia.

C. bonduc (L.)Roxb. [1009]
Not climbing; Shrub/Tree; Perennial.
West Asia: Iran(I), Iraq(I), Oman(I); Africa; South America; South Asia; South-East Asia.
Description: 1009, 1051.
Environmental 1009, 1016; Medicine 1009.
Fever Nut 1009; Indian Nut 1009; Nickar Bean 1009; Physic Nut 1009.
Pantropical; widely introduced; natural range obscure [1009].

C. decapetala (Roth)Alston [1047]
Caesalpinia sepiaria Roxb. [1138].
Climbing; Shrub; Perennial.
West Asia: Iraq(I); South Asia: India(N).
Environmental 1009, 1047.
Mysore Thorn 1009.

C. erianthera Chiov. [1016]
Not climbing; Shrub; Perennial.
West Asia: Oman(N), Saudi Arabia(N), South Yemen(N); Africa.
Description: 1016, 1047; Illustration: 1016.
Medicine 1016; Miscellaneous 1016; Wood 1016.
Used in perfumery [1016].
Brenan (1963), recognizes a var. *pubescens* for South Yemen [1138].

1

C. gilliesii (Hook.)D.Dietr. [1009]
    Not climbing; Shrub/Tree; Perennial.
    West Asia: Afghanistan(I), Iran(I), Iraq(I); South America.
    Description: 1009, 1015, 1051.
    Environmental 1009; Medicine 1016.
    Townsend (1974) records from "Arabia" [1009].

C. mexicana A.Gray [1009]
    Not climbing; Shrub/Tree; Perennial.
    West Asia: Iraq(I); Central America
    Description: 1009.

C. pulcherrima (L.)Sw. [1009]
    Poinciana pulcherrima L. [1047].
    Not climbing; Shrub/Tree; Perennial.
    West Asia: Iraq(I), Saudi Arabia(I), South Yemen(I}.
    Description: 1047.
    Domestic 1016; Environmental 1009, 1018, 1044; Toxins 1016.
    Barbados Pride 1009, 1018, 1047.

DELONIX Raf.

A genus of small trees with large showy flowers. One native species; another is very widely planted, but rare and endangered in its native Madagascar.

D. elata (L.)Gamble [1001]
    Not climbing; Shrub/Tree; Perennial.
    West Asia: North Yemen(N), Oman(N), Saudi Arabia(N), South Yemen(N); Africa.
    Description:1001, 1007, 1016; Illustration: 1001, 1016.
    Domestic 1016; Environmental 1016; Food or drink 1016.
    Forage 1016; Medicine 1016.
    Little Lebi 1055.

D. regia (Hook.)Raf. [1009]
    Not climbing; Tree; Perennial.
    West Asia: Iran(I), Oman(I); Indian Ocean: Madagascar(N).
    Description: 1009.
    Environmental 1016.
    Flamboyant 1016, 1047; Flame of the forest 1009.

GLEDITSIA L.

A small but widely dispersed genus of trees with branched thorns on their trunks and smaller stems.

G. caspica Desf. [1051]
    Gleditsia horrida subsp. *caspica* (Desf.)Paclt [1051]; *Gleditsia horrida* var. *caspica* (Desf.)C.Schneider [1051].
    Not climbing; Tree; Perennial.
    West Asia: Iran(N); U.S.S.R.
    Description: 1015, 1051; Illustration: 1051.
    Forage 1009.
    Caspian Honey-locust 1009.
    Paclt(Taxon 31:337) treats this as a synonym of *G.horrida* (Thunb.)Miq. from Japan [1051];
        Rechinger does not explain why he rejects Paclt's view [1051].

G. triacanthos L.
    Not climbing; Tree; Perennial.
    West Asia: Afghanistan(I), Iran(I), Iraq(I); North America.
    Description: 1009; Illustration: 1009.
    Environmental 1009, 1018; Food or drink 1009; Forage 1009, 1018.
    Common Honey-locust 1009; Sweet Locust 1009; Three-thorned Acacia 1009.

HAEMATOXYLUM L.

A genus of three species, all small trees, two from tropical America and one from southern Africa. The American species provide timber, and the wood contains a red dye.

H. campechianum L. [1007]
Haematoxylon campechianum L.
Not climbing; Tree; Perennial.
West Asia: Saudi Arabia(I); Central America.
Description: 1007.
Environmental 1018.
Logwood 1018, 1047.

PARKINSONIA L.

A genus of small spiny trees, mostly American but with four species in Africa. One species is widely cultivated in dry areas for shade, ornament and fuel.

P. aculeata L. [1051]
Parkinsonia acleata L. [1015].
Not climbing; Shrub/Tree; Perennial.
West Asia: Iran(I), Iraq(I), Oman(I), South Yemen(I); Central America; North America; South America.
Description: 1009, 1051; Illustration: 1009, 1051.
Environmental 1018, 1033.
Jerusalem Thorn 1009, 1018.

PTEROLOBIUM (Vogel) Walp.

A small genus of about 10 species of spiny climbers in the tropical and sub-tropical parts of the Old World.

P. stellatum (Forsskål) Brenan [1251]
Climbing; Shrub; Perennial.
West Asia: North Yemen(N); Africa.
Description: 1047; Illustration: 1047.

CASSIEAE

CASSIA L.

The old genus *Cassia* sens. lat. has been split into three in America, Africa and Australia. The members of *Cassia* sens. strict. are mostly trees with more-or-less cylindrical indehiscent fruits, and three large sigmoidally-curved filaments.

C. fistula L. [1009]
Not climbing; Tree; Perennial.
West Asia: Iran(I), Iraq(I); South-East Asia.
Description: 1009, 1015, 1048.
Domestic 1009; Environmental 1009, 1048, 1049; Forage 1009; Medicine 1048; Wood 1009.
Golden Shower 1009, 1049; Indian Laburnum 1009; Purging Cassia 1009.
Townsend (1974) records from "Arabia" [1009].

CERATONIA L.

Two species, both small trees, found in Oman, Northeast Africa and the Mediterranean.

C. oreothauma Hillc., G.P.Lewis & Verdc. [1050]
Not climbing; Tree; Perennial.
West Asia: Oman(N), Africa.
Description: 1050; Illustration: 1050.
Forage 1050.
Another subspecies in Somalia [1050].

subsp. oreothauma [1050]
Not climbing; Tree; Perennial.
West Asia: Oman(N).
Description: 1050; Illustration: 1050.
Forage 1050.

C. siliqua L. [1051]
Not climbing; Shrub/Tree; Perennial.
West Asia: Iran(U), Iraq(I), North Yemen(U); Africa; Europe; Middle East: Israel(N), Jordan(N), Lebanon(N), Sinai(N), Syria(N), Turkey(N); U.S.S.R.
Description: 1009, 1051.
Food or drink 1009; Forage 1009; Medicine 1009; Wood 1009.
Carob Tree 1009, 1047, 1050; Locust Bean 1047; St. John's Bread 1009, 1050.
Hillcoat,D. et al.(1980) record from "Arabia" [1050].

CHAMAECRISTA Moench

Formerly part of *Cassia* sens. lat., but distinguished by the pod, which dehisces elastically along both sutures, and the short filaments of more-or-less equal length. Most of the Old World species are small slightly woody herbs. Unlike *Senna*, most species that have been studied have root nodules.

C. absus (L.)Irwin & Barneby
Cassia absus L. [1054]
Not climbing; Herb/Shrub; Annual.
West Asia: South Yemen(N); Africa.
Description: 1047, 1048; Illustration: 1047.
Medicine 1053.

C. dimidiata (Roxb.)Lock [1054]
Cassia hochstetteri Ghesq. [1054]; *Senna dimidiata* Roxb. [1054].
Not climbing; Herb; Annual.
West Asia: North Yemen(N); Africa; East Asia: China(N); South Asia: India(N).
Description: 1047, 1048.

C. fallacina (Chiov.)Lock [1054]
Cassia fallacina Chiov. [1054].
Not climbing; Herb/Shrub; Perennial.
West Asia: North Yemen(N); Africa.
Description: 1047.

C. mimosoides (L.)Greene [1054]
Cassia mimosoides L. [1054].
Not climbing; Herb/Shrub; Annual.
West Asia: North Yemen(I), Saudi Arabia(I); Africa.
Description: 1047, 1048.
Cover crop 1048; Medicine 1048.

C. nigricans (Vahl)Greene [1054]
Cassia nigricans Vahl [1054].
Not climbing; Herb; Annual.
West Asia: North Yemen(N), South Yemen(N); Africa; South Asia: India(N).
Description: 1047.

SENNA Miller

Formerly part of *Cassia* sens. lat., but distinguished by the cylindrical or flattened irregularly dehiscent pods, and by the filaments, three of which are longer than the rest but not sigmoidally curved (cf. *Cassia*). Investigated species lack root nodules.

S. alexandrina Miller [1054]
Cassia acutifolia Del. [1007, 1049, 1054; *Cassia alexandrina* (Miller)Thell. [1007]; *Cassia angustifolia* Vahl [1049, 1054]; *Cassia senna* L. [1054]; *Senna acutifolia* (Del.)Batka [1007, 1054].
Not climbing; Herb/Shrub; Perennial.
West Asia: Iraq(I), North Yemen(I), Saudi Arabia(I), South Yemen(I), United Arab Emirates(I); Africa.
Description: 1007, 1009, 1051; Illustration: 1001, 1007.
Food or drink 1048; Medicine 1009, 1047.
Alexandrian Senna 1009.

S. artemisioides (DC.)Randall [1251]
Cassia artemisioides DC. [1251].
Not climbing; Shrub; Perennial.
West Asia: Iraq(I); Australasia.
Description: 1009, 1047.
Environmental 1009

S. auriculata (L.)Roxb. [1054]
Cassia auriculata L. [1054].
Not climbing; Shrub/Tree; Perennial.
West Asia: South Yemen(I); Africa.
Description: 1048.
Domestic 1048; Environmental 1044, 1048.
Tanner's Cassia 1048.

S. corymbosa (Lam.)Irwin & Barneby [1054]
Cassia corymbosa Lam. [1054].
Not climbing; Shrub/Tree; Perennial.
West Asia: Iraq(I); South America.
Description: 1009.
Environmental 1009.
Common Cassia 1009.

S. didymobotrya (Fresen.)Irwin & Barneby [1054]
Cassia didymobotrya Fresen. [1054].
Not climbing; Shrub/Tree; Perennial.
West Asia: Iraq(I), North Yemen(I); Africa.
Description:1009, 1047, 1048; Illustration: 1047.
Environmental 1009, 1048.

S. holosericea (Fresen.)Greuter [1054]
Cassia holosericea Fresen. [1054]; *Cassia pubescens* R.Br. [1007]; *Cassia oocarpa* Baker.
Not climbing; Herb/Shrub; Perennial.
West Asia: Iran(N), Oman(N), Saudi Arabia(N), South Yemen(N); Africa.
Description: 1007, 1016, 1051; Illustration: 1001, 1007, 1016.
Cassia oocarpa Benth., Bull. Misc. Inf., Kew, 1895: 181. Type: South-east Arabia, Merbat, foot of Dhofar mountains, *J.T. Bent* 69 (K!); **synon nov.**

S. hookeriana Batka [1054]
Cassia adenensis Benth. [1054].
Not climbing; Shrub; Perennial.
West Asia: South Yemen(N).
Description: 1140.

S. italica Miller [1054]

Cassia aschrek Forsskal [1047, 1049]; *Cassia italica*
(Miller)Sprengel [1054]; *Cassia obovata* Colladon [1007, 1047, 1049].
Not climbing; Herb/Shrub; Perennial.
West Asia: Bahrain(N), Iran(N), Iraq(N), North Yemen(N), Oman(N), Qatar(N), Saudi Arabia(N), South Yemen(N), United Arab Emirates(N); Africa; Middle East: Israel(N), Sinai(N); South Asia: India(N), Pakistan(N).
Description:1009, 1016, 1051; Illustration: 1001, 1007, 1016.
Medicine 1009, 1014, 1016.
Italian Senna 1009; Mecca Senna 1009.

subsp. italica Miller [1054]

Cassia italica (Miller)Sprengel [1054].
Not climbing; Herb; Perennial.
West Asia: Iran(N), Iraq(N), South Yemen(N); Africa; Middle East: Israel(N); South Asia: Pakistan(N).
Description: 1015, 1051; Illustration: 1051.
Sanna 1056.

S. obtusifolia (L.)Irwin & Barneby [1054]

Cassia obtusifolia L. [1054].
Not climbing; Herb/Shrub; Perennial.
West Asia: Iraq(I), North Yemen(I), Oman(I); Africa; South America.
Description:1009, 1047, 1048.
Food or drink 1048; Medicine 1016, 1048.

S. occidentalis (L.)Link [1054]

Cassia occidentalis L. [1054].
Not climbing; Herb/Shrub
West Asia: Iran(I), Iraq(I), North Yemen(I), Saudi Arabia(I), South Yemen(N); Africa; Australasia; East Asia: China(U); Middle East: Lebanon(U); South America; South Asia: India(U); South-East Asia.
Description: 1007, 1009, 1051; Illustration: 1001, 1007, 1009.
Environmental 1009, 1033; Food or drink 1016, 1048; Medicine 1015, 1016, 1048.
Coffee Senna 1009; Stinking Head 1047.

S. sophera (L.)Roxb. [1054]

Cassia sophera L. [1054].
Not climbing; Shrub; Perennial.
West Asia: Iraq(I), Oman(I); Africa; Australasia; East Asia: China(I); South Asia: India(U).
Description: 1009, 1047, 1048; Illustration: 1009.
Environmental 1009; Medicine 1012, 1016, 1048; Toxins 1048.
S. sophera is very similar and closely related to *S. occidentalis* [1047].
A. Radcliffe-Smith & S.J. Henchie, 4892 from North Yemen is similar to this.

S. surattensis (Burm.f.)Irwin & Barneby [1054]

Cassia glauca Lam. [1054]; *Cassia surattensis* Burm.f. [1054].
Not climbing; Shrub; Perennial.
West Asia: Iran(I); South-East Asia.

S. tora (L.)Roxb. [1054]

Cassia tora L. [1054].
Not climbing; Herb/Shrub; Perennial.
West Asia: Oman(I), Saudi Arabia(I); South Asia: India(N).
Description:1001, 1048, 1051; Illustration: 1001.
Food or drink 1048; Medicine 1016, 1048.

CERCIDEAE

BAUHINIA L.

A large pantropical genus, mainly trees or shrubs but also with many woody climbers. Many have been widely planted for their showy flowers. The genus has often been subdivided but a broad view is taken here.

B. acuminata L. [1015]
Not climbing; Shrub/Tree; Perennial.
West Asia: Iran(I); South-East Asia.
Description: 1015, 1046; Illustration: 1046, 1245.
Environmental 1046; Medicine 1046; Wood 1046.
Mountain Ebony 1046.

B. ellenbeckii Harms
West Asia: South Yemen(N); Africa.
The specimen at Kew is almost certainly this (det. Brenan).

B. purpurea L. [1009]
Not climbing; Shrub/Tree; Perennial.
West Asia: Iraq(I); South-East Asia.
Description: 1009, 1046; Illustration: 1245.
Domestic 1046; Environmental 1009, 1046; Fibre 1046; Forage 1046; Medicine 1009; Wood 1046.
Camel's foot 1009; Horse footprint 1009; Purple Bauhinia 1009.
A white-flowered form also occurs [1009].

B. racemosa Lam. [1009]
Not climbing; Tree; Perennial.
West Asia: Iraq(I); South Asia: India(N).
Description: 1009; Illustration: 1245.
Fibre 1009; Wood 1009.

B. thonningii Schum. [1047]
Piliostigma thonningii (Schum.)Milne-Redh. [1047].
Not climbing; Shrub/Tree; Perennial.
West Asia: North Yemen(N); Africa.

B. tomentosa L. [1038]
B. inermis Forsskål (Doubtful synonym)
Not climbing; Shrub/Tree; Perennial.
West Asia: North Yemen(N); Africa.
Description: 1046, 1047; Illustration: 1245.
Environmental 1046; Food or drink 1046; Medicine 1046; Wood 1046.
Dr B. Verdcourt (K) believes that *B. inermis* belongs here.

B. variegata L. [1009]
Not climbing; Tree; Perennial.
West Asia: Iran(I), Iraq(I); East Asia: China(N); Middle East: Israel(I), Lebanon(I); South Asia: India(N).
Description: 1009, 1015, 1046; Illustration: 1245.
Domestic 1009; Environmental 1009; Fibre 1009; Food or drink 1009; Forage 1009; Medicine 1009; Wood 1009.
Camel's foot 1009; Mountain Ebony 1009; Variegated Bauhinia 1009.
2 varieties, *variegata* & *candida* Voight, the latter white-flowered [1245] occur.
Townsend (1974) records from "Arabia" [1009].

CERCIS L.

A small genus of deciduous trees of temperate regions, mostly flowering from the trunk and main branches (cauliflorous). Sometimes planted for their flowers.

C. griffithii Boiss. [1051]
Not climbing; Shrub; Perennial.
West Asia: Afghanistan(N), Iran(N); U.S.S.R.
Description: 1051; Illustration: 1051.
Environmental 1004.

C. siliquastrum L. [1051]
Not climbing; Tree; Perennial.
West Asia: Afghanistan(N), Iran(N), Iraq(N); Africa; Europe; Middle East: Israel(N), Jordan(N), Lebanon(N), Syria(N), Turkey(N); U.S.S.R.
Description: 1009, 1051; Illustration: 1009, 1051.
Domestic 1009; Environmental 1009; Food or drink 1009; Medicine 1009.
Judas Tree 1009.

subsp. hebecarpa (Bornm.)Yalt. [1009]
Not climbing; Tree; Perennial.
West Asia: Iraq(N).
Description: 1009.

subsp. siliquastrum L. [1009]
Not climbing; Tree; Perennial.
West Asia: Iraq(I).
Description: 1009.
Environmental 1009.

MIMOSOIDEAE

ACACIEAE

ACACIA Miller

A very large pantropical genus with over 1000 species, mainly in dry country. The local species are armed with spines; most of the species introduced from Australia are spineless and have phyllodes instead of leaves.

A. abyssinica Benth. [1001]
Not climbing; Tree; Perennial.
West Asia: North Yemen(N), Saudi Arabia(N); Africa.
Description: 1001, 1003; Illustration: 1001.
Forage 1002.
2 subspecies, *abyssinica* & *calophylla*; only the former occurs in West Asia [1013].

subsp. **abyssinica** Benth. [1001]
Not climbing; Tree; Perennial.
West Asia: North Yemen(N), Saudi Arabia(N); Africa.
Description: 1001, 1003; Illustration: 1001.
West Asian plants are var. *macroloba* Schweinf. [1001].

A. asak (Forsskal)Willd. [1001]
A. glaucophylla A.Rich. [1007]; *A. triacantha* A.Rich. [1013].
Not climbing; Tree; Perennial.
West Asia: North Yemen(N), Oman(N), Saudi Arabia(N), South Yemen(N); Africa.
Description: 1001, 1007; Illustration: 1001, 1007.

A. campoptila Schweinf. [1003]
Not climbing; Tree; Perennial.
West Asia: Saudi Arabia(N), South Yemen(N).
Description: 1003; Illustration: 1003.

A. cornigera (L.)Willd. [1009]
Not climbing; Shrub/Tree; Perennial.
West Asia: Iraq(I); Central America.
Description: 1009.

A. eburnea (L.f.)Willd. [1010]
Not climbing; Shrub/Tree; Perennial.
West Asia: Afghanistan(N), Oman(N); South Asia: India(N), Pakistan(N).
Description: 1010.

A. edgeworthii T.Anderson [1013]
A. erythraea Chiov. [1013]; *A. humifusa* Chiov. [1013]; *A. pseudosocotrana* Chiov. [1013]; *A. socotrana* Balf.f. [1013]; *A. sultani* Chiov. [1013].
Not climbing; Shrub/Tree; Perennial.
West Asia: North Yemen(N), Qatar(U), South Yemen(N); Africa.
Description: 1013; Illustration: 1013.

A. ehrenbergiana Hayne [1014]
A. flava (Forsskal)Schweinf. [1007]; A. flava var. ehrenbergiana (Hayne)Roberty [1013].
Not climbing; Shrub/Tree; Perennial.
West Asia: Iran(N), North Yemen(N), Oman(N), Qatar(N), Saudi Arabia(N), South Yemen(N); Africa.
Description: 1003, 1013, 1014; Illustration: 1001, 1007, 1014.
Seyal 1043.

A. elatior Brenan [1001]
Not climbing; Tree; Perennial.
West Asia: Saudi Arabia(N); Africa.
Description: 1001; Illustration: 1001.
Has affinities with A. etbaica & is also closely akin to A. negrii [1017].

subsp. elatior Brenan [1001]
Not climbing; Tree; Perennial.
West Asia: Saudi Arabia(N); Africa.
Description: 1001; Illustration: 1001.

A. etbaica Schweinf. [1001]
Not climbing; Shrub/Tree; Perennial.
West Asia: North Yemen(N), Oman(N), Saudi Arabia(N), South Yemen(N); Africa.
Description: 1003, 1013, 1016; Illustration: 1001, 1007, 1016].
Domestic 1016; Miscellaneous 1016; Wood 1016.
Dead wood burned and ash taken as snuff [1016].

subsp. etbaica Schweinf. [1013]
Not climbing; Shrub/Tree; Perennial.
West Asia: Saudi Arabia(N); Africa.
Description: 1013.

subsp. uncinata Brenan [1001]
Not climbing; Shrub/Tree; Perennial.
West Asia: North Yemen(N), Oman(N), Saudi Arabia(N), South Yemen(N); Africa.
Description: 1001; Illustration: 1001.

A. farnesiana (L.)Willd. [1009]
Not climbing; Shrub/Tree; Perennial.
West Asia: Afghanistan(I), Iran(I), Iraq(I), Saudi Arabia(I); Central America; Middle East: Israel(I), Lebanon(I), Syria(I); South America.
Description: 1003, 1009, 1010; Illustration: 1009, 1010.
Domestic 1009, 1045; Environmental 1009, 1018, 1045; Forage 1002, 1009; Medicine 1009; Wood 1009.
Acacia 1018; Cassie 1018.

A. greggii A.Gray [1009]
Not climbing; Shrub/Tree; Perennial.
West Asia: Iran(I), Iraq(I); Central America.
Description: 1009; Illustration: 1009.
Environmental 1009.

A. hamulosa Benth. [1001]
Not climbing; Shrub/Tree; Perennial.
West Asia: North Yemen(N), Oman(N), Saudi Arabia(N), South Yemen(N); Africa.
Description: 1001, 1003, 1016; Illustration: 1001, 1003.

A. hockii De Wild. [1003]
A. oerfota sensu Brenan [1008, 1019].
Not climbing; Shrub/Tree; Perennial.
West Asia: North Yemen(N), South Yemen(N); Africa.
Description: 1003, 1013; Illustration: 1003, 1013].
Occupies a wide range of habitats and altitudes, and there may well be variants [1038].

A. hunteri Oliver [1044]
Not climbing; Shrub/Tree; Perennial.
West Asia: South Yemen(U).
Description: 1136; Illustration: 1136.

A. hydaspica R.Parker [1020]
Not climbing; Shrub; Perennial.
West Asia: Iran(N); South Asia: India(N), Pakistan(N).
Description: 1010; Illustration: 1010.
Confused with *A. arabica* (= *A. nilotica* subsp. *nilotica*) when young [1020].

A. iraqensis Rech.f. [1023]
A. gerrardii sensu auctt. [1137]; *A. gerrardii* subsp. *negevensis* Zoh. [1023]; *A. iraquensis* Rech.f. [1001]; *A. negevensis* (Zoh.)Zoh. [1137].
Not climbing; Shrub/Tree; Perennial.
West Asia: Iraq(N), Kuwait(N), North Yemen(N), Oman(N), Saudi Arabia(N), South Yemen(N); Middle East: Jordan(N), Sinai(N).
Description: 1001, 1003, 1009; Illustration: 1001, 1003, 1009.
Fibre 1016; Forage 1016; Wood 1016.

A. jacquemontii Benth. [1015]
West Asia: Iran(N).
Description: 1015.

A. karroo Hayne [1009]
A. dekindtiana A.Chev. [1013]; *A. horrida* (L.)Willd.,p.p. [1013]; *A. natalitia* E.Meyer [1013].
Not climbing; Shrub/Tree; Perennial.
West Asia: Iraq(I); Africa.
Description: 1009.
Domestic 1021; Environmental 1009, 1021; Fibre 1021; Food or drink 1021; Forage 1002, 1021; Wood 1021.
Doorn-boom 1021; Karroo-thorn 1021.

A. laeta Benth. [1009]
Not climbing; Shrub/Tree; Perennial.
West Asia: Iraq(I), North Yemen(N), Oman(N), Saudi Arabia(N), South Yemen(N);Africa; Middle East: Israel(N), Syria(N).
Description: 1003, 1009, 1016; Illustration: 1001, 1016].
Possibly the result of hybridization between *A. mellifera* & some other species [1038].

A. leprosa DC. [1022]
Not climbing; Shrub/Tree; Perennial.
West Asia: Iran(I); Australasia.
Description: 1022.

A. macracantha Willd. [1009]
Not climbing; Shrub/Tree; Perennial.
West Asia: Iraq(I); Central America.
Description: 1009.

A. mellifera (Vahl)Benth. [1001]
Not climbing; Tree; Perennial.
West Asia: North Yemen(N), Saudi Arabia(N), South Yemen(N); Africa.
Description: 1001, 1003, 1007; Illustration: 1001, 1003.

subsp. **mellifera** (Vahl)Benth. [1038]
Not climbing; Shrub/Tree; Perennial.
West Asia: North Yemen(N), Saudi Arabia(N), South Yemen(N); Africa.
Description: 1003; Illustration: 1003.

A. modesta Wall. [1010]
Not climbing; Tree; Perennial.
West Asia: Afghanistan(N); South Asia: India(N), Pakistan(N).
Description: 1010, 1015; Illustration: 1010.
Phula 1018.

A. nilotica (L.)Del. [1010]
Not climbing; Tree; Perennial.
West Asia: Iran(N), Iraq(N), Oman(N); Middle East: Israel(N), Syria(N): South Asia: India(N).
Description: 1009, 1010, 1016; Illustration: 1009, 1016.
Domestic 1016; Environmental 1009; Forage 1002, 1009; Medicine 1016; Wood 1016.

subsp. adstringens (Schum.& Thonn.)Roberty [1010]
A. adansonii Guillemin & Perrottet [1013]; *A. nilotica* var. *adansonii* (Guillemin & Perrottet)A.Chev. [1013].
Not climbing; Tree; Perennial.
West Asia: Iran(N), North Yemen(N), Oman(N); Africa.
Description: 1010; Illustration: 1010.
Food or drink 1002.

subsp. indica (Benth.)Brenan [1010]
A. arabica sensu Brenan [1013]; *A. nilotica* var. *indica* (Benth.)A.F.Hill [1017, 1024].
Not climbing; Tree; Perennial.
West Asia: Iran(N), North Yemen(N), Oman(N), Qatar(N), South Yemen(N); South Asia: India(N), Pakistan(N).
Description: 1010; Illustration: 1010.
The distinction between this and subsp. *tomentosa* not always clear; they may be only varietally distinct.[1024]

subsp. kraussiana (Benth.)Brenan [1016]
A. arabica var. *kraussiana* Benth. [1025, 1026]; *A. nilotica* var. *kraussiana* (Benth.)A.F.Hill [1017].
Not climbing; Tree; Perennial.
West Asia: Iraq(N), North Yemen(N), Oman(N); Africa.
Description: 1016; Illustration: 1016.

subsp. nilotica (L.)Del. [1009]
A. arabica (Lam.)Willd. [1013]; *A. scorpioides* (L.)W.Wight [1013].
Not climbing; Tree; Perennial.
West Asia: Iran(N), Iraq(N), Oman(N), Saudi Arabia(N), South Yemen(N); Africa.
Description: 1009, 1010.
Babul 1018.

subsp. tomentosa (Benth.)Brenan [1007]
Not climbing; Tree; Perennial.
West Asia: Iran(N), Saudi Arabia(N); Africa.
Description: 1007, 1015; Illustration: 1007.
Forage 1015.
Distinction between this and subsp. *indica* not always clear; they may be only varietally distinct [1024].

A. oerfota (Forsskal)Schweinf. [1016]
A. nubica Benth. [1003]; *A. orfota* sensu auctt.; *A. pterygocarpa* Benth. [1007].
Not climbing; Shrub/Tree; Perennial.
West Asia: Iran(N), North Yemen(N), Oman(N), Saudi Arabia(N), South Yemen(N); Africa.
Description: 1003, 1010, 1016; Illustration: 1010, 1016.
Domestic 1016; Food or drink 1002.
When cut or bruised, gives out a rather strong, unpleasant smell [1003].

A. origena A.Hunde [1001]
A. menachensis Schweinf.; *A. negrii* sensu Hepper.
Not climbing; Tree; Perennial.
West Asia: North Yemen(N), Saudi Arabia(N), South Yemen(N); Africa.
Description: 1001, 1003, 1027; Illustration: 1001, 1003, 1027.
A. menachensis is not a published name [1027].

A. pachyceras O. Schwartz [1262]
Not climbing; Tree; Perennial.
West Asia: North Yemen(N).
Description: 1262.
Mandaville (1990 – ref.1252) has suggested that this may be the same as *Acacia iraqensis* (q.v.).

A. paolii Chiov. [1003]
Not climbing; Shrub; Perennial.
West Asia: North Yemen(N); Africa.
Description: 1013, 1038; Illustration: 1013.
Not mentioned for West Asia by Ross (1979) or Brenan (1959) [1013, 1038].

A. saligna (Labill.)Wendl. [1045]
A. ctanophylla Lindley
Not climbing; Shrub/Tree; Perennial.
West Asia: Iraq(I), Saudi Arabia(I); Australasia.
Description: 1007, 1009.
Environmental 1009.

A. senegal (L.)Willd. [1016]
Not climbing; Shrub/Tree; Perennial.
West Asia: Oman(N); Africa; South Asia; India(N), Pakistan(N).
Description: 1010, 1015, 1016; Illustration: 1010, 1016.
Domestic 1016; Fibre 1016; Food or drink 1002, 1016, 1040; Forage 1016; Medicine 1016; Wood 1016.
This species is the most important commercial source of best quality gum arabic [1002]. It is part of a complex of closely related species, including *A. asak* & *A. hamulosa* [1041]; the relationship of the species in the complex is unclear [1041].

A. seyal Del. [1001]
Not climbing; Shrub/Tree; Perennial.
West Asia: North Yemen(N), Saudi Arabia(N); Africa.
Description: 1001, 1003, 1007; Illustration: 1001, 1003, 1007.
Food or drink 1002.
Often confused with *A. hockii* De Wild.[1039]
A. senegal Gum Arabic is of superior quality to that from *A. seyal* [1002].

A. tortilis (Forsskal)Hayne [1016]
Not climbing; Shrub/Tree; Perennial.
West Asia: Iran(I), North Yemen(N), Oman(N), Qatar(N), Saudi Arabia(N), South Yemen(N); Africa.
Description: 1003, 1007, 1016; Illustration: 1001, 1007, 1016.
Environmental 1016; Food or drink 1016; Forage 1002, 1016, 1029; Medicine 1016; Miscellaneous 1016; Wood 1016.
Ash of the dead wood was taken as snuff or added to purchased ground tobacco [1016].

subsp. raddiana (Savi)Brenan [1001]
A. raddiana Savi [1007]; *A. tortilis* var. *pubescens* Aylmer [1029].
Not climbing; Tree; Perennial.
West Asia: Saudi Arabia(N); Africa.
Description: 1001, 1003, 1007; Illustration: 1001, 1003, 1007.
Forage 1002.
Includes vars. *raddiana* and *pubescens* A.Chev.; var. *pubescens* is less frequent than var. *raddiana* [1013].

subsp. spirocarpa (A.Rich.)Brenan [1003]
A. spirocarpa A.Rich. [1013]; *A. spirocarpa* var. *major* Schweinf. [1013].
Not climbing; Tree; Perennial.
West Asia: North Yemen(N), Oman(N), Saudi Arabia(N), South Yemen(N); Africa; Middle East: Israel(N), Syria(N).
Description: 1003.

subsp. tortilis [1016]
A. spirocarpa A.Rich. var. *minor* Schweinf.
Not climbing; Tree; Perennial.
West Asia: North Yemen(N), Oman(N), Qatar(N), Saudi Arabia(N), South Yemen(N), United Arab Emirates(N); Africa; Middle East: Israel(N).
Description: 1016.

FAIDHERBIA A.Chev.

A monotypic genus, until recently regarded as part of *Acacia*. Distinct in its pollen structure, in the basal fusion of the stamen filaments, and its phenology; the tree is usually leafless in the wet season and leafy in the dry.

F. albida (Del.)A.Chev. [1005]
> *A. albida* Del. [1005].
> Not climbing; Shrub/Tree; Perennial.
> West Asia: Iran(N), North Yemen(N), Saudi Arabia(N); Africa; Middle East: Israel(N), Lebanon(N), Syria(N).
> Description: 1001, 1003, 1007; Illustration: 1001, 1003, 1007.
> Domestic 1006; Environmental 1006; Food or drink 1006; Forage 1006; Medicine 1006; Toxins 1006; Wood 1006.
> An indicator of fertile sites [1006].

INGEAE

ALBIZIA Durazz.

A large and mostly tropical genus of woody plants. Some are planted for their timber, and for shade and ornament.

A. julibrissin Durazz. [1009]
> Not climbing; Shrub/Tree; Perennial.
> West Asia: Afghanistan(I), Iran(N), Iraq(I); East Asia: China(N); South Asia: India(N); U.S.S.R.
> Description: 1009, 1010, 1015; Illustration: 1010.
> Environmental 1009; Wood 1009.

A. lebbeck (L.)Benth. [1009]
> *A. lebbec* sensu Migahid; *A. lebbek* sensu auctt. [1016].
> Not climbing; Tree; Perennial.
> West Asia: Afghanistan(I), Iran(I), Iraq(I), North Yemen(I), Oman(I), Saudi Arabia(I), South Yemen(I); East Asia: China(N); South Asia: India(N).
> Description: 1003, 1009, 1010; Illustration: 1003, 1007, 1009.
> Domestic 1009; Environmental 1003, 1009; Forage 1009; Medicine 1009; Wood 1009.
> East Indian Walnut 1009.

PITHECELLOBIUM Martius

A small, mainly neotropical genus. One species is very widely planted as a thorny hedge.

P. dulce (Roxb.)Benth. [1036]
> *Pithecolobium littorale* Rec. [1037].
> Not climbing; Shrub/Tree; Perennial.
> West Asia: Iran(I), Iraq(I), North Yemen(I), Oman(I), Saudi Arabia(I), South Yemen(I); Central America.
> Description: 1003, 1009, 1015; Illustration: 1003.
> Domestic 1009; Environmental 1003, 1009, 1018, 1025; Food or drink 1009; Forage 1009; Wood 1009.

MIMOSEAE

DICHROSTACHYS Wight & Arn.

A small genus, mostly from Africa and Madagascar. The common African species has many subspecies and varieties. An invasive weed in overgrazed grasslands.

D. cinerea (L.)Wight & Arn. [1001]
Not climbing; Shrub/Tree; Perennial.
West Asia: North Yemen(N), Saudi Arabia(N); Africa.
Description: 1001, 1003; Illustration: 1001.

LEUCAENA Benth.

A small neotropical genus, one species of which is very widely planted as a fast-growing fuelwood and pole crop. It can also be used as fodder, but with caution, as many strains are toxic if they form a high proportion of the feed.

L. leucocephala (Lam.)de Wit [1009]
Albizia julibrissin sensu Blakelock [1009, 1025]; *Leucaena glauca* (L.)Benth. [1009].
Not climbing; Tree; Perennial.
West Asia: Iraq(I), North Yemen(I), Saudi Arabia(I); Central America.
Description: 1003, 1009; Illustration: 1003, 1009.
Domestic 1009; Environmental 1009; Food or drink 1009; Forage 1003, 1009; Wood 1009.

MIMOSA L.

A tropical genus of prickly herbs or shrubs, very abundant and diverse in the neotropics, but with some species in the Old World, and others which are pantropical weeds.

M. himalayana Gamble [1010]
Not climbing; Shrub; Perennial.
West Asia: Afghanistan(N); South Asia: India(N), Pakistan(N).
Description: 1010.

M. pudica L. [1015]
Not climbing; Herb/Shrub
West Asia: Iran(I); Africa; South America.
Description: 1015.
Weed 1030.
Sensitive plant 1030.

PROSOPIS L.

A genus of trees, mostly from the drier parts of America, but with a few Old World species. Several of the American species have been very widely planted in dry regions for their timber and pods (fodder); in some areas they show signs of becoming invasive weeds.

P. africana (Guillemin & Perrottet)Taubert [1007]
Prosopis oblonga Benth. [1007].
Not climbing; Tree; Perennial.
West Asia: Saudi Arabia(N); Africa.
Description: 1007.

P. chilensis (Molina)Stuntz emend. Burkart [1031]
Not climbing; Tree; Perennial.
West Asia: Saudi Arabia(I), South Yemen(I); South America
Food or drink 1024; Forage 1024; Wood 1024
Highly variable; other taxa often mis-named as this.

P. cineraria (L.)Druce [1016]
Prosopis spicigera L. [1010, 1032, 1042].
Not climbing; Shrub/Tree; Perennial.
West Asia: Afghanistan(N), Iran(N), Oman(N), Saudi Arabia(N), South Yemen(N), United Arab Emirates(N); South Asia: India(N), Pakistan(N).
Description: 1003, 1010, 1032; Illustration: 1003, 1010.
Domestic 1015; Food or drink 1015; Wood 1031

P. farcta (Banks & Sol.)J.F.Macbr. [1009]
Not climbing; Shrub; Perennial.
West Asia: Afghanistan(N), Bahrain(N), Iran(N), Iraq(N), Kuwait(N), Oman(N); Saudi Arabia(N), South Yemen(N), United Arab Emirates(N); Africa; Middle East: Israel(N), Jordan(N), Lebanon(N), Syria(N), Turkey(N); South Asia: India(N); U.S.S.R.
Description: 1009, 1010, 1032; Illustration: 1001, 1009, 1010.
Domestic 1009, 1031; Forage 1031; Weed 1003; Wood 1009.
Includes two varieties, *farcta* & *glabra*.

P. glandulosa Torrey [1010]
Prosopis juliflora (Sw.)DC. var. *glandulosa* (Torrey)Cockerell [1031].
Not climbing; Tree; Perennial.
West Asia: Iran(I), Iraq(I), South Yemen(I); Central America; South America.
Description: 1010; Illustration: 1010.
Environmental 1009; Forage 1031, 1034; Wood 1034.
Honey Mesquite 1031.
Two varieties, *torreyana* & *prostrata*; var. *torreyana* is the one widely introduced to West Asia [1031].

P. juliflora (Sw.)DC. [1009]
Not climbing; Shrub/Tree; Perennial.
West Asia: Bahrain(I), Iran(I), Iraq(I), Kuwait(I), North Yemen(I), Oman(I), Saudi Arabia(I), South Yemen(I), United Arab Emirates(I); Central America; North America; South America.
Description: 1001, 1003, 1010; Illustration: 1001, 1003, 1010.
Domestic 1009; Environmental 1009, 1018; Forage 1009, 1018, 1034; Wood 1009, 1018, 1034.
Confusion with *P. glandulosa* in the past makes many of the records somewhat doubtful.

P. koelziana Burkart [1031]
Not climbing; Tree; Perennial.
West Asia: Iran(N), Saudi Arabia(N), South Yemen(N).
Description 1031, 1032, 1042.
Includes the varieties *koelziana* & *puberula* [1042].

P. sp. Chaudhary (provisional) [1003]
Not climbing; Shrub; Perennial.
West Asia: Saudi Arabia(I)
Description: 1003; Illustration: 1003.
This may be a hybrid between *P. farcta* & *P. juliflora*.

PAPILIONOIDEAE

ABREAE

ABRUS Adans.

A genus of small woody climbers or woody herbs. There is some controversy about species delimitation within the genus.

A. bottae Defl. (provisional) [1105]
    Climbing; Herb/Shrub; Perennial.
    West Asia: North Yemen(N), Saudi Arabia(N).
    Description: 1001; Illustration:1001.
    Verdcourt (1970) questions whether this is a good species [1105].
    Abrus precatorius has also been recorded (1251). In view of the uncertainty over the status of *A. bottae*, the record should be regarded as provisional.

AESCHYNOMENEAE

AESCHYNOMENE L.

A pantropical genus of woody herbs or shrubs, many of them growing in wet places.

A. indica L. [1058]
    Not climbing; Herb; Annual.
    West Asia: Afghanistan(U), Iran(U); Africa; Australasia; South America; South Asia; South-East Asia.
    Description: 1058; Illustration: 1058.

ARACHIS L.

A South American genus of about 20 species, one of which is very widely cultivated for its edible seeds.

A. hypogaea L. [1009]
    Not climbing; Herb; Perennial.
    West Asia: Iraq(I); South America.
    Description: 1009.
    Cover crop 1009; Domestic 1009; Food or drink 1009; Forage 1009; Medicine 1009.
    Ground-Nut 1009; Monkey-Nut 1009; Peanut 1009.

ORMOCARPUM P.Beauv.

Shrubs or small trees; about 20 species in the Old World tropics in both dry and wet climates.

O. dhofarense Hillc. & J.B.Gillett [1016]
Not climbing; Shrub; Perennial.
West Asia: North Yemen(N), Oman(N).
Description: 1016, 1096; Illustration: 1016.
Forage 1016; Medicine 1016; Wood 1016.

O. yemenense J.B.Gillett [1016]
O. bibracteatum sensu auctt. [1096]
Not climbing; Shrub; Perennial.
West Asia: North Yemen(N), Oman(N), South Yemen(N).
Description: 1096.

STYLOSANTHES Sw.

Herbs; about 25 species in the Old and New World tropics. The specific limits are hard to define. Some species are much planted as fodder.

S. fruticosa (Retz.)Alston
Not climbing; Herb; Perennial.
West Asia: North Yemen(U), Saudi Arabia(U).
Description: 1001; Illustration: 1001.

AMORPHEAE

AMORPHA L.

A North American genus of shrubs or woody herbs; one species is fairly widely planted elsewhere.

A. fruticosa L. [1009]
Not climbing; Shrub; Perennial.
West Asia: Iraq(I); North America.
Description: 1009, 1058.
Environmental 1009.
Bastard Indigo 1009; Indigobush Amorpha 1009.

CICEREAE

CICER L.

A genus of about 40 species, mainly herbs, with its centre of diversity in West Asia. One species is a very important pulse crop.

C. acanthophyllum Boriss. [1129]
C. garanicum Boriss. [1246].
Not climbing; Herb; Perennial.
West Asia: Afghanistan(N); South Asia:Pakistan(N).
Description: 1128, 1129; Distribution Map: 1128; Illustration: 1128, 1129.
Closely related to C. macracanthum [1129].

C. anatolicum Alef. [1009]
C. glutinosum Alef. [1129]; *C. songaricum* sensu Jaub. & Spach [1129].
Not climbing; Herb; Perennial.
West Asia: Iran(N), Iraq(N); Middle East: Turkey(N): U.S.S.R.
Description: 1009, 1128, 1129; Distribution Map: 1128; Illustration: 1009, 1128, 1129.
Food or drink 1128.

C. arietinum L. [1009]
Not climbing; Herb; Annual.
West Asia: Afghanistan(N), Iran(N), Iraq(N), North Yemen(I), Oman(I).
Description: 1009, 1128, 1129; Distribution Map: 1128; Illustration: 1128, 1129.
Food or drink 1009, 1128; Forage 1009, 1128; Medicine 1009, 1128.
Chick Pea 1009; Chick-Pea 1009, 1128; Chickpea 1246.
Probably originating in S.W. Asia, but its native area obscured by cultivation [1009].
'Chickpea' (one word, no hyphen) is the preferred vernacular name spelling [1246].

C. bijugum Rech.f. [1009]
Not climbing; Herb; Annual.
West Asia: Iraq(N); Middle East: Syria(N), Turkey(N).
Description: 1009, 1128, 1129; Distribution Map: 1128; Illustration: 1009, 1128, 1129.

C. chorassanicum (Bunge)Popov [1129]
C. trifoliatum sensu Parsa; *C. trifoliolatum* Bornm. [1128].
Not climbing; Herb; Annual.
West Asia: Afghanistan(N), Iran(N)
Description: 1015, 1128, 1129; Distribution Map: 1128; Illustration: 1052, 1128, 1129.

C. cuneatum A.Rich. [1128]
Climbing; Herb; Annual.
West Asia: Saudi Arabia(N); Africa.
Description: 1001, 1128; Illustration: 1001, 1128.

C. echinospermum P.Davis [1009]
Not climbing; Herb; Annual.
West Asia: Iraq(N);Middle East: Turkey(N).
Description: 1128; Illustration: 1128.
Forage 1128.
No definite record from Iraq [1009].

C. fedtschenkoi Lincz. [1129]
C. songaricum var. *pamiricum* Paulsen [1129]; *C. songaricum* var. *schugnanicum* Popov [1129].
Climbing; Herb; Perennial.
West Asia: Afghanistan(N)
Description: 1085, 1128, 1129; Distribution Map: 1128; Illustration: 1128, 1129.

C. incisum (Willd.)K.Maly [1129]
C. caucasicum Bornm.; *C. ervoides* (Sieber)Fenzl [1246]; *C. incisum* var. *libanoticum* (Boiss.)Bornm.; *C. minutum* Boiss. & Hohen. [1128].
Not climbing; Herb; Perennial.
West Asia: Iran(N); Europe; Middle East: Lebanon(N), Turkey(N).
Description: 1015, 1128, 1129; Distribution Map: 1128; Illustration: 1128, 1129.
Forage 1128.

C. kermanense Bornm. [1129]
Climbing; Herb/Shrub; Perennial.
West Asia: Iran(N)
Description: 1015, 1128, 1129; Distribution Map: 1128; Illustration: 1128, 1129.
Endemic to Iran [1129].

C. macracanthum Popov [1129]
C. songaricum var. *spinosum* Aitch.
Not climbing; Herb/Shrub; Perennial.
West Asia: Afghanistan(N); South Asia: Pakistan(N).
Description: 1128, 1129; Distribution Map: 1128; Illustration: 1128, 1129.
Closely related to *C. acanthophyllum* [1129].

C. microphyllum Benth. [1129]
> *C. jacquemontii* Jaub. & Spach [1128].
> Climbing; Herb/Shrub; Perennial.
> West Asia: Afghanistan(N); East Asia: China(N) (Xizang Zizhiqu); South Asia: Pakistan(N).
> Description: 1128, 1129; Distribution Map: 1128; Illustration: 1128, 1129.
> Forage 1128.

C. multijugum Maesen [1129]
> Not climbing; Herb; Perennial.
> West Asia: Afghanistan(N)
> Description: 1128, 1129; Distribution Map: 1128; Illustration: 1128, 1129.
> Endemic to Afghanistan [1129].

C. nuristanicum Kitam. [1129]
> *C. flexuosum* sensu Koie [1052, 1246]
> Climbing; Herb; Perennial.
> West Asia: Afghanistan(N); South Asia: Pakistan(N).
> Description: 1011, 1128, 1129; Distribution Map: 1128; Illustration: 1011, 1128, 1129.

C. oxyodon Boiss. & Hohen. [1009]
> Climbing; Herb; Perennial.
> West Asia: Afghanistan(N), Iran(N), Iraq(N).
> Description: 1009, 1128, 1129; Distribution Map: 1128; Illustration: 1009, 1128, 1129.

C. pinnatifidum Jaub. & Spach [1009]
> *C. sintenisii* Bornm. [1128].
> Not climbing; Herb; Annual.
> West Asia: Iraq(N); Middle East: Lebanon(N), Syria(N), Turkey(N).
> Description: 1009, 1128, 1129; Distribution Map: 1128; Illustration: 1128, 1129.
> Forage 1128.

C. pungens Boiss. [1129]
> *C. spinosum* Popov [1129].
> Not climbing; Herb; Perennial.
> West Asia: Afghanistan(N); U.S.S.R.
> Description: 1128, 1129; Distribution Map: 1128; Illustration: 1128, 1129.

C. rechingeri Podl. [1129]
> Not climbing; Herb; Perennial.
> West Asia: Afghanistan(N).
> Description: 1128, 1129; Distribution Map: 1128; Illustration: 1128, 1129.
> Endemic to Afghanistan [1129].

C. spiroceras Jaub. & Spach [1129]
> Climbing; Herb; Perennial.
> West Asia: Iran(N).
> Description: 1015, 1128, 1129; Distribution Map: 1128; Illustration: 1128, 1129.
> Endemic to Iran [1129].

C. stapfianum Rech.f. [1129]
> Not climbing; Herb/Shrub; Perennial.
> West Asia: Iran(N).
> Description: 1128, 1129; Distribution Map: 1128; Illustration: 1128, 1129.
> A spiny xerophytic species, related to *C. pungens* [1129].
> Endemic to Iran; known only from the type [1129].

C. subaphyllum Boiss. [1129]
> Climbing; Herb/Shrub; Perennial.
> West Asia: Iran(N).
> Description: 1015, 1128, 1129; Distribution Map: 1128; Illustration: 1128, 1129.
> Endemic to Iran; known only from the type [1129].

C. tragacanthoides Jaub. & Spach [1129]
    C. kopetdaghense Lincz. [1128]; *C. straussii* Bornm. [1128].
    Not climbing; Herb; Perennial.
    West Asia: Iran(N).
    Description: 1015, 1128, 1129; Distribution Map: 1128; Illustration: 1128, 1129.
    Endemic to Iran [1129].
    Includes vars. tragacanthoides & turcomanicum Popov

C. yamashitae Kitam. [1129]
    C. longearistatum Rech.f. [1129].
    Not climbing; Herb; Annual.
    West Asia: Afghanistan(N).
    Description: 1011, 1128, 1129; Distribution Map: 1128; Illustration: 1011, 1128, 1129.
    Endemic to Afghanistan [1246].
    Index Kewensis also records from Iran — an error [1246].

CORONILLEAE

Following advice from Dr Per Lassen (Lund, Sweden), genera formerly placed in Coronilleae have been transferred to Loteae.

CROTALARIEAE

CROTALARIA L.

A very large genus (at least 600 species) of herbs and soft-wooded shrubs. Mainly African (c.500 species), but also in the rest of the Old and New World tropics. Many are poisonous; one (*C.juncea*) is an important source of fibre.

C. aculeata De Wild. [1007]
    Not climbing; Herb/Shrub; Perennial.
    West Asia: Saudi Arabia(U); Africa.
    Description: 1007, 1060; Illustration: 1007.

C. aegyptiaca Benth. [1001]
    Crotalaria wissmannii Schwartz [1060].
    Not climbing; Herb/Shrub; Perennial.
    West Asia: Iran(N), North Yemen(N), Oman(N), Saudi Arabia(N), South Yemen(N), United Arab Emirates(N); Africa; Middle East: Israel(N), Jordan(N); South Asia: India(N).
    Description: 1015, 1060; Illustration: 1001.

C. burhia Benth. [1011]
    Not climbing; Shrub; Perennial.
    West Asia: Afghanistan(N), Iran(N); Africa; South Asia: India(N), Pakistan(N).
    Description: 1015, 1058; Illustration: 1058.
    Includes var. tomentosa Boiss.

C. deflersii Schweinf. [1060]
    Not climbing; Herb; Perennial.
    West Asia: North Yemen(N); Africa.
    Description: 1060; Illustration: 1060.

C. emarginella Vatke [1060]
Crotalaria laxa Franchet [1060]; *Crotalaria rathjensiana* Schwartz [1060].
Not climbing; Herb/Shrub; Perennial.
West Asia: North Yemen(N), Saudi Arabia(N); Africa.
Description: 1001, 1060; Illustration: 1001.

C. hirta Willd. [1062]
Crotalaria mysorensis sensu Pelt. [1060].
Not climbing; Herb; Perennial.
West Asia: North Yemen(I); South Asia: India(N).
Description: 1060.

C. impressa Walp. [1060]
Crotalaria astragalina A.Rich. [1060].
Not climbing; Herb; Annual.
West Asia: North Yemen(N), South Yemen(N); Africa.
Description: 1060.

C. incana L. [1060]
Not climbing; Herb; Perennial.
West Asia: North Yemen(U), Saudi Arabia(U); Africa.
Description: 1060.
Weed 1063.

subsp. incana [1060]
Not climbing; Herb; Perennial.
West Asia: North Yemen(I), Saudi Arabia(I); Africa.
Description: 1060; Illustration: 1060.

subsp. purpurascens (Lam.)Milne-Redh. [1060]
Not climbing; Herb, Shrub; Perennial.
West Asia: North Yemen(N); Africa.
Description: 1060; Illustration: 1060.

C. juncea L. (accepted)
Not climbing; Herb/Shrub; Annual.
West Asia: Afghanistan(I), Iraq(I); Australasia; East Asia; South Asia: India(N), Pakistan(N); South-East Asia.
Description: 1009, 1058, 1060; Illustration: 1009.
Cover crop 1009, 1064; Environmental 1009; Fibre 1009, 1060; Forage 1009, 1060; Weed 1063.
Sunn Crotalaria 1009; Sunn Hemp 1009, 1060, 1063.
Ranks next in order of importance to jute as a fibre crop [1009].
Animals will not eat when green, only when dry [1009].

C. leptocarpa Balf.f. [1001]
Not climbing; Herb; Perennial.
West Asia: North Yemen(N), Oman(N), Saudi Arabia(N), South Yemen(N); Africa.
Description: 1001, 1060; Illustration: 1001.

subsp. leptocarpa [1060]
Not climbing; Herb; Perennial.
West Asia: North Yemen(N), Oman(N), Saudi Arabia(N), South Yemen(N).
Description: 1060.
Subsp. contracta may also occur [1060].

C. medicaginea Lam. [1011]
Not climbing; Herb; Annual.
West Asia: Afghanistan(U), Oman(U); East Asia: China(N); South Asia: India(N), Pakistan(N).
Description: 1058; Illustration: 1058, 1141.

C. microphylla Vahl [1060]
Crotalaria sennii Cuf. [1060].
Not climbing; Herb; Annual.
West Asia: North Yemen(N), Saudi Arabia(N); Africa.
Description: 1001, 1060; Illustration: 1001.

C. natalitia Meissner [1060]
Not climbing; Herb/Shrub; Perennial.
West Asia: North Yemen(N); Africa.
Description: 1060.
Includes var. rutshuruensis De Wild.

C. oocarpa Baker [1060]
Not climbing; Herb; Perennial.
West Asia: North Yemen(N); Africa.
Description: 1060.

subsp. microcarpa Milne-Redh. [1060]
Not climbing; Herb; Perennial.
West Asia: North Yemen(N).
Description: 1060.

C. persica (Burm.f.)Merr. [1012]
Not climbing; Herb/Shrub; Perennial.
West Asia: Iran(N), Oman(N), South Yemen(N), United Arab Emirates(N); Africa; South Asia: Pakistan(N).
Description: 1015, 1058, 1060; Illustration: 1058, 1060.

C. plowdenii Baker [1060]
Not climbing; Herb; Perennial.
West Asia: North Yemen(N); Africa.
Description: 1060.

C. pteropoda Balf.f.
Not climbing; Herb; Perennial.
West Asia: Oman(N).
Description: 1065.

C. pycnostachya Benth. [1060]
Not climbing; Herb; Annual.
West Asia: North Yemen(N), South Yemen(N); Africa.
Description: 1060.

C. quartiniana A.Rich. [1001]
Crotalaria azaisii Sacl. [1060]; *Crotalaria platycalyx* Baker [1060].
Not climbing; Herb/Shrub; Perennial.
West Asia: North Yemen(N), Saudi Arabia(N); Africa.
Description: 1060; Illustration: 1001.

C. retusa L. [1007]
Not climbing; Herb/Shrub; Perennial.
West Asia: Iran(N), North Yemen(N), Oman(N), Saudi Arabia(N), South Yemen(N); Africa.
Description: 1007, 1015, 1060; Illustration: 1141.
Cover crop 1060, 1066; Domestic 1060; Fibre 1060; Weed 1063.
In some instances the stipules may be completely lacking in *C. retusa* [1063].

C. saltiana Andrews [1060]
C. lupinoides Hochst.
Not climbing; Herb; Perennial.
West Asia: Oman(N), South Yemen(N); Africa.
Description: 1060.

C. schweinfurthii Deflers [1251]
Not climbing; Herb; Annual.
West Asia: North Yemen(N).

C. senegalensis (Pers.)DC. [1001]
    Not climbing; Herb; Perennial.
    West Asia: North Yemen(N), Saudi Arabia(N); Africa.
    Description: 1001, 1060; Illustration: 1001.

C. sp. S.Collenette [1001]
    Not climbing; Shrub; Perennial.
    West Asia: Saudi Arabia(N)
    Description: 1001; Illustration: 1001.
    Collenette 4600 & 4716 appear to be the same.

C. spinosa Benth. [1060]
    Not climbing; Herb/Shrub; Perennial.
    West Asia: North Yemen(N), South Yemen(N); Africa.
    Description: 1060.
    Weed 1060.

C. squamigera Deflers [1251]
    Not climbing; Herb; Annual.
    West Asia: North Yemen(N).

C. thebaica (Del.)DC. [1007]
    Not climbing; Herb/Shrub; Perennial.
    West Asia: Saudi Arabia(N); Africa; South Asia: India(N).
    Description: 1060.

LOTONONIS (DC.)Ecklon & Zeyher

A genus of about 100 species, mostly herbs but some woody at the base. Most are South African but a few species extend to the Mediterranean and India.

L. platycarpa (Viv.)Pichi-Serm. [1007]
    Amphinomia platycarpa (Viv.)Cuf. [1014]; *L. dichotoma* (Del.)Boiss. [1007]; *L. leobordea* Benth. [1007].
    Not climbing; Herb; Perennial.
    West Asia: Iran(N), North Yemen(N), Oman(N), Qatar(N), Saudi Arabia(N); Africa; Middle East: Israel(N), Syria(N).
    Description: 1001, 1058; Illustration: 1001, 1058.

DALBERGIEAE

DALBERGIA L.f.

About 100 species of trees, shrubs and woody climbers, mainly in the moist tropics; a few species extend into drier regions, including *D.melanoxylon*, the source of African Blackwood.

D. sissoo Roxb. [1009]
    Not climbing; Tree; Perennial.
    West Asia: Afghanistan(N), Iran(N), Iraq(I), Oman(I); South Asia: India(N), Pakistan(N).
    Description: 1009, 1015, 1058; Illustration: 1009, 1058.
    Wood 1009.
    Sissoo Tree 1009.
    Improves poor soil [1009].

TIPUANA Benth.

The only species, native of South America, is planted as a street tree in many parts of the world.

T. tipu (Benth.)Kuntze [1009]
Not climbing; Tree; Perennial.
West Asia: Iraq(I).
Description: 1009.
Environmental 1009.

DESMODIEAE

ALYSICARPUS Desv.

A herbaceous genus with about 25 species in the Old World tropics. Some of the species are valued as fodder.

A. glumaceus (Vahl)DC.
Not climbing; Herb; Annual.
West Asia: North Yemen(N), Oman(N), Saudi Arabia(N).

A. longifolius Wight & Arn.
Not climbing; Herb; Annual.
West Asia: Oman(U)

A. ovalifolius (Schum.)Leonard [1058]
Not climbing; Herb; Annual.
West Asia: Afghanistan(N), Oman(N); Africa; Indian Ocean; South Asia: India(N), Pakistan(N); South-East Asia.
Description: 1058.

A. rugosus (Willd.)DC. [1064]
Not climbing; Herb; Perennial.
West Asia: Oman(N); Africa.
Description: 1064.

A. vaginalis (L.)DC. [1064]
Not climbing; Herb; Perennial.
West Asia: Oman(N); Africa; South Asia.
Description: 1064.

DESMODIUM Desv.

A large genus of the wetter tropics, with about 300 species. Many have been widely introduced as fodder crops; others have spread through accidental dispersal of their adhesive fruits, so that the native range may not always be clear.

D. elegans DC. [1058]
Desmodium tiliaefolium G.Don [1058]; *Desmodium tiliifolium* G.Don [1058].
Not climbing; Shrub; Perennial.
West Asia: Afghanistan(N); South Asia; India(N), Nepal(N), Pakistan(N).
Description: 1058; Illustration: 1058.

D. gangeticum (L.)DC. [1001]
Not climbing; Herb; Perennial.
West Asia: North Yemen(N), Oman(N), Saudi Arabia(N); Africa; Australasia; South Asia.
Description: 1001, 1064; Illustration: 1001.

D. ospriostreblum Chiov. [1064]
   Not climbing; Herb; Annual.
   West Asia: North Yemen(U), Oman(U); Africa;
   Description: 1064.

D. repandum (Vahl)DC. [1064]
   Not climbing; Herb/Shrub; Perennial.
   West Asia: North Yemen(N).
   Description: 1064.

D. triflorum (L.)DC. [1015]
   Not climbing; Herb; Perennial.
   West Asia: Iran(N), North Yemen(N).
   Description: 1015.

D. triquetrum DC. [1015]
   Not climbing; Herb; Perennial.
   West Asia: Iran(N).
   Description: 1015.

LESPEDEZA Michaux f.

A genus of herbs, often large, occurring in North America, East Asia and Australia. Some yield useful fodder.

L. cuneata (Du Mont.)G.Don
   West Asia: Afghanistan(N); East Asia: China(N); South Asia: Pakistan(N).
   Not mentioned by Rechinger (1984).

L. juncea (L.)Pers. [1058]
   Lespedeza aitchisonii Ricker [1058]; *Lespedeza nuristanica* Rech.f. [1058].
   Not climbing; Shrub; Perennial.
   West Asia: Afghanistan(N); South Asia: India(N), Pakistan(N).
   Description: 1058; Illustration: 1058.

GALEGEAE

ALHAGI Adans.

A small genus of desert shrubs or sub-shrubs. Species delimitation is not easy, and there has been much nomenclatural confusion. Some species are weedy, and some are browsed.

A.canescens(Regel) Keller & Shap. [1058]
   Not climbing; Herb/Shrub; Perennial.
   West Asia: Afghanistan(N); U.S.S.R.
   Description: 1058; Illustration: 1145.

A.graecorum Boiss. [1086]
   A.mannifera Desv. [1086]; *A.mannifera* var. *karduchorum* Boiss. [1086]; *A.maurorum* sensu auctt. [1086]; *A.maurorum* var. *assyriacum*Nab. [1086]; *A.maurorum* var. *karduchorum* Boiss. [1086].
   Not climbing; Herb/Shrub; Perennial.
   West Asia: Bahrain(N), Iran(N), Iraq(N), Kuwait(N), Saudi Arabia(N), Africa; Europe; Middle East.
   Description: 1009, 1086.; Illustration: 1009, 1086.
   Food or drink 1009; Medicine 1009.
   The cited descriptions & illustrations allow *A.graecorum* & *A.maurorum* to be distinguished easily. A source of manna [1009].

A. kirghisorum Schrenk (provisional) [1058]
A. sparsifolium Shap. [1058].
Not climbing; Herb/Shrub; Perennial.
West Asia: Afghanistan(N); U.S.S.R.
Description: 1058.

A. maurorum Medikus [1086]
A. camelorum Fischer [1086]; *A.camelorum* subsp. *turcorum* Sinsk. [1086]; *A.persarum* Boiss. & Buhse [1009]; *A. pseudalhagi* Desv.; *A.turcorum* Boiss.
Not climbing; Herb/Shrub; Perennial.
West Asia: Afghanistan(N), Bahrain(N),Iran(N), Iraq(N); Middle East: Turkey(N); South Asia: Pakistan(N); U.S.S.R.
Description: 1009, 1086; Illustration: 1009, 1086.
Food or drink 1009; Forage 1009; Medicine 1009.
Application of the name much confused; cited literature allows easy distinction of the two taxa. A source of manna [1009].

ASTRACANTHA Podl.

The genus *Astracantha* has been separated from *Astragalus* by Podlech (1983). All members of *Astracantha* have paripinnate leaves with spine-tipped rachises. The 2-15-flowered inflorescences are sessile and usually densely congested in the leaf axils. The sessile fruits are small, unilocular, and 1-2-seeded. The outer tissues of the stem contain gum. The genus is still much in need of revision as a whole.

A. acetabulosa (C.Towns.)Podl. [1146]
Astragalus acetabulosus C.Towns. [1146].
Not climbing; Herb/Shrub; Perennial.
West Asia: Iran(N), Iraq(N).
Description: 1009, 1132; Illustration: 1009, 1132.

A. acmophylla (Bunge)Podl. [1146]
Astragalus acmophyllus Bunge [1146].
Not climbing; Herb; Perennial.
West Asia: Iran(N).
Description: 1015, 1102.
According to Flora of Turkey, this species is endemic to Turkey. No Iran record has been found, apart from Parsa's.

A. adscendens (Boiss. & Hausskn.)Podl. [1146]
Astragalus adscendens Boiss. & Hausskn. [1146]; *Astragalus turrillii* Eig [1009].
Not climbing; Herb/Shrub; Perennial.
West Asia: Iran(N), Iraq(N).
Description: 1009, 1124; Illustration: 1009.
Chemical 1009; Forage 1009.

A. albispina (Sirj. & Bornm.)Podl. [1146]
Astragalus albispinus Sirj. & Bornm. [1146].
Not climbing; Herb; Perennial.
West Asia: Iran(N).
Description: 1015.
Type from Iran (Sirjaev & Bornmuller 1943) [1015]

A. alexandri (Sirj.)Podl. [1146]
Astragalus alexandri Sirj. [1146].
Not climbing; Shrub; Perennial.
West Asia: Iran(N).
Description: 1147.
Type from Iran (Sirjaev 1939).

A. amblolepis (Fischer)Podl. [1146]

Astracantha psilacmos (Bunge)Podl. [1009]; *Astracantha wartoensis* (Boiss. & Kotschy)Podl. [1009]; *Astragalus amadiensis* Eig [1009]; *Astragalus amblolepis* Fischer [1146]; *Astragalus amblolepis* var. *rayatensis* Eig [1009]; *Astragalus psilacmos* Bunge [1009]; *Astragalus wartoensis* Boiss. & Kotschy [1009].
Not climbing; Herb/Shrub; Perennial.
West Asia: Iraq(N); Middle East: Turkey(N).
Description: 1009, 1102.
Chemical 1009; Medicine 1009.
A. amblolepis, A. psilacmos & *A. wartoensis* are retained as three species in a poorly-understood complex [1102].
Townsend, in Flora of Iraq, has Hohenacker (Hoh.) as authority.

A. atropatana (Bunge)Podl. [1146]

Astragalus atropatanus Bunge [1146].
Not climbing; Shrub; Perennial.
West Asia: Iran(N).
Description: 1015.
Type from Iran (Bunge).

A. aurea (Willd.)Podl. [1146]

Astragalus aureus Willd. [1146]; *Astragalus chromolepis* Boiss. & Hohen. [1149].
Not climbing; Herb/Shrub; Perennial.
West Asia: Iran(N); U.S.S.R.
Description: 1015, 1150; Illustration: 1015, 1150, 1151.
Parsa (1948) includes a var.chromolepis, described from Iran.

A. baghensis (Bunge)Podl. [1146]

Astragalus baghensis Bunge [1146].
Not climbing; Herb; Perennial.
West Asia: Iran(N).
Description: 1015.

A. basianica (Boiss. & Hausskn.)Podl. [1146]

Astragalus basianicus Boiss. & Hausskn. [1146].
Not climbing; Herb; Perennial.
West Asia: Iran(N), Iraq(N).
Description: 1009, 1015.
Type from Iran.

A. bienerti (Bunge)Podl. [1146]

Astragalus bienerti Bunge [1146].
Not climbing; Shrub; Perennial.
West Asia: Iran(N).
Description: 1015.
Type from Iran.

A. brachycalyx (Fischer)Podl. [1146]

Astragalus brachycalyx Fischer [1146]; *Astragalus cbraehycalyx* Fischer [1015].
Not climbing; Herb/Shrub; Perennial.
West Asia: Iran(N), Iraq(N); Middle East: Turkey(N).
Description: 1009, 1015; Illustration: 1009.
Parsa's "cbraehycalyx" is a gross misprint!

A. caspica (M.Bieb.)Podl. [1146]

Astragalus caspicus M.Bieb. [1146]; *Astragalus caspius* M.Bieb. [1015]; *Astragalus cretensis* Pallas [1147]; *Astragalus leptodendron* Fischer [1147].
Not climbing; Shrub; Perennial.
West Asia: Iran(N); U.S.S.R.
Description: 1015, 1102, 1150.
Parsa mentions 4 varieties [1015].
Sirjaev(1939) & others spell as 'caspius'; this is wrong.

A. cephalotes (Banks & Sol.)Podl. [1146]

Astragalus andrachnaefolius sensu auctt. [1009]; *Astragalus andrachne* Bunge [1009]; *Astragalus andrachne* var. *sefinensis* (Bornm.)Sirj. [1009]; *Astragalus andrachne* var. *sintenisianus* Sirj. [1009]; *Astragalus cephalotes* Banks & Sol. [1146]; *Astragalus cephalotes* var. *brevicalyx* Eig [1009]; *Astragalus sefinensis* Bornm. [1009].
Not climbing; Herb/Shrub; Perennial.
West Asia: Iran(N), Iraq(N); Middle East: Lebanon(N), Syria(N), Turkey(N).
Description: 1009, 1015.
May produce gum [1009].
Flora of Turkey [1182] recognises 3 varieties, *cephalotes*, *brevicalyx* & *sintenisianus*.
Parsa's record is the only evidence of the occurrence of this taxon in Iran.

A. cerasocrena (Bunge)Podl. [1146]

Astragalus cerasocrenus Bunge [1146]; *Astragalus fragilidens* Freyn & Sint. [1150].
Not climbing; Shrub; Perennial.
West Asia: Iran(N); U.S.S.R.
Description: 1015, 1150; Illustration: 1150.
Type from Iran [1150].

A. cesarensis (Sirj. & Bornm.)Podl. [1146]

Astragalus cesarensis Sirj. & Bornm. [1146].
Not climbing; Herb; Perennial.
West Asia: Iran(N).
Description: 1015.
Parsa mentions a var. *obtuseauriculatus* Sirj. & Bornm [1015].
Type from Iran.

A. chorassanica (Bunge)Podl. [1146]

Astragalus chorassanicus Bunge [1146].
Not climbing; Herb; Perennial.
West Asia: Iran(N).
Description: 1015.
Type from Iran.

A. crassinervia (Boiss.)Podl. [1146]

Astragalus crassinervius Boiss. [1146].
Not climbing; Shrub; Perennial.
West Asia: Iran(N); Middle East: Turkey(N).
Description: 1015, 1102.
Known only from the type, from Kurdistan [1102]; Not recorded from Iran in Fl.Turkey.

A. crenophila (Boiss.)Podl. [1146]

Astragalus crenophilus Boiss. [1146].
Not climbing; Herb/Shrub; Perennial.
West Asia: Iran(N), Iraq(N).
Description: 1009, 1015; Illustration: 1009.
Parsa's record is the only evidence for occurrence in Iran [1015].

A. deinacantha (Boiss.)Podl. [1146]

Astragalus deinacanthus Boiss. [1146].
Not climbing; Herb/Shrub; Perennial.
West Asia: Iran(N); Middle East: Syria(N), Turkey(N).
Description: 1015, 1102.
Parsa's record is the only evidence for occurrence in Iran [1015].
Afghanistan material so named at Kew is probably *A. elisabethae*.

A. diphtherites (Fenzl)Podl. [1146]

Astragalus argyrophyllus Boiss. & Gaill. [1009]; *Astragalus diphtherites* Fenzl [1146]; *Astragalus diphtherites* subsp. *brachyanthus* Eig [1009]
Not climbing; Herb/Shrub; Perennial.
West Asia: Iraq(N); Middle East: Syria(N), Turkey(N).
Description: 1009.
Chemical 1153.

A. divaricata (Bunge)Podl. [1146]
   Astragalus divaricatus Bunge [1146].
   Not climbing; Herb/Shrub; Perennial.
   West Asia: Iran(N).
   Description: 1015.
   Type from Iran [1149].

A. drymophila (Bornm.)Podl. [1146]
   Astragalus drymophilus Bornm. [1146].
   Not climbing; Herb/Shrub; Perennial.
   West Asia: Iran(N).
   Description: 1154; Illustration: 1154.
   Type from Iran [1154].

A. dschuparensis (Freyn & Bornm.)Podl. [1146]
   Astragalus djuparensis Freyn & Bornm. [1015]; *Astragalus dschuparensis* Freyn & Bornm. [1146].
   Not climbing; Shrub; Perennial.
   West Asia: Iran(N).
   Description: 1015.
   Type from Iran [1147].

A. echidna (Bunge)Podl. [1146]
   Astragalus echidna Bunge [1146].
   Not climbing; Shrub; Perennial.
   West Asia: Iran(N).
   Description: 1015.
   Type from Iran [1149].

A. echidnaeformis (Sirj.)Podl. [1146]
   Astragalus echidnaeformis Sirj. [1146].
   Not climbing; Herb/Shrub; Perennial.
   West Asia: Iran(N).
   Description: 1015, 1147; Illustration: 1147.
   Type from Iran [1147].

A. echinus (DC.)Podl. [1155]
   Astragalus echinus DC. [1146].
   Not climbing; Herb/Shrub; Perennial.
   West Asia: Saudi Arabia(N); Middle East: Israel(N), Lebanon(N), Sinai(N), Syria(N).
   Description: 1155.

   subsp. **arabica** Hedge & Podl. [1155]
   Astragalus sp.(sect.Rhacophorus) S.Collenette [1155].
   Not climbing; Herb/Shrub; Perennial.
   West Asia: Saudi Arabia(N).
   Description: 1155; Illustration: 1001.
   Collenette's illustration is of 'Astragalus,Sect.Rhacophorus' (Collenette 4773 [1001]).

A. elisabethae (Sirj. & Rech.f.)Podl. [1146]
   Astragalus elisabethae Sirj. & Rech.f. [1146].
   Not climbing; Shrub; Perennial.
   West Asia: Afghanistan(N).
   Description: 1052; Illustration: 1052.

A. erinacea (Fischer)Podl. [1146]
   Astragalus erinaceus Fischer [1146].
   Not climbing; Shrub; Perennial.
   West Asia: Iran(N): Middle East: Turkey(N). U.S.S.R.
   Description: 1015, 1102, 1150.
   Parsa mentions a var. *sefirudensis* [1015].

A. eriocephala (Willd.)Podl. [1146]
Astragalus breviflorus DC. [1009]; *Astragalus eriocephalus* Willd. [1146].
Not climbing; Herb; Perennial.
West Asia: Iraq(N); Middle East: Turkey(N).
Description: 1009.
Probably also in Iran (but not recorded by Parsa, or later workers) [1009].

A. eriosphaera (Boiss. & Hausskn.)Podl. [1146]
Astragalus eriosphaerus Boiss. & Hausskn. [1146].
Not climbing; Herb/Shrub; Perennial.
West Asia: Iran(N), Iraq(N).
Description: 1009, 1015.

A. eriostyla (Boiss. & Hausskn.)Podl. [1146]
Astragalus eriostylus Boiss. & Hausskn. [1146].
Not climbing; Shrub; Perennial.
West Asia: Iran(N).
Description: 1015.
Type from Iran [1015].

A. eschkerensis (Boiss. & Hausskn.)Podl. [1146]
Astragalus eschkerensis Boiss. & Hausskn. [1146].
Not climbing; Herb/Shrub; Perennial.
West Asia: Iran(N).
Description: 1015.
Type from Iran [1015].

A. esfandiarii (Sirj. & Rech.f.)Podl. [1146]
Astragalus esfandiarii Sirj. & Rech.f. [1146].
Not climbing; Herb/Shrub; Perennial.
West Asia: Iran(N).
Description: 1156.

A. farakulumensis (Sirj. & Rech.f.)Podl. [1146]
Astragalus farakulumensis Sirj. & Rech.f. [1146].
Not climbing; Herb/Shrub; Perennial.
West Asia: Afghanistan(N).
Description: 1052; Illustration: 1052.

A. floccosa (Boiss.)Podl. [1146]
Astragalus floccosus Boiss. [1146].
Not climbing; Shrub; Perennial.
West Asia: Iran(N).
Description: 1015.
Parsa mentions a var. *wafsiensis* Bornm [1015]. Type from Iran.

A. florulenta (Boiss. & Hausskn.)Podl. [1146]
Astragalus florulentus Boiss. & Hausskn. [1146].
Not climbing; Shrub; Perennial.
West Asia: Iran(N).
Description: 1015.

A. garaensis (Sirj.)Podl. [1146]
Astragalus garaensis Sirj. [1146].
Not climbing; Herb/Shrub
Not Threatened; Perennial.
West Asia: Iraq(N); Middle East: Turkey(N).
Description: 1009, 1102; Illustration: 1009.

A. gemiana (Boiss. & Hausskn.)Podl. [1146]
Astragalus gemianus Boiss. & Hausskn. [1146]; *Astragalus geminanus* Boiss. & Hausskn. [1015].
Not climbing; Shrub; Perennial.
West Asia: Iran(N).
Description: 1015.

A. ghilanica (Fischer)Podl. [1146]
Astragalus ghilanicus Fischer [1146].
Not climbing; Shrub; Perennial.
West Asia: Iran(N).
Description: 1015.
Type from Iran [1149].

A. gigantostrobus (Rech.f. & Aellen)Podl. [1146]
Astragalus gigantostrobus Rech.f. & Aellen [1146].
Not climbing; Herb/Shrub; Perennial.
West Asia: Iran(N).
Description: 1157.

A. glabrifolia (Bunge)Podl. [1146]
Astragalus glabrifolius Bunge [1146].
Not climbing; Shrub; Perennial.
West Asia: Afghanistan(N).
Description: 1149.
Type from Afghanistan [1149].

A. glaucopsoides (Bornm.)Podl. [1146]
Astragalus glaucopsoides Bornm. [1146].
Not climbing; Herb; Perennial.
West Asia: Iran(N).
Description: 1015, 1158; Illustration: 1154.
Type from Iran [1158].

A. globiflora (Boiss.)Podl. [1146]
Astragalus andalanicus Boiss. & Hausskn. [1009]; *Astragalus andalanicus* var. *elymaiticus* (Boiss. & Hausskn.)Sirj. [1009]; *Astragalus andalanicus* var. *validior* (Bornm.)Sirj. [1009]; *Astragalus globiflorus* Boiss. [1146]; *Astragalus karadaghicus* Eig [1009].
Not climbing; Herb; Not Threatened.
Perennial.
West Asia: Iran(N), Iraq(N).
Description: 1009, 1015.
Chemical 1009.
Two varieties, *globiflorus* & *andalanicus* [1009].

A. gossypina (Fischer)Podl. [1146]
Astracantha xanthogossypina (Hand.-Mazz.)Podl. [1009]; *Astragalus filagineus* Boiss. [1009]; *Astragalus gossypinoides* Hand.-Mazz. [1009]; *Astragalus gossypinoides* var. *sindscharensis* Sirj. [1009]; *Astragalus gossypinus* Fischer [1146]; *Astragalus xanthogossypinus* Hand.-Mazz. [1009].
Not climbing; Herb/Shrub.
Not Threatened; Perennial.
West Asia: Iran(N), Iraq(N); Middle East: Lebanon(N), Syria(N), Turkey(N).
Description: 1009, 1015.
Note on Chick 1046(K): "..produces the best gum in Fars". (Iran) [1153].

A. gummifera (Labill.)Podl. [1146]
Astracantha rayatensis (Eig)Podl. [1009]; *Astragalus erianthus* Willd. [1147]; *Astragalus gummifer* Labill. [1146]; *Astragalus noemiae* Eig [1009]; *Astragalus rayatensis* Eig [1009].
Not climbing; Herb/Shrub; Perennial.
West Asia: Iraq(N); Middle East: Lebanon(N), Syria(N), Turkey(N).
Description: 1009; Illustration: 1009.
Chemical 1009.
The holotype of *A.rayatensis* is very poor [1009].
Probably also in Iran (but not recorded by Parsa or later workers.)

A. heratensis (Bunge)Podl. [1146]
Astragalus heratensis Bunge [1146].
Not climbing; Shrub; Perennial.
West Asia: Afghanistan(N).
Description: 1015, 1159; Illustration: 1159.
Chemical 1159.
Katira Gum [1159].

A. hypsogeton (Bunge)Podl. [1146]
Astragalus hypsogeton Bunge [1146].
Not climbing; Shrub; Perennial.
West Asia: Iran(N).
Description: 1015.
Type from Iran [1149].

A. karabaghensis (Bunge)Podl. [1146]
Astragalus karabaghensis Bunge [1146]; *Astragalus sivasicus* Bunge [1102].
Not climbing; Shrub; Perennial.
West Asia: Iran(N).
Description: 1015, 1102, 1150.
Type from Iran [1149].

A. kuhistana (Bunge)Podl. [1146]
Astragalus kuhistanus Bunge [1146].
Not climbing; Herb/Shrub; Perennial.
West Asia: Iran(N).
Description: 1015; Illustration: 1015.
Type from Iran [1149].
Trott 1042 (K) — perhaps not this — is recorded as yielding gum [1153].

A. kurdica (Boiss.)Podl. [1146]
Astragalus kurdicus Boiss. [1146]; *Astragalus kurdicus* subsp. *kurdorum* Eig [1009].
Not climbing; Herb/Shrub
Not Threatened; Perennial.
West Asia: Afghanistan(N), Iran(N), Iraq(N); Middle East: Turkey(N).
Description: 1009, 1015, 1102; Illustration: 1009.
Chemical 1009.
Two varieties, *kurdicus* & *muschianus* [1102].
A.muschiana treated as good species (q.v.) by Fl.Iraq [1009].

A. lagonyx (Fischer)Podl. [1146]
Astragalus lagonyx Fischer [1146]; *Astragalus oleifolius* sensu Rawi [1009,1160]; *Astragalus oleifolius* var. *kurdistanicus* Sirj. [1009].
Not climbing; Herb; Perennial.
West Asia: Iran(N), Iraq(N).
Description: 1009, 1015; Illustration: 1009.

A. lateritia (Boiss. & Hausskn.)Podl. [1098]
Astragalus lateritius Boiss. & Hausskn. [1146].
Not climbing; Herb/Shrub; Perennial.
West Asia: Iran(N), Iraq(N).
Description: 1009, 1015.
Iraq record not certain; locality lacks adequate detail [1009].

A. lateritians (Freyn & Bornm.)Podl. [1098]
Astragalus lateritians Freyn & Bornm. [1098].
Not climbing; Shrub; Perennial.
West Asia: Iran(N).
Description: 1015.

A. leioclados (Boiss.)Podl. [1146]
Astragalus belangerianus Fischer [1147]; *Astragalus leioclados* Boiss. [1146]; *Astragalus leiocladus* Boiss. [1015].
Not climbing; Shrub; Perennial.
West Asia: Iran(N).
Description: 1015.
Type from Iran [1149].

A. lepidantha (Boiss.)Podl. [1146]
Astragalus bethlehemiticus subsp. *lepidanthus* (Boiss.)Eig [1009,1146]; *Astragalus lepidanthus* Boiss. [1146].
Not climbing; Herb/Shrub; Perennial.
West Asia: Iraq(N); Middle East: Syria(N).
Description: 1009.
The Iraq record is not definite; the relevant collections are old and poorly-localised [1009].
This taxon may be correctly placed as a subsp. of *A.bethlehemitica* [1009].
There is similar material (e.g. Brown 904 & 3567) from Iran at Kew [1153].

A. leucoptila (Boiss. & Hausskn.)Podl. [1146]
Astragalus leucoptilus Boiss. & Hausskn. [1146].
Not climbing; Herb/Shrub; Perennial.
West Asia: Iran(N), Iraq(N).
Description: 1009, 1015.
The exact collecting locality is uncertain; Iran or Iraq? [1009].
Not seen since it was originally collected, c. 1850 [1009].

A. longifolia (Lam.)Podl. [1146]
Astragalus longifolius Lam. [1146].
Not climbing; Herb; Perennial.
West Asia: Iraq(N); Middle East: Syria(N), Turkey(N).
Description: 1009, 1102.

A. meschhedensis (Bunge)Podl. [1146]
Astragalus meschedensis Bunge [1015]; *Astragalus meschhedensis* Bunge [1146].
Not climbing; Shrub; Perennial.
West Asia: Iran(N).
Description: 1015, 1150.
Type from Iran [1149].

A. mesoleia (Boiss. & Hohen.)Podl. [1146]
Astragalus meseleios Boiss. [1015]; *Astragalus mesoleios* Boiss. & Hohen. [1146].
Not climbing; Shrub; Perennial.
West Asia: Iran(N).
Description: 1015.
Type from Iran [1149].

A. meyeri (Boiss.)Podl. [1146]
Astragalus meyeri Boiss. [1146].
Not climbing; Shrub; Perennial.
West Asia: Iran(N); Middle East: Turkey(N); U.S.S.R.
Description: 1102.

A. michauxiana (Boiss.)Podl. [1146]
Astragalus karabaghensis sensu Rawi [1160]; *Astragalus michauxianus* Boiss. [1146].
Not climbing; Herb; Perennial.
West Asia: Iran(N), Iraq(N).
Description: 1009, 1015; Illustration: 1009.

A. microcephala (Willd.)Podl. [1146]
Astragalus fissilis Freyn & Sint. [1102]; *Astragalus fissus* Freyn & Sint. [1150]; *Astragalus microcephalus* Willd. [1146]; *Astragalus microcephalus* subsp. *pycnocladus* (Boiss. & Hausskn.)Eig [1009]; *Astragalus neglectus* Freyn [1150]; *Astragalus pirimukurunicus* Eig [1009]; *Astragalus pycnocladus* Boiss. & Hausskn. [1009]; *Astragalus pycnophyllus* Steven [1150]; *Astragalus senganensis* Bunge [1009].
Not climbing; Shrub; Perennial.
West Asia: Iran(N), Iraq(N); Middle East: Turkey(N); U.S.S.R.
Description: 1009, 1015, 1150; Illustration: 1151
Chemical 1009.

A. morgani (Freyn)Podl. [1146]
Astragalus morgani Freyn [1146].
Not climbing; Shrub; Perennial.
West Asia: Iran(N).
Description: 1015.
Type from Iran [1015].

A. multispina (Freyn & Bornm.)Podl. [1146]
Astragalus multispinus Freyn & Bornm. [1146].
Not climbing; Shrub; Perennial.
West Asia: Iran(N).
Description: 1015.

A. muschiana (Kotschy & Boiss.)Podl. [1146]
Astragalus kurdicus var. *muschianus* (Kotschy & Boiss.)Chamberlain & V.Matthews [1009]; *Astragalus muschianus* Kotschy & Boiss. [1146].
Not climbing; Herb/Shrub; Perennial.
West Asia: Iraq(N); Middle East: Turkey(N).
Description: 1009; Illustration: 1009.
Treated as a variety of *A.kurdica* in Flora of Turkey [1009].
Likely also to occur in Iran (but not recorded by Parsa or later workers) [1009].

A. myriacantha (Boiss.)Podl. [1098]
Astragalus myriacanthus Boiss. [1098].
Not climbing; Shrub; Perennial.
West Asia: Iran(N).
Description: 1015.
Type from Iran [1149].

A. ochrobia (Bunge)Podl. [1146]
Astragalus ochrobius Bunge [1146].
Not climbing; Shrub; Perennial.
West Asia: Iran(N).
Description: 1015.
Parsa mentions a var. *rechingeri* Sirj. [1015].
Type from Iran [1149].

A. ochtoranensis (Freyn)Podl. [1146]
Astragalus ochtoranensis Freyn [1146].
Not climbing; Shrub; Perennial.
West Asia: Iran(N).
Description: 1015.
Type from Iran [1015].

A. octopus (C.Towns.)Podl. [1146]
Astragalus octopus C.Towns. [1146].
Not climbing; Herb/Shrub; Perennial.
West Asia: Iran(N), Iraq(N).
Description: 1009, 1132; Illustration: 1009, 1132.

A. pachyacantha (Bunge)Podl. [1146]
Astragalus pachyacanthus Bunge [1146].
Not climbing; Shrub; Perennial.
West Asia: Iran(N).
Description: 1015.
Type from Iran [1149].
Parsa mentions a var. *hirtistylus* Boiss [1015].

A. pachystachys (Bunge)Podl. [1146]
Astragalus pachystachys Bunge [1146].
Not climbing; Shrub; Perennial.
West Asia: Iran(N).
Description: 1015.
Type from Iran [1149].

A. parrowiana (Boiss. & Hausskn.)Podl. [1146]
Astragalus parrowianus Boiss. & Hausskn. [1146].
Not climbing; Shrub; Perennial.
West Asia: Iran(N).
Description: 1015.
Two varieties, *intermedius* Sirj. & Bornm., & *gaubae* Sirj. & Bornm [1015].

A. peristerea (Boiss. & Hausskn.)Podl. [1146]
Astragalus peristereus Boiss. & Hausskn. [1146].
Not climbing; Herb/Shrub; Perennial.
West Asia: Iran(N), Iraq(N).
Description: 1009, 1015.
Very similar to *A.globiflora* (q.v.) [1009].
Parsa's record is the only evidence of the occurrence of this taxon in Iran [1152].

A. polyantha (Bunge)Podl. [1146]
Astragalus polyanthus Bunge [1146].
Not climbing; Shrub; Perennial.
West Asia: Iran(N); Middle East: Turkey(N).
Description: 1015, 1102.
Type from Iran [1149].

A. prusiana (Boiss.)Podl. [1146]
Astragalus prusianus Boiss. [1146].
Not climbing; Herb/Shrub; Perennial.
West Asia: Iraq(N); Middle East: Turkey(N).
Description: 1009, 1102.

A. pseudaurea (Sirj. & Rech.f.)Podl. [1146]
Astragalus pseudaureus Sirj. & Rech.f. [1146].
Not climbing; Herb/Shrub; Perennial.
West Asia: Iran(N).
Description: 1156.

A. psilodontia (Boiss.)Podl. [1146]
Astragalus psilodontius Boiss. [1146].
West Asia: Iraq(N); Middle East: Lebanon(N), Syria(N).
The Bornmuller specimen cited by Sirjaev is from Syria [1009].
No Iraq material has been seen; the species probably occurs only in Syria and Lebanon [1009].

A. psilostyla (Bunge)Podl. [1146]
Astragalus psilostylus Bunge [1146].
Not climbing; Herb; Perennial.
West Asia: Iran(N).
Description: 1015.
Type from Iran [1149].

A. ptilocephala (Baker)Podl. [1146]
Astragalus ptilocephalus Baker [1146].
Not climbing; Herb/Shrub; Perennial.
West Asia: Afghanistan(N); South Asia: Pakistan(N).
Description: 1162.

A. pulvinata (Bunge)Podl. [1146]
Astragalus pulvinatus Bunge [1146].
Not climbing; Shrub; Perennial.
West Asia: Iran(N); U.S.S.R.
Description: 1015, 1150.
Type from Iran [1149].

A. pycnocephala (Fischer)Podl. [1146]
   Astragalus guestii Eig [1009]; *Astragalus pycnocephalus* Fischer [1146]; *Astragalus pycnocephalus* var. *nabelekii* Sirj. [1009].
   Not climbing; Herb/Shrub
   Not Threatened; Perennial.
   West Asia: Iraq(N); Middle East: Turkey(N).
   Description: 1009, 1102; Illustration: 1009.
   Two varieties, *pycnocephalus* & *seytunensis* [1102].

A. radschirdensis (Sirj. & Bornm.)Podl. [1146]
   Astragalus radschirdensis Sirj. & Bornm. [1146].
   Not climbing; Herb; Perennial.
   West Asia: Iran(N).
   Description: 1015.

A. rhodochroa (Boiss. & Hausskn.)Podl. [1146]
   Astragalus rhodochrous Boiss. & Hausskn. [1146].
   Not climbing; Herb/Shrub; Perennial.
   West Asia: Iran(N), Iraq(N).
   Description: 1009, 1015; Illustration: 1009.

A. semipellita (Bunge)Podl. [1146]
   Astragalus semipellitus Bunge [1146].
   Not climbing; Shrub; Perennial.
   West Asia: Iran(N).
   Description: 1015.
   Type from Iran [1149].

A. serpentinica (Sirj. & Rech.f.)Podl. [1146]
   Astragalus serpentinicus Sirj. & Rech.f. [1146].
   Not climbing; Shrub; Perennial.
   West Asia: Iran(N).
   Description: 1156.

A. sofica (Bunge)Podl. [1146]
   Astragalus soficus Bunge [1146].
   Not climbing; Herb/Shrub; Perennial.
   West Asia: Iran(N).
   Description: 1015.
   Type from Iran [1149]

A. splendidissima (Sirj. & Rech.f.)Podl. [1146]
   Astragalus splendidissimus Sirj. & Rech.f. [1146].
   Not climbing; Herb/Shrub; Perennial.
   West Asia: Afghanistan(N).
   Description: 1052, 1156; Illustration: 1052.

A. stenonychioides (Freyn & Bornm.)Podl. [1146]
   Astragalus stenonychioides Freyn & Bornm. [1146]; *Astragalus strictifolius* Boiss. [1146].
   Not climbing; Herb/Shrub; Perennial.
   West Asia: Iran(N), Iraq(N); Middle East: Turkey(N); U.S.S.R.
   Description: 1009, 1102, 1150.
   May produce gum [1009].

A. strobilifera (Benth.)Podl. [1146]
   Astragalus strobiliferus Benth. [1146].
   Not climbing; Herb; Perennial.
   West Asia: Afghanistan(N); South Asia: India(N), Pakistan(N).
   Description: 1015.
   Chemical 1153.

A. talarensis (Sirj. & Rech.f.)Podl. [1146]
    Astragalus talarensis Sirj. & Rech.f. [1146].
    Not climbing; Herb/Shrub; Perennial.
    West Asia: Iran(N).
    Description: 1156.

A. totschalensis (Bornm.)Podl. [1146]
    Astragalus totchalensis Bornm. [1015]; *Astragalus totschalensis* Bornm. [1146].
    Not climbing; Herb; Perennial.
    West Asia: Iran(N).
    Description: 1015.

A. zoharyi (Eig)Podl. [1146]
    Astragalus zoharyi Eig [1146].
    Not climbing; Herb/Shrub; Perennial.
    West Asia: Iraq(N).
    Description: 1009.
    Type very poor; not otherwise recorded [1009].

ASTRAGALUS L. (cf. ASTRACANTHA)

As stated above, the genus *Astracantha* is badly in need of revision as whole. The following species are considered by Professor Podlech to belong to *Astracantha*, but all are of doubtful taxonomic status. No combinations have been made in *Astracantha*, and it is most undesirable that such combinations should be made without a thorough revision of the whole group.

A. aeluropus Bunge [1015]
    Not climbing; Herb/Shrub; Perennial.
    West Asia: Iran(N).
    Description: 1015.
    Type from Iran [1149].
    In the Flora of the USSR, this taxon is treated as a synonym of *A.oleifolius* DC. [1150].

A. aphthonus Rech.f. & Aellen [1157]
    Not climbing; Herb; Perennial.
    West Asia: Iran(N).
    Description: 1157.

A. behboudii Sirj. & Rech.f. [1156]
    Not climbing; Herb/Shrub; Perennial.
    West Asia: Iran(N).
    Description: 1156.

A. bibracteolatus Sirj. & Rech.f. [1156]
    Not climbing; Shrub; Perennial.
    West Asia: Iran(N).
    Description: 1156.

A. brachycentrus Fischer [1015]
    Not climbing; Shrub; Perennial.
    West Asia: Iran(N).
    Description: 1015; Illustration: 1015.
    Parsa mentions a var. *koieanus* Sirj. [1015].
    Type from Iran [1149].

A. constrictus Turrill [1164]
    Not climbing; Shrub; Perennial.
    West Asia: Iran(N).
    Description: 1015, 1164.
    A var. *tomentosus* was described at the same time [1164].

A. distantior Turrill [1165]
Not climbing; Shrub; Perennial.
West Asia: Iran(N).
Description: 1015, 1165.

A. dolius Boiss. & Hausskn. (provisional) [1009].
Not climbing; Herb; Perennial.
West Asia: Iran(N), Iraq(N).
Description: 1009.
Very similar to *Astracantha globiflora* (Boiss.)Podl. (q.v.) [1009].
Exact locality uncertain; Iran or Iraq [1009].

A. elymaiticus Boiss. & Hausskn. [1015]
A. elymiaticus Boiss. & Hausskn. [1015].
Not climbing; Shrub; Perennial.
West Asia: Iran(N).
Description: 1015.
Chemical 1153.
Various collections by Chick from Iran at Kew, all named as this, describe different gum qualities, and are given different vernacular names - Katira, Kash Kar, Gumur. They may not all belong to the same taxon [1153].

A. etezadianus Parsa (provisional) [1015].
Not climbing; Shrub; Perennial.
West Asia: Iran(N).
Description: 1015.
The Kew sheet (a drawing) is labelled *A.caspicus* var.*platonychioides* Sirj. & Bornm. [1152].

A. ferociformis Sirj.,Rech.f. & Aellen [1157]
Not climbing; Shrub; Perennial.
West Asia: Iran(N).
Description: 1157.

A. gerruensis Bornm. & Sirj. [1015]
Not climbing; Herb; Perennial.
West Asia: Iran(N).
Description: 1015.

A. getschesarensis Sirj. & Bornm. [1015]
Not climbing; Herb; Perennial.
West Asia: Iran(N).
Description: 1015.
Close to *A. arnacantha* [which is an *Astracantha*] [1015].

A. ghaznianus Sirj. & Rech.f. [1052]
Not climbing; Herb/Shrub; Perennial.
West Asia: Afghanistan(N).
Description: 1052; Illustration: 1052.

A. gillii Sirj. [1166]
Not climbing; Shrub; Perennial.
West Asia: Iran(N).
Description: 1166.

A. glabristylus Turrill [1165]
Not climbing; Shrub; Perennial.
West Asia: Iran(N).
Description: 1015, 1165.

A. glaucops Bornm. [1015]
Not climbing; Shrub; Perennial.
West Asia: Iran(N).
Description: 1015, 1104.

A. haematosemius Sirj. & Bornm. [1015]
   A. haematoemius Sirj. & Bornm. [1015].
   Not climbing; Herb; Perennial.
   West Asia: Iran(N).
   Description: 1015.

A. horridissimus Sirj. & Bornm. [1167]
   A. horrictissimus Sirj. & Bornm. [1015].
   Not climbing; Herb/Shrub; Perennial.
   West Asia: Iran(N).
   Description: 1015, 1167.

A. janthinus Boiss. & Hausskn. [1147]
   Not climbing; Herb/Shrub; Perennial.
   West Asia: Iran(N).

A. keyserlingii Bunge [1015]
   Not climbing; Shrub; Perennial.
   West Asia: Iran(N).
   Description: 1015.
   Type from Iran [1149].

A. knappii Bornm. [1158]
   Not climbing; Herb/Shrub; Perennial.
   West Asia: Iran(N).
   Description: 1015.
   Type from Iran [1158].

A. litostachys Boiss. & Hausskn. [1147]
   A. litostachyus Boiss. & Hausskn. [1015].
   Not climbing; Shrub; Perennial.
   West Asia: Iran(N).
   Description: 1015.
   Type from Iran.

A. longimucronulatus Sirj. & Rech.f. [1156]
   A. longemucronulatus Sirj. & Rech.f. [1156].
   Not climbing; Herb/Shrub; Perennial.
   West Asia: Iran(N).
   Description: 1156.

A. longistylus Bunge [1015]
   Not climbing; Shrub; Perennial.
   West Asia: Iran(N).
   Description: 1015.

A. luristanicus Freyn [1015]
   Not climbing; Herb/Shrub; Perennial.
   West Asia: Iran(N).
   Description: 1015.
   Type from Iran [1015].

A. magnifolius Parsa (provisional) [1015].
   Not climbing; Herb; Perennial.
   West Asia: Iran.
   Description: 1015.
   There is no evidence that this name was ever published [1152].
   The type locality is not that of any other name published by Parsa [1152].

A. medorum Bornm. [1015]
   Not climbing; Herb; Perennial.
   West Asia: Iran(N).
   Description: 1015, 1158.
   Type from Iran [1158].

A. medorum Bornm. [1015]
Not climbing; Herb; Perennial.
West Asia: Iran(N).
Description: 1015, 1158.
Type from Iran [1158].

A. mishouensis Turrill [1015]
Not climbing; Shrub; Perennial.
West Asia: Iran(N).
Description: 1015, 1165.

A. myrianthus G.Beck [1015]
Not climbing; Herb/Shrub; Perennial.
West Asia: Iran(N).
Description: 1015.

A. paralipomenus Bunge [1015]
Not climbing; Shrub; Perennial.
West Asia: Iran(N).
Description: 1015.
Type from Iran [1149].

A. phyllostachys Boiss. & Hausskn. [1015]
Not climbing; Shrub; Perennial.
West Asia: Iran(N).
Description: 1015.

A. pichleri G.Beck [1015]
Not climbing; Shrub; Perennial.
West Asia: Iran(N).
Description: 1015.

A. piletocladus Freyn & Sint. [1150]
Not climbing; Herb/Shrub; Perennial.
West Asia: Iran(N); U.S.S.R.
Description: 1150; Illustration: 1150.

A. piptocephalus Boiss. & Hausskn. [1009]
Not climbing; Herb/Shrub; Perennial.
West Asia: Iran(N), Iraq(N).
Description: 1009, 1015; Illustration: 1009.
The record from Iraq is more than 100 years old, and the locality doubtful; it could well be in Iran [1009].

A. pseudoparrowianus Sirj. & Rech.f. [1015]
Not climbing; Shrub; Perennial.
West Asia: Iran(N).
Description: 1015, 1168.

A. pushtashanicus C.Towns. [1009]
Not climbing; Herb/Shrub; Perennial.
West Asia: Iraq(N).
Description: 1009, 1132; Illustration: 1009, 1132.

A. radkanensis Bunge [1015]
Not climbing; Herb,Shrub; Perennial.
West Asia: Iran(N).
Description: 1015.
Type from Iran [1149].

A. rahensis Sirj. & Rech.f. [1156]
Not climbing; Herb/Shrub; Perennial.
West Asia: Iran(N).
Description: 1156.
A forma *villosa* and a var. *dehbakriensis* have been distinguished [1156].

A. rhodosemius Boiss. & Hausskn. [1015]
   Astragalus rhodosenus Boiss. & Hausskn. [1169].
   Not climbing; Herb/Shrub; Perennial.
   West Asia: Iran(N).
   Description: 1015.
   Parsa mentions a var. *glabrescens* Boiss [1015].

A. rosellus Sirj. & Rech.f. [1170]
   Not climbing; Herb; Perennial.
   West Asia: Iran(N).
   Description: 1170.

A. sarcocolla Dymock (provisional) [1241]
   Not climbing; Shrub; Perennial.
   West Asia: Iran(N).
   Description: 1241.
   Poorly-known; described from fragments in sacks of gum! [1241].

A. segazicus Parsa [1103]
   Not climbing; Shrub; Perennial.
   West Asia: Iran(N).
   Description: 1103; Illustration: 1103.
   The type (Parsa 20024) is at Kew [1153].

A. sharifii Sirj. & Rech.f. [1156]
   Not climbing; Herb/Shrub; Perennial.
   West Asia: Iran(N).
   Description: 1156.

A. stapfii Sirj. [1015]
   Not climbing; Shrub; Perennial.
   West Asia: Iran(N).
   Description: 1015, 1147; Illustration: 1147.

A. stenolepis Fischer [1015]
   Not climbing; Shrub; Perennial.
   West Asia: Iran(N).
   Description: 1015.
   Type from Iran [1149].

A. steppicola Sirj.,Rech.f. & Aellen [1157]
   Not climbing; Shrub; Perennial.
   West Asia: Iran(N).
   Description: 1157.

A. talagonicus Boiss. [1015]
   Not climbing; Herb; Perennial.
   West Asia: Iran(N).
   Description: 1015.
   Type from Iran [1121].

A. trachyacanthus Fischer [1015]
   Not climbing; Herb/Shrub; Perennial.
   West Asia: Iran(N).
   Description: 1015.
   Type from Iran [1149].

A. turrillianus Parsa [1015]
   Not climbing; Shrub; Perennial.
   West Asia: Iran(N).
   Description: 1015, 1124.

A. verus Olivier (provisional) [1149].
Not climbing; Shrub; Perennial.
West Asia: Iran(N).
Description: 1015.
The type is from Iran, but is poor and immature [1149].

A. wrangelii Sirj. [1015]
Not climbing; Shrub; Perennial.
West Asia: Iran(N).
Description: 1015, 1147; Illustration: 1147.
The type is from the Iran/USSR border. The taxon is not mentioned in the Flora of the U.S.S.R. [1147].
There is also a var. *memorabilis* Sirj. [1147].

ASTRAGALUS L.

The genus *Astragalus* is probably the largest genus of flowering plants; there may well be 3000 species. All are herbs, mostly perennial; none climb. Their centre of diversity lies within our area and there may well be more than 1000 species in Iran alone. A great deal of valuable work has been done on the genus by Podlech and his students, but much remains to be done. The genus is divided into sections; those used here are currently recognized by Podlech, and his latest usage of sectional names has been followed (Podlech 1990, ref.1261). At the time of writing there is no modern key to all the sections occurring in the area; those in Davis (1970 — ref.1102) and Goncharov (1946 — ref.1150) are useful.

ASTRAGALUS Section **Acantherioceras**
A small section, confined to Afghanistan. All are perennial spiny shrubs.

A.acantherioceras Rech.f. & Koie [1052]
Not climbing; Shrub; Perennial.
West Asia: Afghanistan(N).
Description: 1052; Illustration: 1052.

A.lugubris Rech.f. & Koie [1052]
Not climbing; Shrub; Perennial.
West Asia: Afghanistan(N).
Description: 1052; Illustration: 1052.

ASTRAGALUS Section **Acanthophace**
Shrubs with paripinnate leaves and spinescent rachides. Flowers sessile in clusters in the upper leaf axils. Pod unilocular, many-seeded. See Deml (1972 —ref. 1171).

A. chionobiiformis C.Towns. [1009]
A. aff.chionobius sensu Rawi [1009,1160].
Not climbing; Herb; Perennial.
West Asia: Iraq(N).
Description: 1009, 1132; Illustration: 1009, 1132.
May well also occur in Iran [1009].

A. farmanfarmajani Sirj. & Rech.f. [1156]
Not climbing; Herb/Shrub; Perennial.
West Asia: Iran(N).
Description: 1156.
There is also a var. *flavus* Sirj. [1156].

A. helgurdensis C.Towns. [1009]
Not climbing; Herb/Shrub; Perennial.
West Asia: Iraq(N).
Description: 1009, 1132; Illustration: 1009, 1132.
Townsend [1009] suggested that it might well occur in Iran, but there have been no records by recent workers.

A. hemsleyi Aitch. & Baker [1171]
A. canispinus Boiss. [1171].
Not climbing; Shrub; Perennial.
West Asia: Afghanistan(N); South Asia: Pakistan(N).
Description: 1015, 1052, 1171; Distribution Map: 1171; Illustration: 1052, 1171.

A. horridus Boiss. [1171]
A. chionobius Bunge [1171]; *A. chionobius* var. *hirtus* Boiss. [1171].
Not climbing; Herb/Shrub; Perennial.
West Asia: Iran(N).
Description: 1015, 1171; Distribution Map: 1171; Illustration: 1171.

A. khaneradarensis Sirj. & Rech.f. [1172]
West Asia: Iran(N).

A. leiophyllus Freyn & Bornm. [1009]
Astracantha leiophylla (Freyn & Bornm.)Greuter [1173]
Not climbing; Herb; Perennial.
West Asia: Iran(N), Iraq(N); Middle East: Turkey(N).
Description: 1009, 1015.
Two varieties, *leiophyllus* and *nigropedunculatus* C.Towns. [1009].

A. lycioides Boiss. [1171]
A. dendridium Fischer [1171]; *A. leptacanthus* Boiss. & Buhse [1171].
Not climbing; Herb/Shrub; Perennial.
West Asia: Iran(N).
Description: 1015, 1171; Distribution Map: 1171.

A. ochrochlorus Boiss. & Hohen. [1015]
Astracantha ochrochlora (Boiss. & Hohen.)Greuter [1143].
Not climbing; Shrub; Perennial.
West Asia: Iran(N); Middle East; Turkey(N).
Description: 1015, 1102.

A. ovigerus Boiss. [1171]
Not climbing; Herb/Shrub; Perennial.
West Asia: Iran(N).
Description: 1015, 1171; Distribution Map: 1171.

A. schistocalyx Bunge [1171]
A. pseudoangustifolius Sirj. & Rech.f. [1171]; *A. schistocalyx* var. *bizgimontanus* Sirj. & Rech.f. [1171]; *A. ssyrtchensis* Bunge [1015,1171]; *A. syrtschensis* Bunge [1171].
Not climbing; Herb/Shrub; Perennial.
West Asia: Iran(N).
Description: 1171]; Distribution Map: 1171.
Two subspecies, *schistocalyx* and *sclerocladus* R[1171].

subsp. schistocalyx [1171]
Not climbing; Herb/Shrub; Perennial.
West Asia: Iran(N).
Description: 1171]; Distribution Map: 1171.

subsp. sclerocladus (Bunge)Deml [1171]
Not climbing; Herb/Shrub; Perennial.
West Asia: Iran(N).
Description: 1171]; Distribution Map: 1171.

A. spinellus Boiss. & Hausskn. [1171]
   Not climbing; Herb/Shrub; Perennial.
   West Asia: Iran(N).
   Description: 1009,1015,1171]; Distribution Map: 1171.
   Known only from the type, whose exact locality ('Kurdistan') remains uncertain [1171].

A. stenostegius Boiss. & Hausskn. [1171]
   Not climbing; Herb; Perennial.
   West Asia: Iran(N).
   Description: 1009,1015,1171]; Distribution Map: 1171.
   Townsend considered that this taxon might occur in Iraq [1009], but Deml records it only from Iran [1171].

ASTRAGALUS Section **Acidodes**
Spiny shrubs. Sometimes regarded as part of Section *Acanthophace* but distinct.

A. calcareus Sirj. & Rech.f. [1157]
   Not climbing; Herb/Shrub; Perennial.
   West Asia: Iran(N).
   Description: 1157.
   Originally placed in Section *Acanthophace*; transferred to *Acidodes* by Deml [1171].

A. carduchorum Boiss. [1009]
   A. carduchorum subsp. *mandaliensis* Eig [1009]
   Not climbing; Herb; Perennial.
   West Asia: Iran(N), Iraq(N).
   Description: 1009, 1015.
   Chemical 1009.
   Placed in Section *Acanthophace* by Townsend; tranferred to *Acidodes* by Deml [1171].

A. chartostegius Boiss. & Hausskn. [1015]
   Not climbing; Herb/Shrub; Perennial.
   West Asia: Iran(N).
   Description: 1015.
   Placed in Section *Acanthace* by Parsa, but in *Acidodes* by Deml [1171].

A. icmadophilus Hand.-Mazz. [1009]
   Not climbing; Herb; Perennial.
   West Asia: Iraq(N); Middle East: Turkey(N).
   Description: 1009.
   Placed in Section *Acanthophace* by Townsend [1009]; Deml transfers to *Acidodes* [1171].

A. jodotropis Boiss. [1121]
   A. iodotropis Boiss. [1171]; *A. jodotrophis* Boiss. [1015].
   Not climbing; Shrub; Perennial.
   West Asia: Iran(N).
   Description: 1015.
   Type from Iran [1121].
   Parsa placed this in Section *Acanthace*; Deml puts it in *Acidodes* [1171].

A. macrosemius Boiss. [1015]
   Not climbing; Herb; Perennial.
   West Asia: Iran(N).
   Description: 1015.

A. paraplesius Bunge [1015]
   Not climbing; Shrub; Perennial.
   West Asia: Iran(N).
   Description: 1015.
   Type from Iran.
   Parsa placed this in Section *Acanthace*; Deml moves it to *Acidodes* [1171].

A. sahendi Buhse [1015]
> *A. ssahandi* Fischer [1015]; *A. ssahendi* Fischer [1015].
> Not climbing; Shrub; Perennial.
> West Asia: Iran(N).
> Description: 1015, 1150.

ASTRAGALUS Section **Acmothrix**
Perennial herbs from a woody base, not spiny. Leaves imparipinnate. Hairs bifurcate. Legume exceeding the calyx, several-seeded.

A. fragrans Willd. [1102]
> *A. xanthinus* Freyn & Bornm. [1102].
> Not climbing; Herb; Perennial.
> West Asia: Iran(N); Middle East: Turkey(N); U.S.S.R.
> Description: 1015,1102,1150]; Illustration: 1150.

A. levieri Freyn [1150]
> *A. supinus* C.A.Meyer [1150].
> Not climbing; Herb; Perennial.
> West Asia: Iran(N); U.S.S.R.
> Description: 1150.
> Originally in Section *Onobrychium*; Flora of U.S.S.R. transfers it to *Acmothrix* [1150].

ASTRAGALUS Section **Aegacantha**
Spiny shrubs, rather similar to *Acanthophace* but differing in the few flowered inflorescences and the few-seeded pods. Mainly in Afghanistan. See Deml (1972, ref.1171).

A. ajfreidii Aitch. & Baker [1171]
> Not climbing; Herb/Shrub; Perennial.
> West Asia: Afghanistan(N).
> Description: 1171; Distribution Map: 1171; Illustration: 1052, 1171.
> Two subspecies, ajfreidii & brevivexillatus [1171].

subsp. **ajfreidii** [1171]
> Not climbing; Herb/Shrub; Perennial.
> West Asia: Afghanistan(N).
> Description: 1171; Distribution Map: 1171.

subsp. **brevivexillatus** Deml [1171]
> Not climbing; Herb/Shrub; Perennial.
> West Asia: Afghanistan(N).
> Description: 1171; Distribution Map: 1171.

A. aloisii Deml [1171]
> Not climbing; Shrub; Perennial.
> West Asia: Afghanistan(N).
> Description: 1171; Distribution Map: 1171; Illustration: 1171.

A. altimurensis Deml [1171]
> Not climbing; Shrub; Perennial.
> West Asia: Afghanistan(N).
> Description: 1171; Distribution Map: 1171; Illustration: 1171.

A. antheliophorus Deml [1171]
> Not climbing; Herb/Shrub; Perennial.
> West Asia: Afghanistan(N).
> Description: 1171; Distribution Map: 1171; Illustration: 1171.

A. aphananthos Kress-Deml [1259]
> Not climbing; Shrub; Perennial.
> West Asia: Afghanistan(N)
> Description: 1259; Illustration: 1259.

Galegeae: Astragalus § Aegacantha

A. baghlanensis Deml [1171]
 Not climbing; Herb/Shrub; Perennial.
 West Asia: Afghanistan(N).
 Description: 1171; Distribution Map: 1171; Illustration: 1171.

A. caroli-henrici Deml [1171]
 Not climbing; Herb/Shrub; Perennial.
 West Asia: Afghanistan(N).
 Description: 1171; Distribution Map: 1171; Illustration: 1171.

A. confertissimus Kitam. [1171]
 A. psilacanthus sensu auctt. [1171]
 Not climbing; Herb/Shrub; Perennial.
 West Asia: Afghanistan(N); South Asia: Pakistan(N).
 Description: 1011, 1171; Illustration: 1011, 1052, 1171.

A. decemjugus Bunge (provisional) [1171]
 West Asia: Afghanistan(N).
 The type (Griffith journ.1050, distrib.1538,p.p.) could not be found by Deml [1171].

A. discernendus Sirj. & Rech.f. [1171]
 Not climbing; Herb/Shrub; Perennial.
 West Asia: Afghanistan(N).
 Description: 1052, 1171; Distribution Map: 1171; Illustration: 1052.

A. endytanthus Podl. & Deml [1171]
 Not climbing; Shrub; Perennial.
 West Asia: Afghanistan(N).
 Description: 1085, 1171; Distribution Map: 1171; Illustration: 1085.

A. franziskae Deml [1171]
 Not climbing; Herb/Shrub; Perennial.
 West Asia: Afghanistan(N).
 Description: 1171; Distribution Map: 1171; Illustration: 1171.

A. freitagii Deml [1171]
 Not climbing; Herb/Shrub; Perennial.
 West Asia: Afghanistan(N).
 Description: 1171; Distribution Map: 1171; Illustration: 1171.

A. gregarius Deml [1171]
 Not climbing; Herb/Shrub; Perennial.
 West Asia: Afghanistan(N).
 Description: 1171; Distribution Map: 1171; Illustration: 1171.

A. hadroacanthus Rech.f. & Gilli [1171]
 Not climbing; Herb/Shrub; Perennial.
 West Asia: Afghanistan(N).
 Description: 1052, 1171; Distribution Map: 1171; Illustration: 1052.

A. hololeios Bornm. [1011,1173]
 Not climbing; Shrub; Perennial.
 West Asia: Afghanistan(N).
 Description: 1218; Illustration: 1052.
 Section *Aegacantha*, not *Hololeios*, according to Podlech [1173].

A. infestus Boiss. [1171]
 Not climbing; Herb/Shrub; Perennial.
 West Asia: Afghanistan(N).
 Description: 1052, 1171; Distribution Map: 1171; Illustration: 1052.

A. interiectus Kress-Deml [1259]
 Not climbing; Shrub; Perennial.
 West Asia: Afghanistan(N).
 Description: 1259; Illustration: 1259.

A. lasiosemius Boiss. [1171]
A. genistoides Boiss. [1171]; *A. lasiosericeus* Hook. et al. [1171]; *A. latistylus* Freyn [1171]; *A. leptus* var. *bamianicus* Sirj. & Rech.f. [1171]; *A. psilopterus* Bunge [1171].
Not climbing; Herb/Shrub; Perennial.
West Asia: Afghanistan(N); South Asia: Pakistan(N); U.S.S.R.
Description: 1052, 1171; Distribution Map: 1171; Illustration: 1052, 1145.
Deml discusses the variation of this taxon. He did not see the type of Kitamura's var. *gracilior* & does not deal with it [1171].

A. leiosemius (Lipsky)Popov [1171]
A. flexirachis Rech.f. & Edelb. [1171]; *A. lasiosemius* var. *leiosemius* Lipsky [1171].
Not climbing; Herb/Shrub; Perennial.
West Asia: Afghanistan(N); U.S.S.R.
Description: 1052, 1150, 1171; Distribution Map: 1171; Illustration: 1171.

A. leptus Boiss. [1171]
Not climbing; Herb/Shrub; Perennial.
West Asia: Afghanistan(N); South Asia: Pakistan(N).
Description: 1171; Distribution Map: 1171; Illustration: 1052, 1171.

subsp. ghazniensis Deml [1171]
Not climbing; Herb/Shrub; Perennial.
West Asia: Afghanistan(N).
Description: 1171; Distribution Map: 1171.

subsp. leptus [1171]
A. spinellifer Rech.f. & Gilli [1171].
Not climbing; Herb/Shrub; Perennial.
West Asia: Afghanistan(N); South Asia: Pakistan(N).
Description: 1171; Distribution Map: 1171; Illustration: 1052.

subsp. longipedunculatus Deml [1171]
Not climbing; Herb/Shrub; Perennial.
West Asia: Afghanistan(N).
Description: 1171; Distribution Map: 1171.

A. melanochiton Deml [1171]
Not climbing; Shrub; Perennial.
West Asia: Afghanistan(N).
Description: 1171; Distribution Map: 1171; Illustration: 1171.

A. molestus Rech.f. [1171]
A. infestus var. *glabrostipulatus* Sirj. & Rech.f. [1171].
Not climbing; Herb/Shrub; Perennial.
West Asia: Afghanistan(N).
Description: 1052, 1171; Distribution Map: 1171; Illustration: 1052.

A. nigrivestitus Podl. & Deml [1171]
Not climbing; Herb/Shrub; Perennial.
West Asia: Afghanistan(N).
Description: 1085, 1171; Distribution Map: 1171; Illustration: 1085, 1171.

A. parwanicus Podl. & Deml [1171]
Not climbing; Shrub; Perennial.
West Asia: Afghanistan(N).
Description: 1085, 1171; Distribution Map: 1171; Illustration: 1085, 1171.

A. pecten-erinis Kress-Deml [1259]
Not climbing; Shrub; Perennial.
West Asia: Afghanistan(N).
Description: 1259; Illustration: 1259.

A. pecten-hystricis Kress-Deml [1259]
Not climbing; Shrub; Perennial.
West Asia: Afghanistan(N).
Description: 1259; Illustration: 1259.

A. phalacrophyton Deml [1171]
Not climbing; Shrub; Perennial.
West Asia: Afghanistan(N).
Description: 1171; Distribution Map: 1171; Illustration: 1171.

A. podlechii Deml [1171]
Not climbing; Shrub; Perennial.
West Asia: Afghanistan(N).
Description: 1171; Distribution Map: 1171; Illustration: 1171.

A. pseudopsilacanthus Ali [1171]
Not climbing; Shrub; Perennial.
West Asia: Afghanistan(N); South Asia: Pakistan(N).
Description: 1171; Distribution Map: 1171; Illustration: 1171.

subsp. **polyneurus** Deml [1171]
Not climbing; Shrub; Perennial.
West Asia: Afghanistan(N).
Description: 1171; Illustration: 1171.

subsp. **pseudopsilacanthus** [1171]
Not climbing; Shrub; Perennial.
West Asia: Afghanistan(N); South Asia: Pakistan(N).
Description: 1171; Illustration: 1171.

A. psilacanthus Boiss. [1171]
Not climbing; Herb/Shrub; Perennial.
West Asia: Afghanistan(N); South Asia: Pakistan(N).
Description: 1052, 1171; Distribution Map: 1171; Illustration: 1052, 1171.
Only the typical subspecies occurs in West Asia [1171].

subsp. **psilacanthus** [1171]
A. chaworth-musteri Sirj. & Rech.f. [1171]; *A. grahamianus* sensu auctt. [1171]; *A. leiosemius* sensu Rech.f. [1052,1171]; *A. myriocladus* Sirj. & Rech.f. [1171]; *A. paghmanensis* Sirj. & Rech.f. [1171]; *A. polyacanthus* sensu Baker [1122,1171].
Not climbing; Herb/Shrub; Perennial.
West Asia: Afghanistan(N); South Asia: Pakistan(N).
Description: 1171; Distribution Map: 1171; Illustration: 1052, 1171.

A. psilocentros Fischer [1171]
A. polemius Boiss. [1171]
Not climbing; Shrub; Perennial.
West Asia: Afghanistan(N); South Asia: India(N), Pakistan(N).
Description: 1171; Distribution Map: 1171; Illustration: 1052, 1171.
Two varieties, *psilocentros* and *pilosus* [1171].

A. raphiodontus Boiss. [1171]
Not climbing; Shrub; Perennial.
West Asia: Afghanistan(N).
Description: 1171; Distribution Map: 1171; Illustration: 1052.

A. sharestanicus Podl. & Deml [1171]
Not climbing; Shrub; Perennial.
West Asia: Afghanistan(N).
Description: 1085, 1171; Distribution Map: 1171; Illustration: 1085, 1171.
Known only from the type [1171].

A. stenopterus Sirj. & Rech.f. [1171]
    Not climbing; Herb/Shrub; Perennial.
    West Asia: Afghanistan(N).
    Description: 1171; Distribution Map: 1171; Illustration: 1052, 1171.
    Known only from the type [1171].

A. strigosostipulatus Rech.f. & Koie emend. Deml [1171]
    Not climbing; Herb/Shrub; Perennial.
    West Asia: Afghanistan(N).
    Description: 1171; Distribution Map: 1171; Illustration: 1171.
    Deml excludes Koie & Rechinger's (1957) Figs.125 & 126 [1171].

A. terrestris Kitam. [1143]
    A. brecklei Deml
    Not climbing; Herb/Shrub; Perennial.
    West Asia: Afghanistan(N).
    Description: 1171; Illustration: 1011.

A. wendelboi Deml [1171]
    Not climbing; Shrub; Perennial.
    West Asia: Afghanistan(N).
    Description: 1171; Distribution Map: 1171; Illustration: 1171.
    Known only from the type [1171].

A. zanskarensis Bunge [1171]
    Not climbing; Shrub; Perennial.
    West Asia: Afghanistan(N), South Asia India(N), Pakistan(N).
    Description: 1171; Distribution Map: 1171; Illustration: 1171.
    Only the typical subspecies occurs in West Asia.

   subsp. **zanskarensis** [1171]
        A. cicerifolius Bunge [1171]; *A. oplites* Parker [1171].
        Not climbing; Shrub; Perennial.
        West Asia: Afghanistan(N); South Asia: India(N).
        Description: 1171; Distribution Map: 1171; Illustration: 1171.

ASTRAGALUS Section **Alopecuroidei**
    Robust herbaceous perennials, not spinous. Stems often long and rambling. Inflorescences many-flowered, axillary. Pods small, bilocular. See Becht (1978—ref.1174).

A. ajubensis Bunge [1174]
    A. durhamii Turrill [1174].
    Not climbing; Herb; Perennial.
    West Asia: Iran(N); Middle East: Turkey(N).
    Description: 1015, 1174; Distribution Map: 1174; Illustration: 1174.

A. alopecias Pallas [1174]
    A. leucospermus Bunge [1174].
    Not climbing; Herb; Perennial.
    West Asia: Afghanistan(N), Iran(N); East Asia: China(N); U.S.S.R.
    Description: 1015, 1150, 1174; Distribution Map: 1174; Illustration: 1145, 1150, 1174.

A. anacardius Bunge [1174]
    A. buschirensis Sirj. & Rech.f. [1174]
    Not climbing; Herb; Perennial.
    West Asia: Iran(N).
    Description: 1015, 1174; Illustration: 1174.
    Townsend [1009] regarded this taxon as synonymous with *A.obtusifolius*, but Becht [1174] keeps the two separate.

A. arbelicus Bornm. (provisional) [1174]
Not climbing; Herb; Perennial.
West Asia: Iran(N), Iraq(N).
Description: 1015, 1174.
Townsend [1009] regarded this as probably synonymous with *A. macrocephalus* ssp. *finitimus*, but Becht [1174] keeps it provisionally distinct. The locality ('Kurdistan') is very imprecise and cannot definitely be placed in a country.

A. decurrens Boiss. [1174]
A. dichroanthus Freyn & Sint. [1174]; *A. pectinatus* Boiss. [1174].
Not climbing; Herb; Perennial.
West Asia: Iraq(N); Middle East: Turkey(N).
Description: 1009, 1174; Distribution Map: 1174; Illustration: 1009, 1174.
The record in Flora of Iraq [1009] is 125 years old. Becht records the taxon only from Turkey [1174].

A. echinops Boiss. [1174]
A. regelii Trautv. [1174]; *A. superbus* Bunge [1174]; *A. superbus* var. *ovalifoliolatus* Sirj. & Rech.f. [1174].
Not climbing; Herb; Perennial.
West Asia: Iran(N), Iraq(N); Middle East: Israel(N), Jordan(N), Lebanon(N), Turkey(N); U.S.S.R.
Description: 1009, 1015, 1174; Distribution Map: 1174; Illustration: 1174.

A. eximius Bunge [1174]
A. ourmitanensis Franchet [1174]
Not climbing; Herb; Perennial.
West Asia: Iran(N); U.S.S.R.
Description: 1015, 1150, 1174]; Distribution Map: 1174; Illustration: 1145, 1174.
Parsa lists this taxon but without a definite Iran record [1015]; Becht does not record from Iran, nor from nearby [1174].

A. ferdovsicus Parsa [1103]
Not climbing; Herb; Perennial.
West Asia: Iran(N).
Description: 1103; Illustration: 1103.
Parsa originally placed this in Section *Erionotus*. The type (Parsa 20022) at Kew [1103]. Podlech tranfers it to Section *Alopecuroidei* [1175].

A. globiceps Bunge [1174]
Not climbing; Herb; Perennial.
West Asia: Afghanistan(N). U.S.S.R.
Description: 1150, 1174; Distribution Map: 1174; Illustration: 1145, 1150, 1174.
Only the nominate subspecies occurs in West Asia [1174].

subsp. **globiceps** Bunge [1174]
A. czuiliensis Golosk. [1174]; *A. flavicomus* Bunge [1174]; *A. jakkabagi* Lipsky [1174]; *A. timuranus* Franchet [1174].
Not climbing; Herb; Perennial.
West Asia: Afghanistan(N); U.S.S.R.
Description: 1174; Distribution Map: 1174; Illustration: 1174.

A. hamadanus Boiss. [1174]
A. sessiliceps Bornm. [1174]
Not climbing; Herb; Perennial.
West Asia: Iran(N).
Description: 1174; Distribution Map: 1174; Illustration: 1174.

A. hymenocalyx Boiss. [1174]
Not climbing; Herb; Perennial.
West Asia: Iran(N).
Description: 1015, 1174; Distribution Map: 1174; Illustration: 1174.

A. jessenii Bunge [1174]
A. ashuricus Parsa [1174]; *A. djadjerudensis* Sirj. & Rech.f. [1174].
Not climbing; Herb; Perennial.
West Asia: Iran(N).
Description: 1015, 1174; Distribution Map: 1174; Illustration: 1174.

A. kirrindicus Boiss. [1174]
A. meridionalis sensu auctt. [1009]; *A. rudehendicus* Sirj., Rech.f. & Aellen [1174].
Not climbing; Herb; Perennial.
West Asia: Iran(N), Iraq(N).
Description: 1009, 1015, 1174; Distribution Map: 1174; Illustration: 1009, 1174.

A. kulabensis Lipsky [1174]
A. faizabadensis Sirj. & Rech.f. [1174].
Not climbing; Herb; Perennial.
West Asia: Afghanistan(N); U.S.S.R.
Description: 1052, 1150, 1174; Distribution Map: 1174; Illustration: 1052, 1145, 1150, 1174.

A. macrocephalus Willd. [1174]
Not climbing; Herb; Perennial.
West Asia: Iran(N), Iraq(N); Middle East: Syria(N), Turkey(N); U.S.S.R.
Description: 1009, 1150, 1174; Distribution Map: 1174; Illustration: 1174.

subsp. **cucullaris** (Boiss.)Chamberlain [1174]
A. cucullaris Boiss. [1174]
Not climbing; Herb; Perennial.
West Asia: Iran(N); Middle East: Turkey(N).
Description: 1015, 1174; Distribution Map[1174.

subsp. **finitimus** (Bunge)Chamberlain [1174]
A. finitimus Bunge [1174]; *A. oloricus* Manden. [1174]; *A. sphaerocephalus* Steven [1174].
Not climbing; Herb; Perennial.
West Asia: Iran(N), Iraq(N); Middle East: Turkey(N); U.S.S.R.
Description: 1009, 1015, 1174; Distribution Map: 1174; Illustration: 1009, 1176.

subsp. **macrocephalus** [1174]
A. megalacmus Freyn & Sint. [1174].
Not climbing; Herb; Perennial.
Middle East:Turkey(N).
Description: 1009, 1174; Distribution Map: 1174; Illustration: 1174.
An old record from 'Mesopotamia' is not in Iraq [1009].

A. maximus Willd. [1174]
Not climbing; Herb; Perennial.
West Asia: Iran(N); Middle East: Turkey(N); U.S.S.R.
Description: 1102, 1150, 1174; Distribution Map: 1174; Illustration: 1150, 1174.
Two varieties, *maximus* & *dasysemius* [1102]; the Iranian record is of var. *maximus* [1176].

A. megalotropis Bunge [1174]
A. julii-gautieri Sirj. & Rech.f. [1174]; *A. melaleucus* Bunge [1174]; *A. nadjafabadensis* Parsa [1174]; *A. nedjefabadensis* Parsa [1015].
Not climbing; Herb; Perennial.
West Asia: Iran(N); U.S.S.R.
Description: 1150, 1174; Distribution Map: 1174; Illustration: 1174.

A. obtusifolius DC. [1174]
A. meridionalis Bunge [1174]; *A. obcordatus* Boiss. [1174].
Not climbing; Herb; Perennial.
West Asia: Iran(N), Iraq(N).
Description: 1009, 1174; Distribution Map: 1174; Illustration: 1009, 1174.
Townsend [1009] placed *A. anacardius* Bunge (q.v.) as a synonym of this, but Becht maintains the two taxa at specific level [1174].

A. oocephalus Boiss. [1174]
A. oocephalus subsp. *stachyophorus* Huber-Mor. & Chamberlain [1174]; *A. oocephalus* var. *stachyophorus* (Huber-Mor. & Chamberlain)Ponert [1174]; *A. stojani* Nab. [1174].
Not climbing; Herb; Perennial.
West Asia: Iran(N); Middle East: Israel(N), Jordan(N), Syria(N), Turkey(N).
Description: 1009, 1174; Distribution Map: 1174; Illustration: 1174.
Parsa records *A.stojani* from Iran, but there is no other evidence for its occurrence there [1015].

A. ponticus Pallas [1174]
A. chartaceus Ledeb. [1174]; *A. chlorotaenius* Freyn & Bornm. [1174].
Not climbing; Herb; Perennial.
West Asia: Iran(N); Europe; Middle East: Turkey(N); U.S.S.R.
Description: 1124, 1150, 1174; Distribution Map: 1174; Illustration: 1174.

A. saetiger R.Becht [1174]
Not climbing; Herb; Perennial.
West Asia: Iran(N).
Description: 1174; Distribution Map: 1174; Illustration: 1174.

A. schahrudensis Bunge [1174]
A. schahrudicus Bunge [1177]; *A. scharudensis* Bunge [1015].
Not climbing; Herb; Perennial.
West Asia: Iran(N); U.S.S.R.
Description: 1015, 1150, 1174; Distribution Map: 1174; Illustration: 1174.

A. sericostachys Stocks [1174]
Not climbing; Herb; Perennial.
West Asia: Iran(N); South Asia: Pakistan(N).
Description: 1015.
Ali records this taxon from Iran; Becht does not [1163]. Ali's locality, 'Mianeh' is in Iranian Azerbaijan, which seems unlikely [1163].

A. speciosus Boiss. & Hohen. [1174]
Not climbing; Herb; Perennial.
West Asia: Iran(N).
Description: 1015, 1174; Distribution Map: 1174; Illustration: 1174.

A. stepporum Podl. [1178]
Not climbing; Herb; Perennial.
West Asia: Iran(N).
Description: 1178.

A. turbinatus Bunge [1174]
A. christophi Trautv. [1174]; *A. grisebachianus* Aitch. & Baker [1174].
Not climbing; Herb; Perennial.
West Asia: Afghanistan(N), Iran(N); U.S.S.R.
Description: 1015, 1150, 1174; Distribution Map: 1174; Illustration: 1174.

ASTRAGALUS Section **Ammodendron**
Semi-woody perennials, often becoming more or less leafless with age. Inflorescence a pedunculate spike. Pod bilocular, slightly longer than the calyx. Plants of sandy areas, mainly in central Asia and Iran. Much in need of revision.

A. acutifolius Bunge [1179]
Not climbing; Herb/Shrub; Perennial.
West Asia: Iran(N).
Description: 1015.
Parsa mentions a var.elegans Bornm. & Gauba [1015].
The name is probably illegitimate as there is also an *A. acutifolius* DC. ex Steud. (= *A.stella*) [1102].
The type is from Iran [1149].

A. ahmed-adlii Bornm. & Gauba [1179]
A. ahmad-adlii Bornm. & Gauba [1015]
Not climbing; Herb/Shrub; Perennial.
West Asia: Iran(N).
Description: 1015, 1180; Illustration: 1180.
The type is from Iran [1180].

A. aiwadzhii B.Fedtsch. [1181]
Not climbing; Herb/Shrub; Perennial.
West Asia: Afghanistan(N); U.S.S.R.
Description: 1150; Illustration: 1145.

A. ammodendroides Bornm. [1179]
Not climbing; Shrub; Perennial.
West Asia: Iran(N).
Description: 1180.

A. anachoreticus Podl. [1181]
Not climbing; Herb/Shrub; Perennial.
West Asia: Afghanistan(N).
Description: 1181; Illustration: 1181.

A. brevipetiolatus Sirj. & Rech.f. [1179]
Not climbing; Herb; Perennial.
West Asia: Iran(N).
Description: 1157.

A. camelorum Barbey [1182]
A. macrobotrys var. *camelorum* Velen. [1182]
Not climbing; Shrub; Perennial.
West Asia: Saudi Arabia(N); Middle East: Sinai(N).
Description: 1183; Illustration: 1183.
Barbey's locality appears to be in Saudi Arabia, but this may be incorrect; Danin [1183] regards this taxon as endemic to Sinai.

A. collenettiae Hedge & Podl. [1155]
Not climbing; Shrub; Perennial.
West Asia: Saudi Arabia(N).
Description: 1155.

A. erwinii-gaubae Sirj. & Rech.f. [1184]
A. turcomanicus var. *elegans* Bornm. & Gauba [1184].
Not climbing; Shrub; Perennial.
West Asia: Iran(N).
Description: 1184.

A. gemellus Podl. [1181]
Not climbing; Herb/Shrub; Perennial.
West Asia: Afghanistan(N).
Description: 1181; Illustration: 1181.

A. ghoratensis Podl. [1185]
Not climbing; Herb; Perennial.
West Asia: Afghanistan(N).
Description: 1185; Illustration: 1185.

A. glabellus Podl. [1181]
Not climbing; Shrub; Perennial.
West Asia: Afghanistan(N).
Description: 1181; Illustration: 1181.

A. goreanus Aitch. & Baker [1011]
Not climbing; Shrub; Perennial.
West Asia: Afghanistan(N).
Description: 1159.
Not mentioned by Rechinger et al.(1961) [1179].

A. griffithii Bunge [1011]
Not climbing; Herb/Shrub; Perennial.
West Asia: Afghanistan(N).
Description: 1149.

A. hyrcanus Pallas [1179]
A. karakugensis Grossh. [1179]
Not climbing; Herb/Shrub; Perennial.
West Asia: Afghanistan(N); South Asia: Pakistan(N); U.S.S.R.
Description: 1150.

A. kavirensis Freitag [1186]
Not climbing; Shrub; Perennial.
West Asia: Iran(N).
Description: 1186; Illustration: 1186.

A. macrobotrys Bunge [1179]
Not climbing; Herb/Shrub; Perennial.
West Asia: Iran(N).
Description: 1015, 1150.

A. nigricans Barneby [1181]
A. nigrescens Popov [1181].
Not climbing; Herb/Shrub; Perennial.
West Asia: Afghanistan(N).
Description: 1150.

A. oligophyllus Boiss. [1179]
West Asia: Afghanistan(N), Iran(N).
Rechinger et al. mention a var. *nigrocalycinus* Sirj. & Rech.f. [1179].

A. podolobus Boiss. [1179]
Not climbing; Herb/Shrub; Perennial.
West Asia: Iran(N).
Description: 1015, 1150.
Rechinger et al. describe a forma *albovillosus* [1179].

A. quinquejugus Sirj. & Rech.f. [1179]
Not climbing; Herb; Perennial.
West Asia: Afghanistan(N).
Description: 1052; Illustration: 1052.

A. squarrosus Bunge [1187]
A. biabanensis Sirj. & Rech.f. [1187]; *A. farsicus* Sirj. & Rech.f. [1187]; *A. pseudosquarrosus* Sirj. & Rech.f. [1163]; *A. sangonensis* Sirj. & Rech.f. [1163].
Not climbing; Herb/Shrub; Perennial.
West Asia: Afghanistan(N), Iran(N), United Arab Emirates(N); South Asia: Pakistan(N); U.S.S.R.
Description: 1187; Illustration: 1187.
Leonard distinguishes 2 varieties, *squarrosus* & *sangonensis* [1187].

A. talbotianus Aitch. & Baker [1179]
Not climbing; Shrub; Perennial.
West Asia: Afghanistan(N).
Description: 1162.
Forage 1162.
Kitamura mentions a var. *condensatus* Aitch. & Baker [1011].

A. tarumensis Sirj. & Rech.f. [1179]
>Not climbing; Shrub; Perennial.
>West Asia: Iran(N).
>Description: 1184.

A. urgunensis Podl. [1185]
>Not climbing; Herb; Perennial.
>West Asia: Afghanistan(N).
>Description: 1185; Illustration: 1185.

ASTRAGALUS Section Ankylotus
Annuals with simple hairs. Leaves imparipinnate. Pods semibilocular, many-seeded.

A. affghanus Boiss.
>*A. affganus* Boiss. [1015]; *A. afganicus* Bornm. [1015]; *A. afghanicus* Bornm. [1188]; *A. brahuicus* Boiss. [1188].
>Not climbing; Herb; Annual.
>West Asia: Iran(N).
>Description: 1015.
>*A. brahuicus* Boiss. is a later homonym of *A. brahuicus* Bunge [1152]; Boissier noticed this and published a new name, *A.affghanus*, in an addendum [1152]. Bornmuller overlooked the addendum and published *A. afghanicus* [1152]. Parsa retained *A. brahuicus* (wrongly), and mis-spelled both replacement names.

A. ankylotus Fischer & Meyer [1189]
>Not climbing; Herb; Annual.
>West Asia: Iran(N).
>Description: 1015, 1150.

A. commixtus Bunge [1189]
>*A. intermedius* Boiss. [1189]; *A. karelinii* Bunge [1189].
>Not climbing; Herb; Annual.
>West Asia: Afghanistan(N), Iran(N): Middle East: Turkey(N); U.S.S.R.
>Description: 1015, 1102, 1150; Illustration: 1145, 1150.

A. gracilipes Bunge [1190]
>*A. ninae* Gontsch. [1190].
>Not climbing; Herb; Annual.
>West Asia: Afghanistan(N); U.S.S.R.
>Description: 1150.

A. hamrinensis Bornm. [1009]
>*A. duplostrigosus* Dinsm. [1009]
>Not climbing; Herb; Annual.
>West Asia: Iraq(N); Middle East: Syria(N).
>Description: 1009; Illustration: 1009.

A. maharluensis Bornm. & Gauba [1188]
>Not climbing; Herb; Annual.
>West Asia: Iran(N).
>Description: 1015, 1188.

A. stalinskyi Sirj. [1085]
>Not climbing; Herb; Annual.
>West Asia: Afghanistan(N); U.S.S.R.
>Description: 1194.

ASTRAGALUS Section **Annulares**
Formerly Section *Haematodes*.

A. annularis Forsskal [1009]
A. maculatus Lam. [1009].
Not climbing; Herb; Annual.
West Asia: Bahrain(N), Iran(N), Iraq(N), Kuwait(N), Qatar(N), Saudi Arabia(N), United Arab Emirates(N); Middle East: Israel(N), Sinai(N).
Description: 1009, 1015; Illustration: 1001.
Forage 1009.

ASTRAGALUS Section **Astragalus**
Erect perennial herbs. Hairs simple, basifixed. Leaves imparipinnate. Inflorescences axillary, short. Pod ovoid to subglobose, tardily dehiscent. Formerly referred to as section *Christiana*. See Agerer-Kirchhoff (1976, ref.1191).

A. acanthochristianopsis Rech.f. & Koie [1191]
Not climbing; Herb; Perennial.
West Asia: Afghanistan(N).
Description: 1052, 1191; Distribution Map: 1191; Illustration: 1052.

A. aleppicus Boiss. [1191]
Not climbing; Herb; Perennial.
West Asia: Iraq(N); Middle East: Israel(N), Jordan(N), Lebanon(N), Syria(N), Turkey(N).
Description: 1102, 1191; Distribution Map: 1191; Illustration: 1191.
Only one subspecies of two occurs in West Asia [1191].

subsp. megaloceras (Eig)C.Agerer-Kirchhoff [1191]
A. megaloceras Eig [1191].
Not climbing; Herb; Perennial.
West Asia: Iraq(N); Middle East: Israel(N), Jordan(N), Syria(N), Turkey(N).
Description: 1191; Illustration: 1192, 1193.

A. anthosphaerus Rech.f. & Gilli [1191]
A. pseudosulfuratus Podl. [1191]; *A. yawnuensis* Podl. [1191].
Not climbing; Herb; Perennial.
West Asia: Afghanistan(N).
Description: 1052, 1098, 1191; Distribution Map: 1191; Illustration: 1052, 1098.

A. baissunensis Lipsky [1191]
Not climbing; Herb; Perennial.
West Asia: Afghanistan(N); U.S.S.R.
Description: 1150, 1191; Distribution Map: 1191; Illustration: 1191.

A. basineri Trautv. [1191]
A. ekatherinae B.Fedtsch. [1191]; *A. michelsoni* B.Fedtsch. [1191]; *A. stephanianus* Aitch. & Baker [1011]; *A. stephenianus* Aitch. & Baker [1191].
Not climbing; Herb; Perennial.
West Asia: Afghanistan(N), Iran(N); U.S.S.R.
Description: 1015, 1150, 1191; Distribution Map: 1191; Illustration: 1191.

A. bezudensis Sirj. & Rech.f. [1191]
A. albovillosus Kitam. [1191]; *A. neoalbovillosus* Kitam. [1191]; *A. zarghumensis* Rech.f. [1191].
Not climbing; Herb; Perennial.
West Asia: Afghanistan(N); South Asia: Pakistan(N).
Description: 1011, 1052, 1191; Distribution Map: 1191; Illustration: 1011, 1052.

A. borraginaceus Rech.f. [1191]
A. albosetulosus Sirj. & Rech.f. [1191].
Not climbing; Herb; Perennial.
West Asia: Afghanistan(N).
Description: 1052, 1191; Distribution Map: 1191; Illustration: 1052.

A. caraganae Hohen. [1191]
A. bardsiricus Parsa [1191]; *A. caraganae* var. *brevicalyx* Eig [1191]; *A. nachitschevanicus* Rzazade [1191]; *A. warackensis* Eig [1191].
Not climbing; Herb; Perennial.
West Asia: Iran(N); Middle East: Turkey(N); U.S.S.R.
Description: 1015, 1191; Illustration: 1102.
The type of Parsa's *A. bardsiricus* (Azarhoush 10046) is cited as this by Agerer-Kirchhoff [1191], although she does not list the name as a synonym.

A. caryolobus Bunge [1191]
A. aleppicus sensu Rawi [1009,1160]; *A. assyriacus* Freyn & Bornm. [1191.
Not climbing; Herb; Perennial.
West Asia: Iran(N), Iraq(N).
Description: 1009, 1015, 1191; Distribution Map: 1191.
Food or drink 1009.

A. dietrichii C.Agerer-Kirchhoff [1191]
Not climbing; Herb; Perennial.
West Asia: Afghanistan(N).
Description: 1191; Distribution Map: 1191; Illustration: 1191.

A. eigii C.Agerer-Kirchhoff [1191]
Not climbing; Herb; Perennial.
West Asia: Afghanistan(N); South Asia: Pakistan(N).
Description: 1191; Distribution Map: 1191; Illustration: 1191.

A. elatior Kitam. [1191]
Not climbing; Herb; Perennial.
West Asia: Afghanistan(N).
Description: 1011, 1191; Distribution Map: 1191; Illustration: 1011.

A. harpocarpus Meff. [1191]
A. taluquanensis Podl. [1191].
Not climbing; Herb; Perennial.
West Asia: Afghanistan(N); U.S.S.R.
Description: 1098, 1150, 1191; Distribution Map: 1191; Illustration: 1098, 1145, 1150.

A. hedgei C.Agerer-Kirchhoff [1191]
Not climbing; Herb; Perennial.
West Asia: Afghanistan(N).
Description: 1191; Distribution Map: 1191; Illustration: 1191.

A. orthocarpoides Sirj. & Rech.f. [1191]
Not climbing; Herb; Perennial.
West Asia: Iran(N).
Description: 1191; Distribution Map: 1191; Illustration: 1191.

A. orthocarpus Boiss. [1191]
A. virgaurea Kitam. [1191].
Not climbing; Herb; Perennial.
West Asia: Afghanistan(N); South Asia: Pakistan(N).
Description: 1011, 1191; Distribution Map: 1191; Illustration: 1011.

A. retamocarpus Boiss. [1191]
A. askabadensis Kuntze [1191]; *A. leucomelas* Bunge [1191]; *A. retamocarpus* subsp. *albiflorus* Freyn & Sint. [1191]; *A. spongocarpus* Meff. [1191].
Not climbing; Herb; Perennial.
West Asia: Afghanistan(N), Iran(N); South Asia:Pakistan(N); U.S.S.R.
Description: 1015, 1150, 1191; Distribution Map: 1191; Illustration: 1145, 1150, 1191.

A. rosae C.Agerer-Kirchhoff [1191]
Not climbing; Herb; Perennial.
West Asia: Afghanistan(N).
Description: 1191; Distribution Map: 1191; Illustration: 1191.

A. sabzakensis C.Agerer-Kirchhoff [1191]
    Not climbing; Herb; Perennial.
    West Asia: Afghanistan(N).
    Description: 1191; Distribution Map: 1191; Illustration: 1191.

A. siahderrensis Sirj. & Rech.f. [1191]
    Not climbing; Herb; Perennial.
    West Asia: Afghanistan(N).
    Description: 1052, 1191; Distribution Map: 1191; Illustration: 1052.

A. sieversianus Pallas [1191]
    Not climbing; Herb; Perennial.
    West Asia: Afghanistan(N), Iran(N); U.S.S.R.
    Description: 1015, 1150, 1191; Distribution Map: 1191; Illustration: 1145, 1150, 1191.

A. sulfuratus Rech.f. & Gilli [1191]
    A. tscharikarensis Sirj. & Rech.f. [1191].
    Not climbing; Herb; Perennial.
    West Asia: Afghanistan(N).
    Description: 1052, 1191; Distribution Map: 1191; Illustration: 1052.

A. tephrosioides Boiss. [1191]
    A. tephrosoides Boiss. [1163].
    Not climbing; Herb; Perennial.
    West Asia: Afghanistan(N).
    Description: 1015, 1191; Distribution Map: 1191; Illustration: 1191.

A. turkestanus Bunge [1191]
    A. kullmannii Podl. [1191]
    Not climbing; Herb; Perennial.
    West Asia: Afghanistan(N); U.S.S.R.
    Description: 1098, 1150, 1191; Distribution Map: 1191; Illustration: 1098, 1144, 1145.

ASTRAGALUS Section **Aulacolobus**

Almost glabrous annuals. Leaves imparipinnate. Inflorescences axillary, pedunculate, few-flowered. Pod large, bilocular.

A. guttatus Banks & Sol. [1009]
    A. aulacolobus Boiss. [1009]; *A. conduplicatus* Bertol. [1009]; *A. phaulanthus* Turrill [1009]; *A. pictus* Boiss. [1009]; *A. striatellus* M.Bieb. [1009].
    Not climbing; Herb; Annual.
    West Asia: Iran(N), Iraq(N); Middle East: Israel(N), Lebanon(N), Syria(N), Turkey(N); U.S.S.R.
    Description: 1009, 1015, 1102; Illustration: 1009, 1145.
    Sirjaev regarded all synonyms except *A. pictus* as good species, and does not mention *A. guttatus* [1189].

ASTRAGALUS Section **Brachycarpus**

Low-growing herbaceous perennials. Flowers in a short dense spike at the end of a long peduncle. Pods short, sessile, bilocular or unilocular.

A. bashgalensis Podl. [1185]
    Not climbing; Herb; Perennial.
    West Asia: Afghanistan(N).
    Description: 1185; Illustration: 1185.

A. jagnobicus Lipsky [1011]
    Not climbing; Herb; Perennial.
    West Asia: Afghanistan(N); U.S.S.R.
    Description: 1150; Illustration: 1145, 1150.
    Kitamura lists a var. *shitvensis* Sirj. & Rech.f. (a spelling mistake for *shtivensis*) [1011].

A. melanostachys Bunge [1126]
   Not climbing; Herb; Perennial.
   West Asia: Afghanistan(N); U.S.S.R.
   Description: 1150; Illustration: 1145.

A. miseriflorus Sirj. & Rech.f. [1011]
   Not climbing; Herb; Perennial.
   West Asia: Afghanistan(N).
   Description: 1052; Illustration: 1052.

A. schugnanicus B.Fedtsch. [1126]
   West Asia: Afghanistan(N).

A. subscaposus Boriss. [1126]
   Not climbing; Herb; Perennial.
   West Asia: Afghanistan(N); U.S.S.R.
   Description: 1150; Illustration: 1145.

A. supraglaber Kitam. [1011,1173]
   A. macrostegius Rech.f. [1011].
   Not climbing; Herb; Perennial.
   West Asia: Afghanistan(N); South Asia: Pakistan(N).
   Description: 1011, 1052; Illustration: 1011, 1052.
   Podlech tranfers from Section *Hypoglottoidei* to Section *Brachycarpus* [1173].

ASTRAGALUS Section **Buceras**
Annuals. Leaves imparipinnate. Inflorescence long-pedunculate, few-flowered. Pod linear, curved, sub-bilocular.

A. hamosus L. [1009]
   A. arnoceras Bunge [1173]; *A. brachyceras* Ledeb. [1009]; *A. dorcoceras* Bunge [1009]; *A. hamosus* var. *brachyceras* (Ledeb.)Ledeb. [1009].
   Not climbing; Herb; Annual.
   West Asia: Afghanistan(N), Iran(N), Iraq(N), Kuwait(N), Qatar(N), Saudi Arabia(N), United Arab Emirates(N); Middle East: Israel(N), Jordan(N), Lebanon(N), Sinai(N), Syria(N), Turkey(N); South Asia: Pakistan(N); U.S.S.R.
   Description: 1009, 1102, 1150; Illustration: 1009.
   Forage 1009; Medicine 1009.
   Podlech has examined the type of *A. arnoceras* [1173].

ASTRAGALUS Section **Campylanthus**
Low subshrubs. Leaves imparipinnate with spinous leaf-rhachides. Flowers in dense sessile or shortly pedunculate heads in the leaf axils. Calyx slightly inflated in fruit. Pod small, unilocular. Confined to Iran. See Tietz (1988 — ref. 1195).

A. argyrostachys Boiss. [1195]
   Not climbing; Herb/Shrub; Perennial.
   West Asia: Iran(N).
   Description: 1015, 1195; Distribution Map: 1195; Illustration: 1195.

A. campylanthus Boiss. [1195]
   A. campylanthus var. *ebenidioides* Bornm. [1195]; *A. campylanthus* var. *subglobosus* Bornm. [1195].
   Not climbing; Herb/Shrub; Perennial.
   West Asia: Iran(N).
   Description: 1015, 1195; Distribution Map: 1195; Illustration: 1195.
   Intermediates with *A. susianus* occur; these have been called *A. susianus* subsp. *stapfianus* [1195].

A. chalaranthus Boiss. & Hausskn. [1195]
   Not climbing; Herb/Shrub; Perennial.
   West Asia: Iran(N).
   Description: 1015, 1195; Distribution Map: 1195; Illustration: 1195.
   Some intergradation with *A. susianus* [1195].

A. ecbatanus Bunge [1195]
   A. racemulosus Boiss. & Hausskn. [1195]; *A. racemulosus* var. *leptorhachis* Rech.f. [1195].
   Not climbing; Herb/Shrub; Perennial.
   West Asia: Iran(N).
   Description: 1009, 1015, 1195; Distribution Map: 1195; Illustration: 1195.
   Townsend was doubtful of the original locality; Iran or Iraq?[1009]; he also suggested that the taxon was based on a mixture.
   Tietz regards it as a good species, not occurring in Iraq [1195].

A. erinifolius Pau [1195]
   Not climbing; Herb/Shrub; Perennial.
   West Asia: Iran(N).
   Description: 1015, 1195; Illustration: 1195.
   Known only from the type, collected in 1899 [1195].

A. susianus Boiss. [1195]
   Not climbing; Herb/Shrub; Perennial.
   West Asia: Iran(N).
   Description: 1015, 1195; Distribution Map: 1195; Illustration: 1195.
   Two subspecies, *susianus* and *sericeus* [1195].
   Intermediates occur with *A. campylanthus* and *A. chlaranthus*; those with *A. campylanthus* have been called *A. susianus* subsp. *stapfianus* [1195].

   subsp. **sericeus** S.Tietz [1195]
      Not climbing; Herb/Shrub; Perennial.
      West Asia: Iran(N).
      Description: 1195; Distribution Map: 1195; Illustration: 1195.

   subsp. **susianus** [1195]
      A. acutus Bunge [1195].
      Not climbing; Herb/Shrub; Perennial.
      West Asia: Iran(N).
      Description: 1195; Distribution Map: 1195; Illustration: 1195.

ASTRAGALUS Section **Campylotrichon**
Annual. Racemes short, on long axillary peduncles. Pods sessile, spreading, curved.

A. campylotrichus Bunge [1011]
   Not climbing; Herb; Annual.
   West Asia: Afghanistan(N), U.S.S.R.
   Description: 1150; Illustration: 1144, 1145, 1150.

A. oncotrichus Bunge [1173,1189]
   A. jijaensis Sirj. & Rech.f.
   Not climbing; Herb; Annual.
   West Asia: Afghanistan(N), Iran(N).
   Description: 1015.
   The Bunge type material at Kew is very small and lacks mature fruits [1153].
   Podlech tranfers from Section *Oxyglottis* to *Campylotrichon* [1173].
   The types of *A. oncotrichus* and *A. jijaensis* are identical.

ASTRAGALUS Section **Caprini**
Perennials, mostly from a woody rootstock and with the aerial stem short or absent. Hairs simple. Leaves imparipinnate. Racemes lax. Pod sessile or stipitate. A large and complex section with many closely-related species.

Formerly known as Section *Myobroma*. See Podlech (1988 — ref. 1175).

A. absentivus A.A.R.Maassoumi [1196]
>Not climbing; Herb; Perennial.
>West Asia: Iran(N). Description: 1196; Illustration: 1172, 1196.

A. adpressipilosus Gontsch. [1175]
>*A. adpresse-pilosus* Gontsch. [1011].
>Not climbing; Herb; Perennial.
>West Asia: Afghanistan(N); U.S.S.R.
>Description: 1150, 1175; Distribution Map: 1175; Illustration: 1145, 1175.
>Podlech & Anders [1126] record this taxon for Afghanistan, but the material was redetermined as *A. toppinianus* in Podlech (1988) [1175]. The taxon occurs on the U.S.S.R.-Afghanistan border.

A. adulterinus Podl. [1178]
>Not climbing; Herb; Perennial.
>West Asia: Iran(N).
>Description: 1178.

A. aegobromus Boiss. & Hohen. [1175]
>*A. aegobromus* var. *hirsutus* Boiss. [1175]; *A. aegobromus* var. *longiscapus* Bornm. [1175]; *A. argillosus* Manden. [1175]; *A. derbendicus* Bunge [1175]; *A. kuhikakaschanicus* Sirj. & Rech.f. [1175]; *A. ramiaensis* Sirj. & Rech.f. [1175]; *A. ramianensis* Sirj. & Rech.f. [1175]; *A. rugosus* var. *humilis* Parsa [1175]; *A. sultanabadensis* Sirj. & Rech.f. [1175]; *A. torrentum* Bunge [1175].
>Not climbing; Herb; Perennial.
>West Asia: Iran(N), Iraq(N); Middle East: Turkey(N); U.S.S.R.
>Description: 1009, 1150, 1175; Distribution Map: 1175; Illustration: 1175.

A. aharicus A.A.R.Maassoumi & Podl. [1175]
>Not climbing; Herb; Perennial.
>West Asia: Iran(N).
>Description: 1175, 1197; Distribution Map: 1175; Illustration: 1172, 1175, 1197.

A. aktauensis Gontsch. [1175]
>Not climbing; Herb; Perennial.
>West Asia: Afghanistan(N); U.S.S.R.
>Description: 1150, 1175; Distribution Map: 1175; Illustration: 1145, 1175.

A. alienus Podl. [1175]
>Not climbing; Herb; Perennial.
>West Asia: Iran(N).
>Description: 1175; Distribution Map: 1175; Illustration: 1175.

A. aligudarzicus Parsa (provisional) [1103]
>Not climbing; Herb; Perennial.
>West Asia: Iran(N).
>Description: 1103; Illustration: 1103.
>The type (Hb.Parsa 20004) is at Kew, and probably does not belong to Section *Caprini* [1153].

A. angustiflorus C.Koch [1175]
>Not climbing; Herb; Perennial.
>West Asia: Iran(N), Iraq(N); Europe; Middle East: Turkey(N); U.S.S.R.
>Description: 1015, 1150, 1175; Distribution Map: 1175; Illustration: 1175.
>2 subspecies; only the nominate one occurs in West Asia [1175].

>subsp. **angustiflorus** [1175]
>>*A. fieldianus* Huber-Mor. [1175]; *A. fraxinella* Bunge [1175]; *A. hymenochlaenus* Bunge [1175]; *A. silachorensis* Bornm. [1175].
>>Not climbing; Herb; Perennial.
>>West Asia: Iran(N), Iraq(N); Middle East: Turkey(N); U.S.S.R.
>>Description: 1009, 1015, 1175; Distribution Map: 1175; Illustration: 1175.

A. apricus Bunge [1175]
>*A. polyphyllus* Bunge [1175]].
>Not climbing; Herb; Perennial.
>West Asia: Iran(N); U.S.S.R.
>Description: 1015, 1150, 1175; Distribution Map: 1175; Illustration: 1175.

A. aqrabatensis Podl. [1175]
Not climbing; Herb; Perennial.
West Asia: Afghanistan(N).
Description: 1175, 1185; Distribution Map: 1175; Illustration: 1175, 1185.

A. archibaldii Podl. [1175]
Not climbing; Herb; Perennial.
West Asia: Iran(N).
Description: 1175; Distribution Map: 1175; Illustration: 1175.
Known only from the type [1175].

A. aspreticola Podl. [1175]
Not climbing; Herb; Perennial.
West Asia: Iran(N).
Description: 1175; Distribution Map: 1175; Illustration: 1175.
Known only from the type, coll. 1975 [1198].

A. assadii A.A.R.Maassoumi & Podl. [1175]
Not climbing; Herb; Perennial.
West Asia: Iran(N).
Description: 1175, 1197; Distribution Map: 1175; Illustration: 1172, 1175, 1197.

A. auganus Bunge [1175]
Not climbing; Herb; Perennial.
West Asia: Afghanistan(N); South Asia: Pakistan(N).
Description: 1015, 1175; Distribution Map: 1175; Illustration: 1175.

A. avicennicus Parsa [1175]
Not climbing; Herb; Perennial.
West Asia: Iran(N).
Description: 1103, 1175; Distribution Map: 1175; Illustration: 1103, 1175.
Parsa's figure is erroneously labelled as *A. hamadanicus* [1175].

A. babatagii Popov [1175]
A. babatagi Popov [1175].
Not climbing; Herb; Perennial.
West Asia: Afghanistan(N); U.S.S.R.
Description: 1150, 1175; Distribution Map: 1175; Illustration: 1145, 1175.
A subspecies *indurescens* Rassul. has been described from Tadjikistan (U.S.S.R.) [1175].

A. bamianicus Podl. [1175]
Not climbing; Herb; Perennial.
West Asia: Afghanistan(N).
Description: 1175, 1185; Distribution Map: 1175; Illustration: 1175, 1185.
Fig.82,p.77 (right-hand plant) in Koie & Rechinger 1957 is also this [1175].

A. basilicus Podl. & A.A.R.Maassoumi [1175]
Not climbing; Herb; Perennial.
West Asia: Iran(N).
Description: 1175; Distribution Map: 1175; Illustration: 1175.

A. belcheraghensis Podl. [1175]
Not climbing; Herb; Perennial.
West Asia: Afghanistan(N).
Description: 1175; Distribution Map: 1175; Illustration: 1175.

A. brachystachys DC. [1175]
A. aaronsohnianus Eig [1175]; *A. ascophorus* Fischer [1175]; *A. baibakht* Eig [1175]; *A. brachystachys* var. *argentatus* Eig [1175]; *A. dinsmorei* Mout. [1175]; *A. maaratensis* Eig [1175]; *A. pinetorum* var. *drusorum* Eig [1175]; *A. selemiensis* Mout. [1175].
Not climbing; Herb; Perennial.
West Asia: Iran(N), Iraq(N); Middle East: Israel(N), Jordan(N), Syria(N), Turkey(N).
Description: 1009, 1175; Distribution Map: 1175; Illustration: 1009, 1175.

A. brotherusii Podl. [1178]
Not climbing; Herb; Perennial.
U.S.S.R.
Description: 1178.

A. calamistratus Podl. [1175]
Not climbing; Herb; Perennial.
West Asia: Afghanistan(N).
Description: 1175; Illustration: 1175.
Known only from the type, collected in 1969 [1175].

A. callainus Podl. [1175]
Not climbing; Herb; Perennial.
West Asia: Iran(N).
Description: 1175; Distribution Map: 1175; Illustration: 1175.

A. caprinus L. [1175]
Not climbing; Herb; Perennial.
West Asia: Iran(N), Iraq(N), Saudi Arabia(N); Middle Eas:Cyprus(N), Israel(N), Jordan(N), Lebanon(N), Syria(N), Turkey(N).
Description: 1175; Distribution Map: 1175; Illustration: 1175.
3 subspecies, *caprinus, glaber* & *huetii*, of which only *caprinus* occurs in West Asia [1175].

subsp. caprinus [1175]
A. alexandrinus Boiss. [1175]; *A. beersabeensis* Eig [1175]; *A. beershabensis* Rech.f. [1175]; *A. caprinus* subsp. *lanigerus* (Desf.)Maire [1175]; *A. deserti-syriaci* Eig [1175]; *A. hormozabadensis* Sirj. Rech.f. [1175]; *A. khuzistanicus* Sirj. & Rech.f. [1175]; *A. lanigerus* Desf. [1175]; *A. platyraphis* Fischer [1175].
Not climbing; Herb; Perennial.
West Asia: Iran(N), Iraq(N), Saudi Arabia(N); Middle East: Cyprus(N), Israel(N), Jordan(N), Lebanon(N), Syria(N), Turkey(N).
Description: 1009, 1175; Distribution Map: 1175; Illustration: 1001, 1175.

A. cartilagineus Gontsch. [1175]
Not climbing; Herb; Perennial.
West Asia: Afghanistan(N), Iran(N); U.S.S.R.
Description: 1175; Distribution Map: 1175; Illustration: 1052, 1175.
Two subspecies, *cartilagineus* & *honigbergeri* [1175].

subsp. cartilagineus [1175]
Not climbing; Herb; Perennial.
West Asia: Afghanistan(N), Iran(N).
Description: 1175; Distribution Map: 1175; Illustration: 1052.
Koie & Rechinger's (1957) figure of 'A. honigbergeri' is this [1175].

subsp. honigbergeri (Sirj. & Rech.f.) Podl. [1175]
A. honigbergeri Sirj. & Rech.f. [1175].
Not climbing; Herb; Perennial.
West Asia: Afghanistan(N).
Description: 1052, 1175; Distribution Map: 1175; Illustration: 1052, 1175.
Koie & Rechinger's (1957) figure of 'Ahonigbergeri' is not this, but ssp. *cartilagineus* [1175]

A. catabostrychos Kress-Deml & Podl. [1175]
Not climbing; Shrub; Perennial.
West Asia: Afghanistan(N).
Description: 1175; Distribution Map: 1175; Illustration: 1175.

A. charguschanus Freyn [1175]
A. pamiricus B.Fedtsch. [1175]; *A. staintonianus* Ali [1175]; *A. tianschanicus* var. *pamiricus* B.Fedtsch. [1175].
Not climbing; Herb; Perennial.
West Asia: Afghanistan(N); South Asia: Pakistan(N); U.S.S.R.
Description: 1150, 1175; Distribution Map: 1175; Illustration: 1145, 1175.

A. chrysanthus Boiss. & Hohen. [1175]
A. chrysanthus var. *elburzensis* Parsa [1175]; *A. hadjarianus* Parsa [1015].
Not climbing; Herb; Perennial.
West Asia: Iran(N).
Description: 1175; Distribution Map: 1175; Illustration: 1175.
A. hadjarianus may not be a validly published name.

A. cisoxanus Podl. [1175]
Not climbing; Herb; Perennial.
West Asia: Afghanistan(N).
Description: 1098, 1175; Distribution Map: 1175; Illustration: 1098, 1175.
Known only from the type, collected in 1965 [1175].

A. citrinus Bunge [1175]
Not climbing; Herb; Perennial.
West Asia: Afghanistan(N), Iran(N); U.S.S.R.
Description: 1175; Distribution Map: 1175; Illustration: 1175.
3 subspecies, *citrinus, barrowianus* & *khorasanicus* [1175].

subsp. barrowianus (Aitch. & Baker) Podl. [1175]
A. barrowianus Aitch. & Baker [1175].
Not climbing; Herb; Perennial.
West Asia: Afghanistan(N); U.S.S.R.
Description: 1015, 1175; Distribution Map: 1175; Illustration: 1052, 1175.

subsp. citrinus [1175]
A. angustidens Freyn & Sint. [1175]; *A. angustidens* var. *retusefoliolatus* Sirj. & Rech.f. [1175]; *A. subangustidens* V.V.Nikitin [1175].
Not climbing; Herb; Perennial.
West Asia: Afghanistan(N), Iran(N); U.S.S.R.
Description: 1175; Distribution Map: 1175; Illustration: 1175.

subsp. khorasanicus Podl. [1175]
A. angustidens var. *recterostratus* Sirj. & Rech.f. [1175]; *A. angustidens* var. *sessilis* Sirj. & Rech.f. [1175]; *A. angustidens* var. *strictifolius* Sirj. & Rech.f. [1175].
Not climbing; Herb; Perennial.
West Asia: Afghanistan(N), Iran(N).
Description: 1175; Distribution Map: 1175; Illustration: 1175.

A. concinnus Bunge [1175]
A. concinnus Boiss. [1175]; *A. concinus* Bunge [1175].
Not climbing; Herb; Perennial.
West Asia: Afghanistan(N); South Asia: Pakistan(N).
Description: 1015, 1175; Distribution Map: 1175; Illustration: 1175.
The Afghanistan record is old and poorly localised, and there are no records this century; the occurrence of the taxon must be in doubt [1175].

A. connectens Podl. [1175]
Not climbing; Herb; Perennial.
West Asia: Afghanistan(N).
Description: 1175, 1185; Distribution Map: 1175; Illustration: 1175, 1185.

A. controversus A.A.R.Maassoumi & Podl. [1175]
Not climbing; Herb; Perennial.
West Asia: Iran(N).
Description: 1175, 1197; Distribution Map: 1175; Illustration: 1172, 1175, 1197.

A. curvipes Trautv. [1175]
A. curvipes var. *transitorius* Sirj. & Rech.f. [1175]; *A. supralanatus* Freyn & Sint. [1175].
Not climbing; Herb; Perennial.
West Asia: Iran(N); U.S.S.R.
Description: 1150, 1175; Distribution Map: 1175; Illustration: 1175.

A. darlingtonii Podl. [1175]
>Not climbing; Herb; Perennial.
>West Asia: Iran(N).
>Description: 1175; Distribution Map: 1175; Illustration: 1175.

A. damardanicus Podl. [1175]
>Not climbing; Herb; Perennial.
>West Asia: Afghanistan(N).
>Description: 1175; Distribution Map: 1175; Illustration: 1175.

A. delicatulus Podl. [1175]
>Not climbing; Herb; Perennial.
>West Asia: Afghanistan(N).
>Description: 1175, 1185; Distribution Map: 1175; Illustration: 1175, 1185.

A. denticulatus Podl. [1175]
>Not climbing; Herb; Perennial.
>West Asia: Iran(N).
>Description: 1175; Distribution Map: 1175; Illustration: 1175.
>Known only from the type, collected in 1973 [1175].

A. dignus Boriss. [1175]
>Not climbing; Herb; Perennial.
>West Asia: Afghanistan(N); South Asia: Pakistan(N); U.S.S.R.
>Description: 1150, 1175; Distribution Map: 1175; Illustration: 1145, 1175.
>Intermediates (perhaps hybrids) with *A.adpressipilosus* occur in the Pamirs [1175].

A. echanensis Podl. [1175]
>Not climbing; Herb; Perennial.
>West Asia: Afghanistan(N).
>Description: 1098, 1175; Distribution Map: 1175; Illustration: 1098, 1175.

A. edelbergianus Sirj. & Rech.f. [1175]
>Not climbing; Herb; Perennial.
>West Asia: Afghanistan(N).
>Description: 1052, 1175; Distribution Map: 1175; Illustration: 1052, 1175.

A. edmondsonii Podl. [1175]
>Not climbing; Herb; Perennial.
>West Asia: Iran(N).
>Description: 1175; Illustration: 1175.

A. ekbergii Podl. [1175]
>Not climbing; Herb; Perennial.
>West Asia: Afghanistan(N).
>Description: 1175; Distribution Map: 1175; Illustration: 1175.
>Known only from the type, collected in 1969 [1175].

A. elwendicus Bornm. [1175]
>Not climbing; Herb; Perennial.
>West Asia: Iran(N).
>Description: 1175; Distribution Map: 1175; Illustration: 1175.

A. erionotus Bunge [1175]
>*A. andarabicus* Podl. [1175]; *A. chitralensis* Ali [1175].
>Not climbing; Herb; Perennial.
>West Asia: Afghanistan(N); South Asia: Pakistan(N).
>Description: 1175; Distribution Map: 1175; Illustration: 1175.

A. erubescens Podl. [1175]
>Not climbing; Herb; Perennial.
>West Asia: Iran(N).
>Description: 1175; Distribution Map: 1175; Illustration: 1175.

A. erythrosemius Boiss. [1175]
A. erythroseminus Boiss. [1011].
Not climbing; Herb; Perennial.
West Asia: Afghanistan(N); South Asia: Pakistan(N).
Description: 1175; Illustration: 1175.

A. esferayanicus Podl. & A.A.R.Maassoumi [1175]
Not climbing; Herb; Perennial.
West Asia: Iran(N).
Description: 1175, 1198; Distribution Map: 1175; Illustration: 1172, 1175, 1198.

A. eusarathron Kress-Deml & Podl. [1175]
Not climbing; Herb; Perennial.
West Asia: Iran(N).
Description: 1175; Distribution Map: 1175; Illustration: 1175.

A. evanensis A.A.R.Maassoumi & Podl. [1175]
Not climbing; Herb; Perennial.
West Asia: Iran(N).
Description: 1175, 1198; Distribution Map: 1175; Illustration: 1172, 1175, 1198.

A. fabaceus M.Bieb. [1175]
A. tumidus M.Bieb. [1175].
Not climbing; Herb; Perennial.
West Asia: Iran(N); Middle East: Turkey(N); U.S.S.R.
Description: 1015, 1150, 1175; Distribution Map: 1175; Illustration: 1175.

A. farkharensis Podl. [1175]
Not climbing; Herb; Perennial.
West Asia: Afghanistan(N).
Description: 1098, 1175; Distribution Map: 1175; Illustration: 1098, 1175.

A. filamentosus Bunge (provisional) [1175]
Not climbing; Herb; Perennial.
Description: 1009, 1015.
The material is very poor; the name could be a synonym of either *A. ovinus* or *A. angustiflorus* subsp. *angustiflorus* [1175].
The collecting locality 'Kurdistan' - could be in Iran, Iraq or Turkey [1175].

A. firuzkuhensis Podl. [1175]
Not climbing; Herb; Perennial.
West Asia: Iran(N).
Description: 1175; Distribution Map: 1175; Illustration: 1175.

A. flemingii Ali [1175]
A. bukuensis sensu Baker [1122, 1175.
Not climbing; Herb; Perennial.
West Asia: Afghanistan(N); South Asia: Pakistan(N).
Description: 1175; Distribution Map: 1175; Illustration: 1175.

A. flexus Fischer [1175]
A. aquae-rubrae B.Fedtsch. [1175]; *A. pentapetaloides* Bunge [1175]; *A. pentapetaloides* var. *blepharophyllus* Bunge [1175]; *A. remotijugus* var. *pumilus* Parsa [1175]; *A. stenanthus* Freyn [1175].
Not climbing; Herb; Perennial.
West Asia: Iran(N); U.S.S.R.
Description: 1015, 1150, 1175; Distribution Map: 1175; Illustration: 1145, 1150, 1175.

A. floccosifolius Sumn. [1175]
A. ephemeretorum Gontsch. [1175]; *A. ephemeretorum* subsp. *bilobulatus* Rassul. [1175]; *A. pseudolanuginosus* Gontsch. [1175].
Not climbing; Herb; Perennial.
West Asia: Afghanistan(N); U.S.S.R.
Description: 1150, 1175; Distribution Map: 1175; Illustration: 1145, 1175.

A. gagnieui A.A.R.Maassoumi & Podl. [1175]
  Not climbing; Herb; Perennial.
  West Asia: Iran(N).
  Description: 1175, 1198; Distribution Map: 1175; Illustration: 1172, 1175, 1198.

A. galiifolius Podl. [1175]
  Not climbing; Herb; Perennial.
  West Asia: Afghanistan(N).
  Description: 1098, 1175; Distribution Map: 1175; Illustration: 1098, 1175.

A. gaubae Bornm. [1175]
  A. mobayenicus Parsa [1175].
  Not climbing; Herb; Perennial.
  West Asia: Iran(N).
  Description: 1015, 1175; Distribution Map: 1175; Illustration: 1175.
  Known only from the type; the types of the two names are duplicates of one collection [1175].

A. gaudanensis B.Fedtsch. [1175]
  Not climbing; Herb; Perennial.
  West Asia: Iran(N); U.S.S.R.
  Description: 1150, 1175; Distribution Map: 1175; Illustration: 1175.

A. ghorbandicus Podl. [1175]
  Not climbing; Herb; Perennial.
  West Asia: Afghanistan(N).
  Description: 1175, 1185; Distribution Map: 1175; Illustration: 1175, 1185.

A. gifanicus A.A.R.Maassoumi & Podl. [1175]
  Not climbing; Herb; Perennial.
  West Asia: Iran(N).
  Description: 1175, 1198; Distribution Map: 1175; Illustration: 1172, 1175, 1198.
  Known only from the type, collected in 1984 [1175].

A. gilgitensis Ali [1175]
  Not climbing; Herb; Perennial.
  West Asia: Afghanistan(N); South Asia: Pakistan(N).
  Description: 1175; Distribution Map: 1175; Illustration: 1175.

A. golestanicus A.A.R.Maassoumi & Podl. [1175]
  Not climbing; Herb; Perennial.
  West Asia: Iran(N).
  Description: 1175, 1197; Distribution Map: 1175; Illustration: 1172, 1175, 1197.

A. gompholobium Bunge [1175]
  A. adraskanensis Podl. [1175]; *A. bizgensis* Rech.f. [1175]; *A. magnus* Sirj. & Rech.f. [1175]; *A. pseudoasterothrix* Hand.-Mazz. [1175]; *A. sultani* Ali
  Not climbing; Herb; Perennial.
  West Asia: Afghanistan(N), Iran(N); South Asia: Pakistan(N).
  Description: 1015, 1052, 1175; Distribution Map: 1175; Illustration: 1052, 1175, 1185.

A. grey-wilsonianus Podl. [1175]
  Not climbing; Herb; Perennial.
  West Asia: Afghanistan(N).
  Description: 1175; Distribution Map: 1175; Illustration: 1175.

A. gypsaceus G.Beck [1175]
  Not climbing; Herb; Perennial.
  West Asia: Iran(N).
  Description: 1175; Distribution Map: 1175; Illustration: 1175.

A. hedgeanus Podl. [1175]
  Not climbing; Herb; Perennial.
  West Asia: Afghanistan(N).
  Description: 1175; Distribution Map: 1175; Illustration: 1175.
  Known only from the type, collected in 1969 [1175

A. hermannii Freitag & Podl. [1175]
Not climbing; Herb; Perennial.
West Asia: Iran(N).
Description: 1175, 1199; Distribution Map: 1175; Illustration: 1175, 1199.
Known only from the type, collected in 1978 [1175].

A. ibicinus Boiss. & Hausskn. [1175]
Not climbing; Herb; Perennial.
West Asia: Iran(N).
Description: 1015, 1175; Distribution Map: 1175; Illustration: 1175.

A. imbecillus A.A.R.Maassoumi & Podl. [1175]
Not climbing; Herb; Perennial.
West Asia: Iran(N).
Description: 1175, 1198; Illustration: 1175.
Known only from the type, collected in 1981 [1175].

A. impexus Podl. [1175]
Not climbing; Herb; Perennial.
West Asia: Iran(N).
Description: 1175; Distribution Map: 1175; Illustration: 1175.

A. indistinctus Podl. & A.A.R.Maassoumi [1175]
Not climbing; Herb; Perennial.
West Asia: Iran(N).
Description: 1175, 1198; Distribution Map: 1175; Illustration: 1175.

A. inquilinus A.A.R.Maassoumi [1196]
Not climbing; Herb; Perennial.
West Asia: Iran(N).
Description: 1196; Illustration: 1172, 1196.

A. iranshahrii A.A.R.Maassoumi & Podl. [1200]
Not climbing; Herb; Perennial.
West Asia: Iran(N).
Description: 1200; Illustration: 1172, 1200.

A. ischredensis Bunge [1175]
Not climbing; Herb; Perennial.
West Asia: Iran(N).
Description: 1015, 1175; Distribution Map: 1175; Illustration: 1175.

A. ishkamishensis Podl. [1175]
Not climbing; Herb; Perennial.
West Asia: Afghanistan(N); U.S.S.R.
Description: 1098, 1175; Distribution Map: 1175; Illustration: 1098, 1175.

A. jarmolenkoi Gontsch. [1175]
Not climbing; Herb; Perennial.
West Asia: Iran(N).
Description: 1150, 1175; Distribution Map: 1175; Illustration: 1150, 1175.

A. johannis Boiss. [1175]
Not climbing; Herb; Perennial.
West Asia: Iran(N).
Description: 1015, 1175; Distribution Map: 1175; Illustration: 1175.

A. karateginii Gontsch. [1175]
A. chamaerionotus Rech.f. [1175].
Not climbing; Herb; Perennial.
West Asia: Afghanistan(N); U.S.S.R.
Description: 1052, 1150, 1175; Distribution Map: 1175; Illustration: 1052, 1145, 1175.

A. kashafensis Podl. [1178]
   Not climbing; Herb; Perennial.
   West Asia: Iran(N).
   Description: 1178.

A. kashmarensis A.A.R.Maassoumi & Podl. [1175]
   Not climbing; Herb; Perennial.
   West Asia: Iran(N).
   Description: 1175, 1197; Distribution Map: 1175; Illustration: 1172, 1175, 1197.

A. kaswinensis Bornm. [1175]
   Not climbing; Herb; Perennial.
   West Asia: Iran(N).
   Description: 1015, 1175; Distribution Map: 1175; Illustration: 1175.
   Known only from the type, collected in 1902 [1175].

A. kermanschahensis Bornm. [1175]
   Not climbing; Herb; Perennial.
   West Asia: Iran(N).
   Description: 1175; Distribution Map: 1175; Illustration: 1175.
   Known only from the type, collected in 1904 [1175].

A. khwaja-muhammadensis Podl. [1175]
   Not climbing; Herb; Perennial.
   West Asia: Afghanistan(N).
   Description: 1098, 1175; Illustration: 1098, 1175.
   Known only from the type, collected in 1965 [1175].

A. kirpicznikovii Grossh. [1175]
   Not climbing; Herb; Perennial.
   West Asia: Iran(N); U.S.S.R.
   Description: 1175; Distribution Map: 1175; Illustration: 1175.

A. kopetdaghi Boriss. [1175]
   A. glabriusculus Gontsch. [1175].
   Not climbing; Herb; Perennial.
   West Asia: Iran(N); U.S.S.R.
   Description: 1150, 1175; Distribution Map: 1175; Illustration: 1175.
   Two vars., *kopetdaghi* and *orientikopetdaghensis* [1175].

A. kukkonenii Podl. [1175]
   Not climbing; Herb; Perennial.
   West Asia: Afghanistan(N).
   Description: 1175; Distribution Map: 1175; Illustration: 1175.

A. kunarensis Podl. [1178]
   Not climbing; Herb; Perennial.
   West Asia: Afghanistan(N).
   Description: 1178.

A. kuschkensis Boriss. [1175]
   A. nudus Gontsch. [1175].
   Not climbing; Herb; Perennial.
   West Asia: Afghanistan(N); U.S.S.R.
   Description: 1150, 1175; Illustration: 1175.

A. laetus Bunge [1175]
   A. grantii Bunge [1175]; *A. leiocalyx* Bunge [1175].
   Not climbing; Herb; Perennial.
   West Asia: Afghanistan(N).
   Description: 1175; Distribution Map: 1175; Illustration: 1052, 1175.

A. lalandei Podl. [1175]
Not climbing; Herb; Perennial.
West Asia: Afghanistan(N).
Description: 1175; Distribution Map: 1175; Illustration: 1175.

A. lambinonii Podl. [1175]
Not climbing; Herb; Perennial.
West Asia: Iran(N).
Description: 1175; Distribution Map: 1175; Illustration: 1175.
Two collections only, in 1890 & 1897 [1175].

A. lanceolatus Bunge [1175]
Not climbing; Herb; Perennial.
West Asia: Afghanistan(N).
Description: 1175; Distribution Map: 1175; Illustration: 1175.

A. leonardii A.A.R.Maassoumi [1196]
Not climbing; Herb; Perennial.
West Asia: Iran(N).
Description: 1196; Illustration: 1172, 1196.

A. lepidus Podl. [1178]
Not climbing; Herb; Perennial.
West Asia: Iran(N).
Description: 1178.

A. maassoumii Podl. [1175]
Not climbing; Herb; Perennial.
West Asia: Iran(N).
Description: 1175; Distribution Map: 1175; Illustration: 1175.

A. macronyx Bunge [1175]
A. samarkandinus Freyn [1175].
Not climbing; Herb; Perennial.
West Asia: Afghanistan(N); U.S.S.R.
Description: 1175; Distribution Map: 1175; Illustration: 1175.

A. macropelmatus Bunge [1175]
A. movlavicus Parsa [1143].
Not climbing; Herb; Perennial.
West Asia: Afghanistan(N), Iran(N), Iraq(N).
Description: 1009, 1015, 1175; Distribution Map: 1175; Illustration: 1009, 1175.
2 subspecies, *macropelmatus* & *pseudobuchtormensis* [1175].
Intermediates occur in S.Iran & SE Afghanistan [1175].
The type of *A. movlavicus* is inadequate for subspecific attribution [1143].

subsp. **macropelmatus** [1175]
A. apricus var. *rarus* (Sirj. & Rech.f.)Parsa [1175]; *A. rarus* Sirj. & Rech.f. [1175].
Not climbing; Herb; Perennial.
West Asia: Iran(N), Iraq(N).
Description: 1009, 1175; Distribution Map: 1175; Illustration: 1009, 1175.

subsp. **pseudobuchtormensis** (Sirj. & Rech.f.)Podl. [1175]
A. aitchisonii Sirj. & Rech.f. [1175]; *A. buchtormensis* var. *pseudobuchtormensis* (Sirj. & Rech.f.)Parsa [1175]; *A. pseudobuchtormensis* Sirj. & Rech.f. [1175]; *A. subconduplicatus* Ali [1175]; *A. turbathaidareinsis* Sirj. & Rech.f. [1175].
Not climbing; Herb; Perennial.
West Asia: Afghanistan(N), Iran(N).
Description: 1175; Distribution Map: 1175; Illustration: 1052, 1175.

A. managettae Sirj. & Rech.f. [1175]
Not climbing; Herb; Perennial.
West Asia: Iran(N).
Description: 1175; Distribution Map: 1175; Illustration: 1175

A. miralamensis Podl. [1175]
Not climbing; Herb; Perennial.
West Asia: Afghanistan(N).
Description: 1175; Distribution Map: 1175; Illustration: 1175.
Known only from the type, collected in 1969 [1175].

A. modestus Boiss. & Hohen. [1175]
Not climbing; Herb; Perennial.
West Asia: Iran(N).
Description: 1015, 1150, 1175; Distribution Map: 1175; Illustration: 1175.
Regarded as a forma *alpina* of *A. aegobromus* (q.v.) by Bornmuller [1175].

A. monanthemus Boiss. [1175]
A. heterochrous Bornm. [1175]; *A. heterochrus* Bornm. [1015].
Not climbing; Herb; Perennial.
West Asia: Iran(N).
Description: 1015, 1175; Distribution Map: 1175; Illustration: 1175, 1201.

A. mozaffarianii A.A.R.Maassoumi [1175]
Not climbing; Herb; Perennial.
West Asia: Iran(N).
Description: 1148, 1175; Illustration: 1148, 1172.

A. multijugus DC. [1175]
A. tavernieri Boiss. [1175].
Not climbing; Herb; Perennial.
West Asia: Iran(N).
Description: 1009, 1015, 1175; Distribution Map: 1175; Illustration: 1175.
Doubtfully recorded for Iraq by Townsend & Guest [1009], but not by Podlech (1988) [1175].

A. neo-mobayenii A.A.R.Maassoumi [1196]
Not climbing; Herb; Perennial.
West Asia: Iran(N).
Description: 1196; Illustration: 1196.

A. neo-podlechii A.A.R.Maassoumi [1175]
Not climbing; Herb; Perennial.
West Asia: Iran(N).
Description: 1148, 1175; Distribution Map: 1175; Illustration: 1148, 1172, 1175.

A. nephtonensis Freyn [1175]
A. diversifolius Trautv. [1175]; *A. faryabensis* Podl. [1175]; *A. maymanensis* Podl. [1175].
Not climbing; Herb; Perennial.
West Asia: Afghanistan(N), Iran(N); U.S.S.R.
Description: 1150, 1175, 1185; Distribution Map: 1175; Illustration: 1175, 1185.

A. neubauerianus Sirj. & Rech.f. [1175]
A. barbicalyx Ali [1175]; *A. kurrumensis* Bunge [1175]; *A. neoverticillatus* Kitam. [1175]; *A. verticillaris* Bunge [1175].
Not climbing; Herb; Perennial.
West Asia: Afghanistan(N), Iran(N); South Asia: Pakistan(N).
Description: 1175; Distribution Map: 1175; Illustration: 1052, 1175.
Podlech does not record this taxon from Iran, or from nearby. It seems likely that Parsa's record is an error [1175].
'Kurrumensis' and 'kuramensis' are regarded as orthographic variants, and therefore homonyms [1175].

A. nurabadensis A.A.R.Maassoumi & Podl. [1175]
Not climbing; Herb; Perennial.
West Asia: Iran(N).
Description: 1175, 1198; Distribution Map: 1175; Illustration: 1172, 1175, 1198.

A. orbiculatus Ledeb. [1175]
Not climbing; Herb; Perennial.
West Asia: Afghanistan(N), Iran(N); East Asia: China(N); South Asia: Pakistan(N); U.S.S.R.
Description: 1175; Distribution Map: 1175; Illustration: 1175.

A. ovinus Boiss. [1175]
A. amanus Boiss. [1175]; *A. angustiflorus* subsp. *amanus* (Boiss.)Chamberlain [1175]; *A. bachtiaricus* Bunge [1175]; *A. bachtiaricus* var. *leiocarpus* Bornm. [1175]; *A. faghirensis* Sirj. & Rech.f. [1175]; *A. lobophorus* Boiss. [1175]; *A. lobophorus* var. *pilosus* (Bornm.)C.Towns. [1175]; *A. piestolobus* Bunge [1175]; *A. rugosus* Boiss. [1175]; *A. rugosus* var. *pilosus* Bornm. [1175]; *A. siliquosus* Boiss. (1846) [1175].
Not climbing; Herb; Perennial.
West Asia: Iran(N), Iraq(N); Middle East: Turkey(N).
Description: 1009, 1015, 1175; Distribution Map: 1175; Illustration: 1175.
Forage 1009.
Variable; see Podlech (1988), p.241 [1175].

A. papillosus Podl. [1178]
Not climbing; Herb; Perennial.
West Asia: Afghanistan(N).
Description: 1178.

A. parvulus Bornm. [1175]
Not climbing; Herb; Perennial.
West Asia: Iran(N).
Description: 1015, 1175; Illustration: 1175.
Known only from the type, collected in 1884 [1175].

A. patentipilosus Kitam. [1175]
A. grandiflorus Bunge [1175]; *A. korovinianus* Barneby [1175].
Not climbing; Herb; Perennial.
West Asia: Afghanistan(N); U.S.S.R.
Description: 1011, 1150, 1175; Distribution Map: 1175; Illustration: 1011, 1150, 1175.

A. pellitus Bunge [1175]
A. kucanensis Rech.f. [1175]; *A. pellitus* var. *orbicularis* Parsa [1175]; *A. pubifolius* V.V.Nikitin [1175].
Not climbing; Herb; Perennial.
West Asia: Afghanistan(N), Iran(N); U.S.S.R.
Description: 1015, 1175; Distribution Map: 1175; Illustration: 1103, 1175, 1202.

A. penicillatus Podl. [1175]
Not climbing; Herb; Perennial.
West Asia: Afghanistan(N).
Description: 1098, 1175; Distribution Map: 1175; Illustration: 1098, 1175.
Podlech (1967) recognizes a var. *glaber* [1098].

A. perdurans Podl. [1175]
Not climbing; Herb; Perennial.
West Asia: Iran(N).
Description: 1175; Distribution Map: 1175; Illustration: 1175.

A. perplexans Podl. [1175]
Not climbing; Herb; Perennial.
West Asia: Afghanistan(N).
Description: 1175; Illustration: 1175.
Known only from the type, collected in 1962 [1175].

A. pinetorum Boiss. [1175]
Not climbing; Herb; Perennial.
West Asia: Iran(N), Iraq(N); Middle East: Lebanon(N), Syria(N), Turkey(N); U.S.S.R.
Description: 1009, 1175; Distribution Map: 1175; Illustration: 1009, 1175.
3 subspecies, *pinetorum, declinatus,* & *alamutensis* S[1175].

subsp. **alamutensis** A.A.R.Maassoumi & Podl. [1175]
Not climbing; Herb; Perennial.
West Asia: Iran(N).
Description: 1175, 1197; Distribution Map: 1175; Illustration: 1175, 1197.

subsp. **declinatus** Podl. [1175]
A. declinatus Willd. [1175]; A. declinatus var. superglaber Freyn & Bornm. [1175]; A. nummularius sensu auctt. [1175]; A. samamensis Boiss. & Buhse [1175].
Not climbing; Herb; Perennial.
West Asia: Iran(N); Middle East: Turkey(N); U.S.S.R.
Description: 1175; Distribution Map: 1175; Illustration: 1175.

subsp. **pinetorum** [1175]
A. declinatus subsp. pinetorum (Boiss.)Ponert [1175]; A. declinatus var. suprahirsutus Freyn [1175]; A. djashmensis Sirj., Rech.f. & Aellen [1175]; A. ramicaudex Chamberlain [1175]; A. seidlitzii Bunge [1175]; A. talyschensis Bunge [1175].
Not climbing; Herb; Perennial.
West Asia: Iran(N), Iraq(N); Middle East:Lebanon(N), Syria(N), Turkey(N); U.S.S.R.
Description: 1009, 1175; Distribution Map: 1175; Illustration: 1009, 1175.

A. **piranshahricus** A.A.R.Maassoumi & Podl. [1175]
Not climbing; Herb; Perennial.
West Asia: Iran(N).
Description: 1175, 1198; Distribution Map: 1175; Illustration: 1175.

A. **polybotrys** Boiss. [1175]
A. recollectus Rech.f. [1175].
Not climbing; Herb; Perennial.
West Asia: Afghanistan(N); South Asia: Pakistan(N).
Description: 1015, 1052, 1175; Distribution Map: 1175; Illustration: 1052, 1175.

A. **pseudobrachystachys** Sirj. & Rech.f. [1175]
Not climbing; Herb; Perennial.
West Asia: Iran(N); U.S.S.R.
Description: 1175; Distribution Map: 1175; Illustration: 1175.

A. **pseudogompholobium** Podl. [1175]
Not climbing; Herb; Perennial.
West Asia: Afghanistan(N).
Description: 1175; Illustration: 1175.
Known only from the type, collected in 1977.

A. **pseudoibicinus** A.A.R.Maassoumi & Podl. [1175]
Not climbing; Herb; Perennial.
West Asia: Iran(N).
Description: 1175, 1198; Distribution Map: 1175; Illustration: 1175, 1198.
2 subspecies, *pseudoibicinus* & *kowlikoshensis* [1175].

subsp. **kowlikoshensis** A.A.R.Maassoumi & Podl. [1175]
Not climbing; Herb; Perennial.
West Asia: Iran(N).
Description: 1175, 1198; Distribution Map: 1175; Illustration: 1172, 1175, 1198.

subsp. **pseudoibicinus** [1175]
Not climbing; Herb; Perennial.
West Asia: Iran(N).
Description: 1175, 1198; Distribution Map: 1175; Illustration: 1172, 1175, 1198.

A. **pseudoindurascens** Sirj. & Rech.f. [1175]
A. subinduratus var. pseudoindurascens (Sirj. & Rech.f.)Parsa [1175]; A. subinduratus var. pseudoinduratus (Sirj. & Rech.f.)Parsa [1175].
Not climbing; Herb; Perennial.
West Asia: Iran(N).
Description: 1175; Distribution Map: 1175.

A. pseudojohannis A.A.R.Maassoumi & Podl. [1175]
Not climbing; Herb; Perennial.
West Asia: Iran(N).
Description: 1175, 1197; Distribution Map: 1175; Illustration: 1172, 1175, 1197.
Known only from the type, collected in 1977 [1175].

A. pseudokurrumensis Sirj. & Rech.f. [1175]
Not climbing; Herb; Perennial.
West Asia: Iran(N).
Description: 1175; Distribution Map: 1175; Illustration: 1175.
Known only from the type, collected in 1948 [1175].

A. pseudomultijugus Podl. [1178]
Not climbing; Herb; Perennial.
West Asia: Iran(N).
Description: 1178.

A. pseudopellitus Podl. [1175]
Not climbing; Herb; Perennial.
West Asia: Iran(N).
Description: 1175; Distribution Map: 1175; Illustration: 1175.

A. pseudotomentellus Podl. [1175]
Not climbing; Herb; Perennial.
West Asia: Afghanistan(N).
Description: 1175; Distribution Map: 1175; Illustration: 1175.

A. pseudoutriger Grossh. [1175]
Not climbing; Herb; Perennial.
West Asia: Iran(N); Middle East: Turkey(N); U.S.S.R.
Description: 1150, 1175; Distribution Map: 1175; Illustration: 1175.

A. pseudovinus A.A.R.Maassoumi & Podl. [1198]
Not climbing; Herb; Perennial.
West Asia: Iran(N).
Description: 1197; Illustration: 1172, 1197.

A. pseudozagrosicus A.A.R.Maassoumi & Podl. [1175]
Not climbing; Herb; Perennial.
West Asia: Iran(N).
Description: 1175, 1198; Distribution Map: 1175; Illustration: 1172, 1175, 1198.

A. purpurascens Bunge [1175]
Not climbing; Herb; Perennial.
West Asia: Afghanistan(N); South Asia: Pakistan(N).
Description: 1015, 1175; Distribution Map: 1175; Illustration: 1175.
Fibre 1153.

A. pyrrhotrichus Boiss. [1175]
Not climbing; Herb; Perennial.
West Asia: Afghanistan(N); South Asia: Pakistan(N).
Description: 1175; Distribution Map: 1175; Illustration: 1175.

A. rassoulii Podl. [1175]
Not climbing; Herb; Perennial.
West Asia: Afghanistan(N).
Description: 1175, 1185; Distribution Map: 1175; Illustration: 1175, 1185.

A. rawianus C.Towns. [1175]
Not climbing; Herb; Perennial.
West Asia: Iran(N), Iraq(N).
Description: 1009, 1132, 1175; Distribution Map: 1175; Illustration: 1009, 1132, 1175.

A. remotijugus Boiss. & Hohen. [1175]
Not climbing; Herb; Perennial.
West Asia: Iran(N).
Description: 1015, 1175; Distribution Map: 1175; Illustration: 1175.

A. renzianus Podl. [1175]
Not climbing; Herb; Perennial.
West Asia: Iran(N).
Description: 1175, 1203; Illustration: 1175.

A. reticulato-venosus A.A.R.Maassoumi & Podl. [1175]
Not climbing; Herb; Perennial.
West Asia: Iran(N).
Description: 1175, 1197; Distribution Map: 1175; Illustration: 1172, 1175, 1197.
Known only from the type, collected 1981 [1175].

A. rhabdophorus Bornm. [1175]
Not climbing; Herb; Perennial.
West Asia: Iran(N).
Description: 1015, 1175; Distribution Map: 1175; Illustration: 1175.

A. rhizanthus Benth. [1175]
Not climbing; Herb; Perennial.
West Asia: Afghanistan(N);East Asia: China(N); South Asia: India(N);Nepal(N);Pakistan(N).
Description: 1175]; Distribution Map: 1175; Illustration: 1052, 1175.
2 subspecies, *rhizanthus* & *candolleanus*; only *rhizanthus* in West Asia [1175].

subsp. **rhizanthus** [1175]
A. *anomalus* Bunge [1175]; A. *candolleanus* var. *pindreensis* Baker [1175]; A. *dscheratuensis* var. *viridis* Sirj. & Rech.f. [1175]; A. *malacophyllus* Bunge [1175]; A. *nuristanicus* Sirj. & Rech.f. [1175] A. *nuristanicus* var. *elasoonensis* Sirj. & Rech.f. [1175]; A. *pindreensis* (Baker)Ali [1175].
Not climbing; Herb; Perennial.
West Asia: Afghanistan(N); East Asia: China(N); South Asia: India(N);Pakistan(N)
Description: 1175]; Distribution Map: 1175; Illustration: 1052, 1175.
Two varieties, *rhizanthus* & *pindreensis* [1175].

A. rhizocephalus Baker [1175]
Not climbing; Herb; Perennial.
West Asia: Afghanistan(N); South Asia: Pakistan(N).
Description: 1175]; Distribution Map: 1175; Illustration: 1052, 1175.
2 subspecies, *rhizocephalus* & *hindukushensis* [1175].

subsp. **hindukushensis** (Wendelbo)Podl. [1175]
A. *hindukushensis* Wendelbo [1175].
Not climbing; Herb; Perennial.
West Asia: Afghanistan(N); South Asia: Pakistan(N).
Description: 1175]; Distribution Map: 1175; Illustration: 1175, 1204.

subsp. **rhizocephalus** [1175]
A. *dscheratuensis* Sirj. & Rech.f. [1175].
Not climbing; Herb; Perennial.
West Asia: Afghanistan(N).
Description: 1052, 1175; Distribution Map: 1175; Illustration: 1052, 1175.

A. rubrocalycinus A.A.R.Maassoumi & Podl. [1200]
Not climbing; Herb; Perennial.
West Asia: Iran(N).
Description: 1200; Illustration: 1200.

A. rubromarginatus Czerniak. [1175]
Not climbing; Herb; Perennial.
West Asia: Iran(N); U.S.S.R.
Description: 1150, 1175; Illustration: 1175.
2 subspecies, *rubromarginatus* & *oeroilanicus*; only *rubromarginatus* in West Asia [1175].

subsp. **rubromarginatus** [1175]
Not climbing; Herb; Perennial.
West Asia: Iran(N); U.S.S.R.
Description: 1175; Distribution Map: 1175; Illustration: 1175.

A. rufescens Freyn [1175]
A. variegatus Freyn & Bornm. [1175].
Not climbing; Herb; Perennial.
West Asia: Iran(N).
Description: 1175; Distribution Map: 1175; Illustration: 1175.

A. salangensis Podl. [1175]
Not climbing; Herb; Perennial.
West Asia: Afghanistan(N).
Description: 1175, 1185; Distribution Map: 1175; Illustration: 1175, 1185.

A. sangcharakensis Podl. [1175]
Not climbing; Herb; Perennial.
West Asia: Afghanistan(N).
Description: 1175; Illustration: 1175.

A. savellanicus Podl. [1175]
Not climbing; Herb; Perennial.
West Asia: Iran(N).
Description: 1175; Distribution Map: 1175; Illustration: 1175.
Known only from the type, collected in 1974 [1175].

A. schemachensis Karj. [1175]
Not climbing; Herb; Perennial.
West Asia: Iran(N); U.S.S.R.
Description: 1150, 1175; Distribution Map: 1175; Illustration: 1175.

A. schmidii Podl. [1175]
Not climbing; Herb; Perennial.
West Asia: Iran(N).
Description: 1175; Distribution Map: 1175; Illustration: 1175.

A. semilunatus Podl. [1175]
Not climbing; Herb; Perennial.
West Asia: Iran(N).
Description: 1175; Distribution Map: 1175; Illustration: 1175.

A. shatuensis Podl. [1175]
Not climbing; Herb; Perennial.
West Asia: Afghanistan(N).
Description: 1175; Distribution Map: 1175; Illustration: 1175.
Known only from the type, collected in 1967 [1175].

A. sojakii Podl. [1175]
Not climbing; Herb; Perennial.
West Asia: Iran(N).
Description: 1175; Distribution Map: 1175; Illustration: 1175.

A. subrosulariformis Sirj. & Rech.f. [1175]
Not climbing; Herb; Perennial.
West Asia: Iran(N).
Description: 1175; Distribution Map: 1175; Illustration: 1175.
Known only from the type, collected in 1948 [1175].

A. takharensis Podl. [1175]
Not climbing; Herb; Perennial.
West Asia: Afghanistan(N).
Description: 1175, 1185; Distribution Map: 1175; Illustration: 1175, 1185.

A. terekliensis Gontsch. [1175]
  Not climbing; Herb; Perennial.
  West Asia: Afghanistan(N); U.S.S.R.
  Description: 1150, 1175; Distribution Map: 1175; Illustration: 1150, 1175.

A. tomentellus Podl. [1175]
  A. botschantzevii Kamelin & Rassul. [1175].
  Not climbing; Herb; Perennial.
  West Asia: Afghanistan(N); U.S.S.R.
  Description: 1098, 1175; Distribution Map: 1175; Illustration: 1098, 1145, 1175.

A. toppinianus Ali [1175]
  Not climbing; Herb; Perennial.
  West Asia: Afghanistan(N); South Asia: Pakistan(N).
  Description: 1175, 1205; Distribution Map: 1175; Illustration: 1175, 1205.

A. touranicus Freitag & Podlech [1175]
  Not climbing; Herb; Perennial.
  West Asia: Iran(N).
  Description: 1175, 1199; Distribution Map: 1175; Illustration: 1175, 1199.

A. urbanus Podl. & A.A.R.Maassoumi [1257]
  Not climbing; Herb; Perennial.
  West Asia: Iran(N).
  Description: 1257.

A. urmiensis Bunge [1175]
  Not climbing; Herb; Perennial.
  West Asia: Iran(N); U.S.S.R.
  Description: 1015, 1175; Distribution Map: 1175; Illustration: 1175.

A. vereskensis A.A.R.Maassoumi & Podl. [1175]
  Not climbing; Herb; Perennial.
  West Asia: Iran(N).
  Description: 1175, 1198; Distribution Map: 1175; Illustration: 1172, 1175, 1198.

A. vexans Rech.f. & Koie [1175]
  Not climbing; Herb; Perennial.
  West Asia: Afghanistan(N).
  Description: 1052, 1175; Distribution Map: 1175; Illustration: 1052, 1175.

A. volkii Rech.f. [1175]
  Not climbing; Herb; Perennial.
  West Asia: Afghanistan(N).
  Description: 1175; Distribution Map: 1175; Illustration: 1052, 1175;

A. vulcanicus Bornm. [1175]
  A. longepedunculatus Sirj., Rech.f. & Aellen [1175]; *A. nicharensis* var.*longepedunculatus* (Sirj., Rech.f. & Aellen)Parsa [1175].
  Not climbing; Herb; Perennial.
  West Asia: Iran(N).
  Description: 1015, 1175; Distribution Map: 1175; Illustration: 1175.

A. webbianus Benth. [1175]
  A. minutifoliolatus Wendelbo [1175]; *A. minuto-foliolatus* Wendelbo [1011].
  Not climbing; Herb; Perennial.
  West Asia: Afghanistan(N); East Asia: China(N); South Asia: India(N);Pakistan(N).
  Description: 1175; Distribution Map: 1175; Illustration: 1175, 1204.

A. zagrosicus Boiss. & Hausskn. [1175]
  Not climbing; Herb; Perennial.
  West Asia: Iran(N).
  Description: 1015, 1175; Distribution Map: 1175; Illustration: 1175.
  Known only from the type, collected in 1867 [1175].
  Townsend & Guest (1974) place this as a synonym of *A. aegobromus*; Podlech disagrees [1175].

A. ziaratensis Podl. [1175]
    Not climbing; Herb; Perennial.
    West Asia: Afghanistan(N).
    Description: 1175; Distribution Map: 1175; Illustration: 1175.
    Known only from the type, collected in 1969 [1175].

ASTRAGALUS Section **Caraganella**
Spiny shrubs with both medifixed and basifixed hairs. Heterophyllous; leaves on long shoots paripinnate with a spinous rachis; those on short shoots imparipinnate, not spinescent. Inflorescences axillary. Pods long-stalked, bilocular. See Podlech (1975 — Ref. 1206).

A. koschukensis Boiss. [1206]
    A. koshukensis Boiss. [1011]; *A. mokurensis* Sirj. & Rech.f. [1206]; *A. stocksii* var. *elongatus* Bunge [1206]; *A. stocksii* var. *honigbergeri* Sirj. [1206].
    Not climbing; Herb/Shrub; Perennial.
    West Asia: Afghanistan(N); South Asia: Pakistan(N).
    Description: 1052, 1206; Distribution Map: 1206; Illustration: 1052.

A. reshadensis Podl. [1206]
    Not climbing; Shrub; Perennial.
    West Asia: Afghanistan(N).
    Description: 1206; Distribution Map:1206.

A. stocksii Bunge [1206]
    A. parvistipulus Rech.f. [1206].
    Not climbing; Shrub; Perennial.
    West Asia: Afghanistan(N), Iran(N); South Asia: Pakistan(N).
    Description: 1052, 1206; Distribution Map: 1206; Illustration: 1052.

ASTRAGALUS Section **Cenantrum**
Herbaceous perennials. Fruits of some species inflated.

A. bahrakianus Grey-Wilson [1207]
    A. dictamnoides sensu Podl. [1098,1126].
    Not climbing; Herb; Perennial.
    West Asia: Afghanistan(N).
    Description: 1207; Illustration: 1207.

A. tecti-mundi Freyn [1126]
    West Asia: Afghanistan(N).
    Description: 1150; Illustration: 1150.

ASTRAGALUS Section **Cercidophaca**
Perennial unarmed undershrubs with bifurcate hairs. Pods linear, compressed, acute at both ends, almost completely bilocular. See Podlech & Maassoumi (1989 — Ref. 1208).

A. cercidophacos Podl. & A.A.R.Maassoumi [1208]
    Not climbing; Herb/Shrub; Perennial.
    West Asia: Iran(N).
    Description: 1208; Illustration: 1208.

A. venustus A.A.R.Maassoumi & Podl. [1208]
    Not climbing; Herb/Shrub; Perennial.
    West Asia: Iran(N).
    Description: 1208; Illustration: 1208.

A. weirianus Aitch. & Baker [1208]
   Not climbing; Shrub; Perennial.
   West Asia: Afghanistan(N).
   Description: 1159, 1208; Illustration: 1052, 1208.
   Forage 1159.
   When originally described this species was said to be heavily grazed. It has not been refound recently and it may have been exterminated by overgrazing [1208].

ASTRAGALUS Section **Chaetodon**
Herbaceous perennials, mostly stemless, with bipartite hairs. Inflorescences many-flowered, dense, long- or short-pedunculate. Calyx accrescent at the base, ruptured later by the expanding few-seeded pod. Mainly central Asian.

A. imitans Podl. [1185]
   Not climbing; Herb; Perennial.
   West Asia: Afghanistan(N).
   Description: 1185; Illustration: 1185.

ASTRAGALUS Section **Chlorostachys**
Herbaceous perennials. Inflorescences many-flowered, often elongate. Close to sections *Cenantrum* and *Diplotheca*.

A. atropilosulus (Hochst.)Bunge [1155]
   Not climbing; Herb; Perennial.
   West Asia: North Yemen(N), Saudi Arabia(N).
   Illustration: 1001.

   subsp. **abyssinicus** (Hochst.)J.B.Gillett [1155]
      Not climbing; Herb; Perennial.
      West Asia: North Yemen(N), Saudi Arabia(N).
      Illustration: 1001.

A. calcicolus Podl. [1181]
   Not climbing; Herb; Perennial.
   West Asia: Afghanistan(N).
   Description: 1181, 1185; Illustration: 1185.
   Originally placed in Section *Komaroviella*; now in Section *Chlorostachys* [1181].

A. chlorostachys Lindley [1209]
   A. altissimus Rech.f. [1011]; *A. nuristanicus* Kitam. [1209].
   Not climbing; Herb; Perennial.
   West Asia: Afghanistan(N); South Asia: Nepal(N).
   Description: 1011, 1052; Illustration: 1011, 1052.

A. coluteocarpus Boiss. [1011]
   A. pseudomongholicus Sirj. & Rech.f. [1209].
   Not climbing; Herb; Perennial.
   West Asia: Afghanistan(N); South Asia: India(N), Pakistan(N); U.S.S.R.
   Description: 1052, 1150, 1181; Illustration: 1052, 1145, 1150.

A. falconeri Bunge [1126]
   A. badachschanicus Boriss. [1126].
   Not climbing; Herb; Perennial.
   West Asia: Afghanistan(N); U.S.S.R.
   Description: 1150; Illustration: 1145.

A. hoffmeisteri (Klotzsch)Ali [1126]
   Not climbing; Herb; Perennial.
   West Asia: Afghanistan(N).

A. kuramensis Baker [1185]
    Not climbing; Herb; Perennial.
    West Asia: Afghanistan(N); South Asia: Pakistan(N).
    Description: 1185; Illustration: 1162, 1185
    There is also *Astracantha kuramensis* (based on *Tragacantha kuramensis* Boriss.) [1152].

A. microdontus Baker [1011]
    Not climbing; Herb; Perennial.
    West Asia: Afghanistan(N); South Asia: Pakistan(N).
    Description: 1162.

ASTRAGALUS Section **Chronopus**
Perennial short-stemmed herbs or subshrubs with basifixed hairs. Leaves imparipinnate, the rachis often persisting as a spine. Inflorescences few-flowered, axillary. Pods sessile, often woody, bilocular. See Ott (1978 — Ref. 1210).

A. bazmanicus Podl. [1178]
    Not climbing; Herb/Shrub; Perennial.
    West Asia: Iran(N).
    Description: 1178.
    Known only from the type, collected in 1977 [1178].

A. dactylocarpus Boiss. [1210]
    Not climbing; Herb; Perennial.
    West Asia: Afghanistan(N), Iran(N), Iraq(N), Kuwait(N), Saudi Arabia(N); Middle East: Jordan(N), Syria(N), Turkey(N).
    Description: 1009, 1210; Distribution Map: 1210; Illustration: 1210.
    2 subspecies, *dactylocarpus* & *acinaciferus* [1210].

   subsp. **acinaciferus** (Boiss.)E.Ott [1210]
        A. acinaciferus Boiss. [1210]; *A. calvescens* Bunge [1210]; *A. crassus* Bunge [1210]; *A. gerensis* Boiss. [1210]; *A. kentrodes* Fischer [1210]; *A. lepidotrichus* Bornm. & Gauba [1210]; *A. spinescens* Bunge [1210]; *A. spinescens* subsp. *aitchisonii* Eig [1210].
        Not climbing; Herb; Perennial.
        West Asia: Afghanistan(N), Iran(N), Iraq(N), Saudi Arabia(N); Middle East: Israel(N), Jordan(N), Sinai(N), Syria(N); U.S.S.R.
        Description: 1009, 1210; Distribution Map: 1210; Illustration: 1210.

   subsp. **dactylocarpus** [1210]
        Not climbing; Herb; Perennial.
        West Asia: Iran(N), Iraq(N); Middle East: Syria(N), Turkey(N).
        Description: 1210; Distribution Map: 1210.

A. ensifer Nab. [1210]
    Not climbing; Herb; Perennial.
    West Asia: Iran(N), Iraq(N).
    Description: 1009,1015,1210]; Distribution Map: 1210; Illustration: 1210.

A. jesdianus Boiss. & Buhse [1210]
    A. saadius Parsa [1143]; *A. sirdjanicus* Parsa [1143].
    Not climbing; Herb; Perennial.
    West Asia: Iran(N).
    Description: 1103, 1187, 1210; Distribution Map: 1210; Illustration: 1103, 1187, 1210.
    The type of *A. sirdjanicus* (Hb.Parsa 20016) is at Kew, as is the type of *A. saadius* (Hb. Parsa 20036) [1153].

A. kermanschahanenicus Sirj. & Rech.f. (provisional) [1210]
    Not climbing; Herb/Shrub; Perennial.
    West Asia: Iran(N).
    Description: 1210.
    The type (Rechinger,Aellen & Esfandiari 2843) appears lost; Ott (1978) did not see it [1210].

A. sieberi DC. [1210]
> A. baghdadensis Rech.f. [1210]; A. zubairensis Eig [1210].
> Not climbing; Herb; Perennial.
> West Asia: Iran(N), Iraq(N), Kuwait(N), North Yemen(N), Qatar(N), Saudi Arabia(N); Middle East: Jordan(N), Syria(N).
> Description: 1009, 1014, 1210; Distribution Map: 1210; Illustration: 1001, 1014. Forage 1014.

A. sparsus Decne. [1210]
> Not climbing; Herb; Perennial.
> West Asia: North Yemen(N), Saudi Arabia(N); Middle East: Israel(N), Sinai(N).
> Description: 1210; Distribution Map: 1210; Illustration: 1001.

A. vanillae Boiss. [1210]
> A. arakensis Parsa [1143].
> Not climbing; Herb; Perennial.
> West Asia: Iran(N).
> Description: 1015, 1210; Distribution Map: 1210; Illustration: 1210.

ASTRAGALUS Section **Corethrum**
Herbaceous perennials, usually stemless. Hairs bifurcate. Leaves imparipinnate. Inflorescences long-pedunculate. Pods bilocular. Mainly central Asian.

A. chamaesarathron Rech.f. [1179]
> Not climbing; Herb; Perennial.
> West Asia: Afghanistan(N).
> Description: 1052; Illustration: 1052.

A. chorizanthus Rech.f. & Gilli [1179]
> Not climbing; Herb; Perennial.
> West Asia: Afghanistan(N).
> Description: 1052; Illustration: 1052.

A. schachdarinus Lipsky [1126]
> West Asia: Afghanistan(N); U.S.S.R.

A. virgaeformis Sirj. & Rech.f. [1179]
> Not climbing; Herb; Perennial.
> West Asia: Afghanistan(N).
> Description: 1052; Illustration: 1052.

ASTRAGALUS Section **Craccina**
Herbaceous perennials, sometimes from a woody base. Hairs bifurcate. Leaves imparipinnate. Inflorescences usually elongate. Pods sessile or nearly so, bilocular.

A. huthianus Freyn & Bornm. [1187]
> Not climbing; Herb; Perennial.
> West Asia: Iran(N).
> Description: 1015.

ASTRAGALUS Section **Cremoceras**
Perennial shrublets. Hairs medifixed. Leaves with few leaflets. Pod long, curved into a semi-circle, bilocular.

A. campylanthoides Bornm. [1015]
> Not climbing; Herb; Perennial.
> West Asia: Iran(N).
> Description: 1015, 1158.
> The type is from Iran [1149].

A. ochreatus Bunge [1015]
> Not climbing; Herb; Perennial
> West Asia: Iran(N).
> Description: 1015.

ASTRAGALUS Section **Cyamodes**
Annual. Leaves imparipinnate, simple-hairy. Inflorescence a spike. Pod longer than the calyx. Monotypic.

A. boeticus L. [1102]
A. *baeticus* L. [1188].
Not climbing; Herb; Annual.
West Asia: Iran(N); Europe; Middle East: Turkey(N).
Description: 1102.

ASTRAGALUS Section **Cycloglottis**
Annual. Leaves paripinnate, hairs simple. Inflorescences short, axillary. Pods falcate or spirally twisted, bilocular. Monotypic.

A. contortuplicatus L. [1015]
Not climbing; Herb; Annual.
West Asia: Iran(N); South Asia: India(N); U.S.S.R.
Description: 1015, 1150; Illustration: 1145, 1150.

ASTRAGALUS Section **Cystium**
Acaulescent perennials. Leaves imparipinnate. Hairs bifurcate. Inflorescences long-pedunculate. Pods inflated, membranous.

A. didymophysus Bunge [1015]
Not climbing; Herb; Perennial.
West Asia: Iran(N).
Description: 1015.
The type is from Iran [1149].

A. masenderanus Bunge [1185]
A. *masanderanus* Bunge [1015].
Not climbing; Herb; Perennial.
West Asia: Afghanistan(N), Iran(N).
Description: 1015.
The type is from Iran [1149].

ASTRAGALUS Section **Dasyphyllium**
Small perennials, stemless or almost so, with simple hairs. Leaves imparipinnate, crowded. Inflorescence a dense head or spike. Pod short, included in the calyx.

A. cretaceus Boiss. & Kotschy [1009]
Not climbing; Herb; Perennial.
West Asia: Iraq(N); Middle East: Syria(N), Turkey(N).
Description: 1009.

A. deickianus Bornm. [1177]
Not climbing; Herb; Perennial.
West Asia: Iran(N).
Description: 1177.

A. emarginatus Labill. [1009]
Not climbing; Herb; Perennial.
West Asia: Iraq(N); Middle East: Lebanon(N), Syria(N), Turkey(N).
Description: 1009.

A. inexpectatus A.A.R.Maassoumi & Podl. [1200]
Not climbing; Herb; Perennial.
West Asia: Iran(N).
Description: 1200; Illustration: 1172, 1200.

A. pulchellus Boiss. [1009]
> *A. singarensis* Boiss. & Hausskn. [1009].
> Not climbing; Herb; Perennial.
> West Asia: Afghanistan(N), Iran(N), Iraq(N); Middle East: Turkey(N).
> Description: 1009, 1015, 1102.

A. senilis Bornm. [1179]
> Not climbing; Herb; Perennial.
> West Asia: Iran(N).
> Description: 1015, 1201; Illustration: 1201.
> The type is from Iran [1201].

ASTRAGALUS Section **Dipelta**
Annual. Hairs simple. Inflorescences long, axillary. Pod sessile, short, separating into two parts at maturity. Monotypic.

A. dipelta Bunge [1015]
> *Didymopelta turkestanica* (Regel & Schmalh.)Regel & Schmalh. [1190]. *Dipelta turkestanica* Regel & Schmalh. [1190].
> Not climbing; Herb; Annual.
> West Asia: Afghanistan(N), Iran(N); U.S.S.R.
> Description: 1015, 1150; Illustration: 1145, 1150

ASTRAGALUS Section **Diplotheca**
Tall perennials with simple hairs. Inflorescences axillary, many-flowered, short-pedunculate. Pods membranous, long-stipitate.

A. graveolens Benth. [1011]
> *A. bodinieri* Levl. [1163]; *A. medullaris* Boiss. [1011]; *A. rotundifolius* Royle [1011].
> Not climbing; Herb; Perennial.
> West Asia: Afghanistan(N); East Asia: China(N); South Asia: India(N), Pakistan(N).
> Kitamura doubts the sectional position of this taxon as it has no bracteoles [1011].

A. gymnopodus Boiss. [1098]
> West Asia: Afghanistan(N).

A. nigritus Sirj. & Rech.f. [1011]
> Not climbing; Herb; Perennial.
> West Asia: Afghanistan(N).
> Description: 1052; Illustration: 1052.

A. vicia Sirj. & Rech.f. [1011]
> Not climbing; Herb; Perennial.
> West Asia: Afghanistan(N).
> Description: 1052; Illustration: 1052.

ASTRAGALUS Section **Drepanodes**
Annuals. Hairs simple, basifixed. Inflorescences on rather long axillary peduncles. Pods short-stalked, falcate, bilocular.

A. laricus Boiss. [1015]
> Not climbing; Herb; Annual.
> West Asia: Iran(N).
> Description: 1015.
> The type is from Iran [1015].

A. parimanicus Parsa [1103]
> Not climbing; Herb; Annual.
> West Asia: Iran(N).
> Description: 1103; Illustration: 1103.
> The type (Herb.Parsa 20026) is at Kew [1153].

A. razicus Parsa [1103]
Not climbing; Herb; Annual.
West Asia: Iran(N).
Description: 1103; Illustration: 1103.
The type (Herb.Parsa 20025) is at Kew [1153].

ASTRAGALUS Section **Eremophysa**
Robust herbaceous perennials with simple basifixed hairs. Leaves imparipinnate. Calyx strongly inflated in fruit; pods included within the calyx.

A. kahiricus D.C. [1009]
A. isopetalus Boiss. [1163]; *A. longiflorus* Del. [1009].
Not climbing; Herb; Perennial.
West Asia: Afghanistan(N), Iran(N), Iraq(N), Kuwait(N), Saudi Arabia(N); Middle East: Israel(N), Jordan(N), Sinai(N); South Asia: Pakistan(N).
Description: 1009, 1015; Illustration: 1001, 1009, 1187.

A. registanicus Rech.f. [1212]
Not climbing; Herb; Perennial.
West Asia: Afghanistan(N), Iran(N).
Description: 1212.

A. tashqurghanicus Rech.f. & Podl. [1185]
Not climbing; Herb; Perennial.
West Asia: Afghanistan(N).
Description: 1185; Illustration: 1185.

ASTRAGALUS Section **Erioceras**
Perennials; stems short; hairs bifurcate. Leaves imparipinnate. Pods elogated, bilocular.

A. anacamptoides Sirj. & Rech.f. [1179]
Not climbing; Herb; Perennial.
West Asia: Afghanistan(N).
Description: 1052; Illustration: 1052.

A. anacamptus Bunge [1179]
Not climbing; Herb; Perennial.
West Asia: Iran(N).
Description: 1015.
The type is from Iran [1149].

A. catacamptus Bunge [1179]
Not climbing; Herb; Perennial.
West Asia: Iran(N).
Description: 1015.
The type is from Iran [1149].

A. djenarensis Sirj. & Rech.f. [1179]
West Asia: Iran(N).
Description: 1213.

A. durandianus Aitch. & Baker [1015]
A. shebarensis Podl. [1143].
Not climbing; Herb; Perennial.
West Asia: Afghanistan(N), Iran(N).
Description: 1015, 1185; Illustration: 1052, 1185.
Originally placed in Caraganella [1152]; transferred to a new Section, *Wettsteiniana*, by Sirjaev & Rechinger [1157]; now to be transferred to Section *Erioceras* (Podlech in prep.) [1143].

A. flabellatus Podl. [1185]
Not climbing; Herb/Shrub; Perennial.
West Asia: Afghanistan(N).
Description: 1185; Illustration: 1185.

A. harsukhianus Rech.f. [1052]
Not climbing; Herb; Perennial.
West Asia: Afghanistan(N).
Description: 1052; Illustration: 1052.

A. pentanthus Boiss. [1179]
A. pendanthus Boiss. [1015].
Not climbing; Herb; Perennial.
West Asia: Iran(N).
Description: 1015.
The type is from Iran [1149].

A. sympileicarpus Rech.f. [1011]
Not climbing; Herb; Perennial.
West Asia: Afghanistan(N).
Description: 1052; Illustration: 1052.

A. xylocladus Rech.f. & Gilli [1011]
Not climbing; Herb; Perennial.
West Asia: Afghanistan(N).
Description: 1052; Illustration: 1052.

ASTRAGALUS Section **Falcinellus**

A. amherstianus Benth. [1011]
Not climbing; Herb; Annual.
West Asia: Afghanistan(N), Iran(N); South Asia: India(N), Pakistan(N).
Description: 1015.

A. bakaliensis Bunge [1187]
Not climbing; Herb; Annual.
West Asia: Afghanistan(N), Iran(N); South Asia: Pakistan(N); U.S.S.R.
Description: 1015, 1150, 1187; Illustration: 1145, 1150, 1187.

A. cyrusianus Parsa [1103]
Not climbing; Herb; Annual.
West Asia: Iran(N).
Description: 1103; Illustration: 1103.

A. eremophilus Boiss. [1014]
A. falcinellus Boiss. [1214]; *A. sp. 2408* of Collenette [1155].
Not climbing; Herb; Annual.
West Asia: Iran(N), Oman(N), Qatar(N), Saudi Arabia(N), United Arab Emirates(N); South Asia: Pakistan(N).
Description: 1007, 1014, 1015; Illustration: 1001, 1014, 1187.
Forage 1014.
Ali recognises a variety *falcinellus* (Boiss.)Sirj. [1214].

A. fraternellus Bornm. [1189]
A. elegantua Bornm. & Gauba [1015]; *A. elegantulus* Bornm. & Gauba [1189].
Not climbing; Herb; Annual.
West Asia: Iran(N).
Description: 1015, 1188; Illustration: 1215.

A. sarobiensis Rech.f. [1011]
Not climbing; Herb; Annual.
West Asia: Afghanistan(N).
Description: 1052; Illustration: 1052.

A. scorpiurus Bunge [1205]
> *A. humifusa* Baker [1205]; *A. lasius* Blatter [1163]; *A. punjabicus* Sirj. [1205].
> West Asia: Afghanistan(N); South Asia: Pakistan(N).

A. subumbellatus Klotzsch [1205]
> *A. camporum* Boiss. [1205]; *A. hippocrepidis* Bunge [1205].
> West Asia: Afghanistan(N), Iran(N); South Asia: Pakistan(N).
> Description: 1015, 1052.
> *A. camporum* & *A. hippocrepidis* listed as good species by Kitamura [1011].

ASTRAGALUS Section **Glycyphyllos**

Robust trailing or erect perennials. Leaves imparipinnate, simple-hairy. Inflorescences axillary, short-pedunculate. Calyx not inflated in fruit. Pod much longer than the calyx.

A. botryophorus A.A.R.Maassoumi & Podl. [1257]
> Not climbing; Herb; Perennial.
> West Asia: Iran(N).
> Description: 1257.

A. glycyphyllos L. [1015]
> *A. glycophyllos* L. [1015].
> Not climbing; Herb; Perennial.
> West Asia: Iran(N); Europe; Middle East: Turkey(N); U.S.S.R.
> Description: 1015, 1102.
> 2 subspecies, *glycyphyllos* and *glycyphylloides*, both in Iran [1102].

subsp. **glycyphylloides** (DC.)V.Matthews [1102]
> *A. glycyphylloides* DC. [1102]; *A. uliginosus* M.Bieb. [1102].
> Not climbing; Herb; Perennial.
> West Asia: Iran(N); Europe; Middle East: Turkey(N); U.S.S.R.
> Description: 1102.

subsp. **glycyphyllos** [1102]
> Not climbing; Herb; Perennial.
> West Asia: Iran(N); Europe; Middle East: Turkey(N); U.S.S.R.
> Description: 1102.

ASTRAGALUS Section **Grammocalyx**

Dwarf perennials, stemless or almost so. Hairs simple, basifixed. Leaves imparipinnate. Calyx inflated later, enclosing the fruit.

A. aspadanus Bunge [1015]
> Not climbing; Herb; Perennial.
> West Asia: Iran(N).
> Description: 1015.
> The type is from Iran [1149].

A. grammocalyx Boiss. & Hohen. [1015]
> *A. coelicolor* Sirj. & Rech.f. [1015]; *A. gorganicus* Sirj. & Rech.f. [1015].
> Not climbing; Herb; Perennial.
> West Asia: Iran(N).
> Description: 1015, 1150, 1213.

A. lineatus Lam. [1009]
> *A. cappadocicus* Boiss. [1009]; *A. chionophilus* Boiss. [1009]; *A. longidens* Freyn [1102].
> Not climbing; Herb; Perennial.
> West Asia: Iran(N), Iraq(N); Middle East: Turkey(N); U.S.S.R.
> Description: 1009, 1015, 1102; Illustration: 1009.
> Occurs only on serpentine rocks.
> Flora of Turkey records three varieties, *lineatus*, *jildisianus* & *longidens* [1102].

A. saccatus Boiss. [1015]
> Not climbing; Herb; Perennial.
> West Asia: Iran(N).
> Description: 1015.
> The type is from Iran [1149].

ASTRAGALUS Section **Halicacabus**

Woody-based perennials. Leaves imparipinnate, rachides sometimes persistent but not spinescent. Inflorescences long-pedunculate. Calyx greatly inflated in fruit.

A. chardinii Boiss. [1015]
> Not climbing; Herb; Perennial.
> West Asia: Iran(N).
> Description: 1015.

A. ebenoides Boiss. [1015]
> Not climbing; Herb; Perennial.
> West Asia: Iran(N).
> Description: 1015.

A. raddei Basilevsk. [1015]
> Not climbing; Herb; Perennial.
> West Asia: Iran(N).
> Description: 1015, 1150; Illustration: 1145, 1150.

A. wagneri Bunge [1015]
> Not climbing; Herb; Perennial.
> West Asia: Iran(N).
> Description: 1015.

ASTRAGALUS Section **Harpilobus**

Annuals with simple basifixed hairs. Leaves imparipinnate. Pod sessile, curved, semibilocular.

A. campylorrhynchus Fischer & Meyer [1009]
> *A. damascenus* Boiss. & Gaill. [1009]; *A. leiolobus* Bunge; *A. nadirius* Parsa; *A. negevensis* Zoh. & Fertig [1009].
> Not climbing; Herb; Annual.
> West Asia: Afghanistan(N), Iran(N), Iraq(N); Middle East: Lebanon(N), Syria(N), Turkey(N); South Asia: Pakistan(N); U.S.S.R.
> Description: 1009, 1102, 1150; Illustration: 1009, 1145, 1150.
> Sirjaev placed this in its own section, *Fedtschenkoana*, and recognised 2 forms, *glaber* & *tenuifolius* [1189].
> Podlech moves this from Sect. *Buceras* to Sect. *Harpilobus* [1173].
> Podlech has examined the type of *A. nadirius* [1173].
> Townsend placed this species in Section *Buceras* because of a misinterpretation of the nature of the indumentum [1173].

A. corrugatus Bertol. [1009]
> *A. corrugatus* subsp. *tenuirugis* (Boiss.)Eig [1009]; *A. cruciatus* Link [1009]; *A. cruentus* Balbis [1009]; *A. quadrisulcatus* Bunge [1009]; *A. tenuijugis* Boiss. [1011]; *A. tenuirugis* Boiss. [1009]; *A. tenuirugis* var. *brevipedunculatus* Parsa [1009].
> Not climbing; Herb; Annual.
> West Asia: Afghanistan(N), Bahrain(N), Iran(N), Iraq(N), Kuwait(N), Oman(N), Qatar(N), Saudi Arabia(N); Middle East: Israel(N), Jordan(N), Sinai(N), Syria(N), Turkey(N); South Asia: Pakistan(N).
> Description: 1009, 1015, 1150; Illustration: 1001, 1009, 1145, 1150.
> Forage 1009.
> Migahid [1007] recognises a var. *brevipes* [1007], and also regards *A. tenuirugis* as a good species [1007].

A. harpilobus Karelin & Kir. [1011]
>*A. gyzensis* Aitch. [1011].
>West Asia: Afghanistan(N).

A. hauarensis Boiss. [1009]
>*A. gysensis* Del. [1015]; *A. gyzensis* Del. [1009].
>Not climbing; Herb; Annual.
>West Asia: Afghanistan(N), Bahrain(N), Iran(N), Iraq(N), Kuwait(N), Oman(N), Saudi Arabia(N), United Arab Emirates(N); Middle East: Israel(N), Jordan(N), Sinai(N).
>Description: 1009; Illustration: 1001, 1187.
>Forage 1009.
>Two varieties, *hauarensis* and *glaber* [1009].
>Townsend regarded this as very close to *A. harpilobus*, but kept it distinct [1009].
>Leonard records a var. *brachycarpus* (Sirj. & Rech.f.)Ali [1187].

A. mareoticus Del. [1155]
>Not climbing; Herb; Annual.
>West Asia: Iran(N), Saudi Arabia(N).
>Description: 1015.
>Further material is needed from Saudi Arabia to confirm its occurrence [1155].
>Parsa's Iran record seems unlikely [1152].

A. oxyrrhynchus Fischer & Meyer [1216]
>West Asia: Iran(N).

A. reticulatus M.Bieb. [1189]
>West Asia: Afghanistan(N).
>Description: 1150.

ASTRAGALUS Section Hemiphaca
Perennials. Stems well-developed. Hairs simple. Leaves imparipinnate. Inflorescences axillary, long-pedunculate. Pods sessile, short. Maassoumi (1989 — ref.1142) deals with the Iranian species.

A. azizii A.A.R.Maassoumi [1142]
>Not climbing; Herb; Perennial.
>West Asia: Iran(N).
>Description: 1142.

A. bashmensis A.A.R.Maassoumi [1142]
>Not climbing; Herb; Perennial.
>West Asia: Iran(N).
>Description: 1142.

A. borujerdicus Parsa [1103] (provisional)
>Not climbing; Herb; Perennial.
>West Asia: Iran(N).
>Description: 1103; Illustration: 1103.
>The status and sectional position of this taxon are uncertain; the type (Hb.Parsa 20003) is at Kew [1153].

A. daenaensis Boiss. [1142]
>*A. daenensis* Boiss. [1015].
>Not climbing; Herb; Perennial.
>West Asia: Iran(N); Middle East: Turkey(N).
>Description: 1015, 1102, 1142.

A. facetus A.A.R.Maassoumi & Podl. [1142]
>Not climbing; Herb; Perennial.
>West Asia: Iran(N).
>Description: 1142, 1200; Illustration: 1172, 1200.

A. macropterus DC. [1126]
    Not climbing; Herb; Perennial.
    West Asia: Afghanistan(N); U.S.S.R.
    Description: 1150; Illustration: 1144, 1145.

A. minutissimus Freyn & Bornm. [1142]
    Not climbing; Herb; Perennial.
    West Asia: Iran(N).
    Description: 1015, 1142.

A. minutulus A.A.R.Maassoumi [1142]
    Not climbing; Herb; Perennial.
    West Asia: Iran(N).
    Description: 1142.

A. penetratus A.A.R.Maassoumi [1142]
    Not climbing; Herb; Perennial.
    West Asia: Iran(N).
    Description: 1142.

A. regestus A.A.R.Maassoumi [1142]
    Not climbing; Herb; Perennial.
    West Asia: Iran(N).
    Description: 1142.

A. zerdanus Boiss. [1142]
    A. confertiformis Sirj. & Rech.f. [1142].
    Not climbing; Herb; Perennial.
    West Asia: Iran(N).
    Description: 1015, 1170.

ASTRAGALUS Section **Hemiphragmium**
Herbaceous perennials, usually caulescent. Hairs simple, basifixed. Leaves imparipinnate. Inflorescences axillary on rather long peduncles. Pods usually unilocular, stipitate, generally pendulous when mature.

A. aksuensis Bunge [1015]
    Not climbing; Herb; Perennial.
    West Asia: Iran(N).
    Description: 1015, 1150; Illustration: 1145.
    Not recorded for Iran in Fl. U.S.S.R. 12 [1150].

A. luteocarpus Baker [1011]
    West Asia: Afghanistan(N).

ASTRAGALUS Section **Herpocaulos**
Prostrate annual with leafy stems. Hairs bifurcate. Leaves imparipinnate. Inflorescences sessile, few- to many-flowered. Pods sessile. Monotypic. See Podlech (1984 — ref. 1217).

A. vogelii (Webb)Bornm. [1217]
    Not climbing; Herb; Annual.
    West Asia: Iran(N), North Yemen(N), Oman(N), Saudi Arabia(N); Africa; South Asia: India(N), Pakistan(N).
    Description: 1007, 1217; Illustration: 1001.
    2 subspecies, *vogelii* & *fatimensis* [1217].

subsp. **fatimensis** Maire [1217]
> *A. arabicus* Bunge [1217]; *A. fatmensis* Chiov. [1217].
> Not climbing; Herb; Annual.
> West Asia: Iran(N), North Yemen(N), Oman(N), Saudi Arabia(N), South Yemen; Africa; South Asia: India(N), Pakistan(N).
> Description: 1015, 1217

subsp. **vogelii** [1217]
> *A. gautieri* Battand. & Trabut [1217]; *A. prolixus* Bunge [1217]; *A. vogelii* subsp. *prolixus* (Bunge)Maire [1217].
> Not climbing; Herb; Annual.
> West Asia: Saudi Arabia(N); Africa.
> Description: 1217.

ASTRAGALUS Section **Hololeuce**

Dwarf perennial herbs with medifixed hairs. Leaves imparipinnate. Flowers sessile in capitate long-pedunculate inflorescences. Pods sessile, short, bilocular or nearly so. Sytin (1986 — ref. 1219) has revised the species from the Caucasus.

A. **alyssoides** Lam. [1009]
> *A. elbrusensis* Boiss. [1009]; *A. elbursensis* Boiss. [1015]; *A. hololeucus* Boiss. & Buhse [1009].
> Not climbing; Herb; Perennial.
> West Asia: Iran(N), Iraq(N); Middle East: Turkey(N); U.S.S.R.
> Description: 1009, 1102, 1150; Illustration: 1009, 1145.
> Sytin upholds *A. elbrusensis*, but did not see the type of *A. alyssoides* [1219].

A. **gjunaicus** Grossh. [1219]
> *A. ketzkhovelianus* Manden. [1219]; *A. mukusiensis* Rech.f. [1219].
> Not climbing; Herb; Perennial.
> West Asia: Iran(N); Middle East: Turkey(N); U.S.S.R.
> Description: 1102, 1219.
> The Flora of Turkey maintains *A. mukusiensis* [1102].

A. **hirsutus** Vahl [1102]
> *A. chlorosphaerus* Boiss. & Noe [1102].
> Not climbing; Herb; Perennial.
> West Asia: Iran(N); Middle East: Turkey(N).
> Description: 1102.
> Iranian material at Kew determined by Townsend. The taxon is not mentioned by Parsa [1153].

ASTRAGALUS Section **Hymenostegis**

Perennial woody-based herbs or shrublets. Leaves paripinnate, simple-hairy, rachis spine-tipped. Inflorescences axillary, usually long-pedunculate. Calyx becoming inflated in fruit. Pod 1-2-seeded. A distinctive section, but species delimitation is difficult.

A. **bounophilus** Boiss. [1220]
> *A. bunophilus* Boiss. [1149].
> Not climbing; Herb; Perennial.
> West Asia: Iran(N).
> Description: 1015.
> The type is from Iran [1149].
> The spelling 'bunophilus' appears on the type label, & in Bunge; but the name was published as 'bounophilus' [1149].

A. **brunsianus** Bornm. [1220]
> Not climbing; Herb,Herb/Shrub; Perennial.
> West Asia: Iran(N).
> Description: 1015, 1216.

A. cordatus Bunge [1220]
>Not climbing; Shrub; Perennial.
>West Asia: Iran(N).
>Description: 1015.
>The type is from Iran [1149].

A. glumaceus Boiss. [1220]
>Not climbing; Shrub; Perennial.
>West Asia: Iran(N).
>Description: 1015.
>The type is from Iran [1149].

A. hirticalyx Boiss. [1220]
>Not climbing; Herb/Shrub; Perennial.
>West Asia: Iran(N); Middle East: Turkey(N).
>Description: 1015, 1102.

A. hymenocystis Fischer & Meyer [1220]
>Not climbing; Shrub; Perennial.
>West Asia: Iran(N); Middle East: Turkey(N).
>Description: 1015, 1102.

A. hymenostegis Fischer [1220]
>Not climbing; Shrub; Perennial.
>West Asia: Iran(N).
>Description: 1015.
>The type is from Iran [1149].

A. kohrudicus Bunge [1220]
>Not climbing; Shrub; Perennial.
>West Asia: Iran(N).
>Description: 1015.
>The type is from Iran [1149].

A. lagopoides Lam. [1221]
>*A. brachypodus* Boiss. [1220]; *A. chrysostachys* Boiss. [1009]; *A. chrysostachys* var. *villosus* Bornm. [1009]; *A. lagurus* Willd. [1221]; *A. melanostictus* Freyn [1009]; *A. melanosticus* Freyn [1015]; *A. nervistipulus* Boiss. & Hausskn. [1009].
>Not climbing; Herb/Shrub; Perennial.
>West Asia: Iran(N), Iraq(N); Middle East: Turkey(N); U.S.S.R.
>Description: 1009, 1102, 1150.
>Two varieties, *lagurus* and *villosus* [1009].

A. laguriformis Freyn [1220]
>*A. laguroides* Freyn [1009].
>Not climbing; Herb/Shrub; Perennial.
>West Asia: Iraq(N).
>Description: 1009, 1015; Illustration: 1009.
>Parsa lists this taxon for Iran, but without a definite record [1015].
>Assadi lists and illustrates material from Iran as 'A.aff.laguriformis' [1176].

A. leucargyreus Bornm. [1220]
>Not climbing; Herb/Shrub; Perennial.
>West Asia: Iran(N).
>Description: 1015, 1158.

A. manucehrii Sirj. & Rech.f. [1220]
>Not climbing; Herb/Shrub; Perennial.
>West Asia: Iran(N).
>Description: 1156.

A. naftabensis Sirj. & Rech.f. [1220]
Not climbing; Herb/Shrub; Perennial.
West Asia: Iran(N).
Description: 1156.
Rechinger et al. recognise a var. *brevipedunculatus* Sirj. & Rech.f. [1220].

A. paralurges Bunge [1220]
Not climbing; Herb; Perennial.
West Asia: Iran(N).
Description: 1015.
The type is from Iran [1149].

A. persicus (DC.)Fischer & Meyer [1009]
A. kapherrianus Fischer [1102]; *A. mesopotamicus* Boiss. [1009]; *A. olivieri* Bunge [1009]
Not climbing; Herb/Shrub; Perennial.
West Asia: Iran(N), Iraq(N); Middle East: Turkey(N); U.S.S.R.
Description: 1009, 1102, 1150; Illustration: 1150.
Very similar to *A. lagurus* (now *A. lagopoides*), and probably conspecific [1009].
The record from Iraq is very doubtful [1009].

A. porphyrodon C.Towns. [1009]
Not climbing; Herb/Shrub; Perennial.
West Asia: Iraq(N).
Description: 1009, 1132; Illustration: 1009, 1132.

A. recognitus Fischer [1220]
Not climbing; Herb/Shrub; Perennial.
West Asia: Iran(N).
Description: 1015.
The type is from Iran[1149]

A. rubriflorus Bunge [1220]
Not climbing; Shrub; Perennial.
West Asia: Iran(N).
Description: 1015.
The type is from Iran [1149].

A. rubrostriatus Bunge [1220]
Not climbing; Shrub; Perennial.
West Asia: Iran(N).
Description: 1015.
Parsa mentions a var. *longepedunculatus* Parsa [1015].
The type is from Iran [1149].

A. sciureus Boiss. & Hohen. [1220]
Not climbing; Shrub; Perennial.
West Asia: Iran(N).
Description: 1015.
Parsa mentions a var. *subsessilis* Bornm. (as "subsesselis") [1015].

A. seidabadensis Bunge [1220]
Not climbing; Shrub; Perennial.
West Asia: Iran(N).
Description: 1015.
The type is from Iran [1149].

A. sirensis Turrill [1220]
Not climbing; Herb/Shrub; Perennial.
West Asia: Iran(N).
Description: 1165.

A. straussii Bornm. [1220]
Not climbing; Shrub; Perennial.
West Asia: Iran(N).
Description: 1015, 1104
Parsa mentions a var. *albiflora* Bornm [1015].

A. tabrisianus Boiss. & Buhse [1015]
Not climbing; Herb/Shrub; Perennial.
West Asia: Iran(N).
Description: 1015.
Parsa mentions a var. *pedunculatus* Turrill [1015].

A. tenax Bunge [1015]
Not climbing; Shrub; Perennial.
West Asia: Iran(N).
Description: 1015.
The type is from Iran [1149].

A. tephreschensis Rech.f. et al. [1220] (Provisional)
West Asia: Iran(N).
Description: 1220.
Bornmuller used this name (as 'tefreschensis') but did not validly publish it [1104].
Rechinger et al. used the name in their key and probably validated it [1220].

A. thyrsiflorus Sirj. & Rech.f. [1220]
Not climbing; Herb; Perennial.
West Asia: Iran(N).
Description: 1015, 1168; Illustration: 1168.

A. uranolimneus Boiss. [1220]
Not climbing; Herb/Shrub; Perennial.
West Asia: Iran(N); Middle East: Turkey(N); U.S.S.R.
Description: 1015, 1150.
Close to *A. lagopodioides* (now *A. zohrabi*), and may be synonymous [1102].
There is no record for Iran in Flora of Turkey (or in Fl. U.S.S.R. 12).
MedChecklist considers this to be a doubtful synonym of *A. zohrabi* [1221].

A. velenowskyi Nab. [1102]
A. velenovskyi Nab. [1015].
Not climbing; Shrub; Perennial.
West Asia: Iran(N); Middle East: Turkey(N).
Description: 1015, 1102.
Parsa queries the locality; the Iran record is, at best, doubtful [1015].
Not recorded for Iran by Rechinger et al. [1220].

A. zohrabi Bunge [1221]
A. lagopodioides Willd. [1221].
Not climbing; Herb/Shrub; Perennial.
West Asia: Iran(N); Middle East: Turkey(N).
Description: 1102.
Treated as synonymous with *A. lagopodioides* in Flora of Turkey [1102].

ASTRAGALUS Section Hypoglottoidei
Perennials, acaulescent or shortly caulescent, with basifixed hairs. Leaves imparipinnate. Inflorescences pedunculate. Pods short or elongated, sometimes inflated, almost bilocular. See Maassoumi (1988 — ref.1222) for the Iranian species.

A. arenarius Pallas [1143]
A. hypoglottis sensu Parsa [1015, 1143].
Not climbing; Herb; Perennial.
West Asia: Iran(N).
Description: 1015, 1150; Illustration: 1150.
Fl. U.S.S.R. 12 makes *A. arenarius* a synonym of *A. danicus* Retz [1150].
The application of the name *A. hypoglottis* to Iranian plants is risky pending a full revision of the section [1143].

A. atricapillus Bornm. [1222]
Not climbing; Herb; Perennial.
West Asia: Iran(N).
Description: 1015, 1201, 1222; Illustration: 1015, 1201.

A. brachypetalus Trautv. [1222]
A. brachyanthus Freyn & Sint. [1102].
Not climbing; Herb; Perennial.
West Asia: Iran(N); Middle East: Turkey(N); U.S.S.R.
Description: 1102, 1150, 1222; Illustration: 1150.

A. haematinus Sirj. & Rech.f. [1222]
A. astarabadensis Sirj. & Rech.f. [1222]; *A. asterabadensis* Sirj. & Rech.f. [1103]; *A. viciaefolius* var. *haematinus* (Sirj. & Rech.f.)Parsa [1222].
Not climbing; Herb; Perennial.
West Asia: Iran(N).
Description: 1170, 1222.

A. herbertii A.A.R.Maassoumi [1222]
Not climbing; Herb; Perennial.
West Asia: Iran(N).
Description: 1222.

A. nurensis Boiss. & Buhse [1222]
Not climbing; Herb; Perennial.
West Asia: Iran(N).
Description: 1015, 1222.
Parsa mentions a var. *melanotrichus* Bornm. & Gauba [1015].

A. perpexus A.A.R.Maassoumi [1222]
Not climbing; Herb; Perennial.
West Asia: Iran(N).
Description: 1222.

A. pish-chakensis A.A.R.Maassoumi [1222]
Not climbing; Herb; Perennial.
West Asia: Iran(N).
Description: 1222.

A. pseudofragrans C.Towns. [1009]
Not climbing; Herb; Perennial.
West Asia: Iraq(N).
Description: 1009, 1132; Illustration: 1009, 1132.

A. rimarum Bornm. [1222]
Not climbing; Herb; Perennial.
West Asia: Iran(N).
Description: 1015, 1201, 1222; Illustration: 1201.
Parsa placed this taxon in Section *Tapinodes* [1015].

A. tibetanus Bunge [1126]
A. chadjanensis Franchet [1190]; *A. olufsenii* Freyn [1190].
Not climbing; Herb; Perennial.
West Asia: Afghanistan(N), Iran(N); East Asia:China(N) (Xizang Zizhiqu); South Asia: Pakistan(N); U.S.S.R.
Description: 1015, 1150; Illustration: 1144, 1145.

A. vexillaris Boiss. [1009]
Not climbing; Herb; Perennial.
West Asia: Iraq(N); Middle East: Syria(N), Turkey(N).
Description: 1009, 1102.
The Iraq record is doubtful; the locality is given only as 'Mesopotamia' [1009].

A. viciifolius DC. [1102]
> *A. saxatilis* Freyn & Bornm. [1102]; *A. viciaefolius* M.Bieb. [1015].
> Not climbing; Herb; Perennial.
> West Asia: Iran(N); Middle East: Turkey(N).
> Description: 1015, 1102, 1150.

ASTRAGALUS Section **Hypsophilus**
Caulescent herbaceous perennials with medifixed hairs. Inflorescence long-pedunculate, few-flowered, capitate. Calyx inflated in fruit. Pods compressed, sub-bilocular.

A. nivalis Karelin & Kir. [1190]
> *A. orthanthoides* Boriss. [1190]; *A. orthanthus* Freyn [1190].
> Not climbing; Herb; Perennial.
> West Asia: Afghanistan(N); U.S.S.R.
> Description: 1150.

ASTRAGALUS Section **Hystrix**
Dwarf perennial cushion-forming shrubs.

A. hystrix Fischer & Meyer [1015]
> Not climbing; Shrub; Perennial.
> West Asia: Iran(N).
> Description: 1015.
> Parsa(1948) placed this species in Section *Acanthace*; Deml transfers it to *Hystrix* [1171].

ASTRAGALUS Section **Incani**
Formerly section *Proselius*. Perennial herbs, acaulescent or almost so. Hairs medifixed. Leaves imparipinnate, tri- or unifoliolate in a few species. Inflorescences long-pedunculate, axillary. Calyx not inflated, often splitting later. Pods sessile or stipitate, very variable in form, virtually bilocular. A difficult section, much in need of revision.

A. abnormalis Rech.f. [1157]
> Not climbing; Herb; Perennial.
> West Asia: Iran(N).
> Description: 1157.

A. ackerbergensis Freyn & Sint. [1233]
> Not climbing; Herb; Perennial.
> West Asia: Iran(N); U.S.S.R.
> Description: 1015, 1150.

A. ancistrocarpus Boiss. & Hausskn. [1009]
> *A. nitidulus* Hand.-Mazz. [1009].
> Not climbing; Herb; Perennial.
> West Asia: Iran(N), Iraq(N); Middle East: Jordan(N), Syria(N), Turkey(N).
> Description: 1009, 1102; Illustration: 1009.
> Eig recognised three varieties, *brachycarpus, stepposus, nitidulus*. Townsend, in Flora of Iraq [1009], does not [1192].

A. askius Bunge [1233]
> Not climbing; Herb; Perennial.
> West Asia: Iran(N).
> Description: 1015.
> Parsa mentions a var. "busheanus" Boiss. [1015].
> Rechinger & Dulfer treat this as a species, *A. buhseanus* (q.v.) [1233].

A. bayattii Bornm. & Gauba [1233]
Not climbing; Herb; Perennial.
West Asia: Iran(N).
Description: 1015, 1180; Illustration: 1180.

A. bordschensis Bornm. [1233]
Not climbing; Herb; Perennial.
West Asia: Iran(N).
Description: 1015, 1188.

A. brahuicus Boiss. [1233]
Not climbing; Herb; Perennial.
South Asia: Pakistan(N).
Description: 1015.
Parsa's locality is in Pakistan, as is the type locality. The taxon does not occur in West Asia [1015].

A. buhseanus Bunge [1233]
A. askius var. *buhseanus* Boiss. [1233]; *A. refractus* Boiss. & Buhse [1233].
Not climbing; Herb; Perennial.
West Asia: Iran(N).
Description: 1015.

A. campylosema Boiss. [1015]
Not climbing; Herb; Perennial.
West Asia: Iran(N); Middle East: Turkey(N).
Description: 1015, 1102, 1150.
3 subspecies, *campylosema, atropurpureus & nigripilis* [1102]; none of these is recorded from Iran in Flora of Turkey, Flora of U.S.S.R. or Rechinger & Dulfer (1969) [1102].
Archibald 1831, 2168 & 2068 (K) are all named as this species; the subspecies is uncertain [1153].

subsp. atropurpureus (Boiss.)Chamberlain [1102]
A. atropurpureus Boiss. [1102]; *A. pithyusarum* Bornm. [1102].
Not climbing; Herb; Perennial.
West Asia: Iran(N); Middle East: Turkey(N).
Description: 1102.
Parsa lists this taxon, but without a definite Iran record [1015]. He also mentions var. *rewaduzicus* Nab. [1015], of which the status is uncertain according to the Flora of Turkey [1102].

subsp. campylosema [1102]
A. feddei Sirj. [1102].
Not climbing; Herb; Perennial.
Middle East: Turkey(N).
Description: 1102.

A. cerasinus Baker [1011]
Not climbing; Herb; Perennial.
West Asia: Afghanistan(N); South Asia: Pakistan(N).
Description: 1162.

A. chaetopodus Bunge [1233]
Not climbing; Herb; Perennial.
West Asia: Iran(N).
Description: 1015.
The type is from Iran [1149].

A. cinereus Willd. [1233]
Not climbing; Herb; Perennial.
West Asia: Iran(N); U.S.S.R.
Description: 1150.

A. coeruleus Parsa (provisional) [1103]
Not climbing; Herb; Perennial.
West Asia: Iran(N).
Description: 1015.
This does not appear to be a published name, and the epithet *coeruleus* is also preoccupied at least once [1152]. The name is not mentioned by Rechinger & Dulfer [1233].

A. confiniorum Boriss. [1233]
Not climbing; Herb; Perennial.
West Asia: Iran(N); U.S.S.R.

A. confusus Bunge [1233]
Not climbing; Herb; Perennial.
West Asia: Iran(N).
Description: 1015.
The type is from Iran [1149].

A. cottonianus Aitch. & Baker [1233]
A. dilankuri Lipsky [1233]; *A. involucratus* Lipsky [1233]; *Chondrocarpus dilankuri* (Lipsky)Nevski [1190].
Not climbing; Herb; Perennial.
West Asia: Afghanistan(N).
Description: 1015, 1162; Illustration: 1145.

A. curvirostris Boiss. [1233]
Not climbing; Herb; Perennial.
West Asia: Iran(N), Iraq(N).
Description: 1009, 1015.
2 subspecies, *curvirostris* & *psilocarpus* [1009]; Bornmuller described another subspecies, *oligozygon*, from Iran [1188].

subsp. **curvirostris** [1009]
Not climbing; Herb; Perennial.
West Asia: Iran(N).
Description: 1009.

subsp. **psilocarpus** C.Towns. [1009]
Not climbing; Herb; Perennial.
West Asia: Iran(N), Iraq(N).
Description: 1009, 1132.

A. cuscutae Bunge [1233]
Not climbing; Herb; Perennial.
West Asia: Iran(N); U.S.S.R.
Description: 1015, 1150.
Parsa mentions a var. *pulcher* Stapf from Iran [1015].

A. cyclophyllon G.Beck [1233]
Not climbing; Herb; Perennial.
West Asia: Iran(N).
Description: 1015.

A. dekazygus Sirj. & Rech.f. [1233]
Not climbing; Herb; Perennial.
West Asia: Afghanistan(N); South Asia: Pakistan(N).
Description: 1052; Illustration: 1052.
Kitamura placed this as a synonym of *A. cerasinus* [1011].

A. demavendicus Boiss. & Buhse [1233]
A. demawendicus Boiss. & Buhse [1180]; *A. rudbaricus* Bunge [1233].
Not climbing; Herb; Perennial.
West Asia: Iran(N).
Description: 1015.

A. drepanophorus Bornm. [1233]
A. drepanophora Bornm. [1188].
Not climbing; Herb; Perennial.
West Asia: Iran(N).
Description: 1015, 1188.

A. faridanicus Parsa [1103]
  Not climbing; Herb; Perennial.
  West Asia: Iran(N).
  Description: 1103; Illustration: 1103.
  Ali has determined the type sheet (Gentry 14894 - Kew) as *A. micrancistrus* [1153].

A. fridae Rech.f. [1233]
  Not climbing; Herb; Perennial.
  West Asia: Iran(N).
  Description: 1015, 1168; Illustration: 1168.

A. fuliginosus G.Beck [1233]
  Not climbing; Herb; Perennial.
  West Asia: Iran(N).
  Description: 1015, 1150.

A. gillettii C.Towns. [1009]
  Not climbing; Herb; Perennial.
  West Asia: Iraq(N).
  Description: 1009, 1132; Illustration: 1009, 1132.

A. groetzbachii Podl. [1098]
  Not climbing; Herb; Perennial.
  West Asia: Afghanistan(N).
  Description: 1098; Illustration: 1098.

A. gudrunensis Boiss. [1009]
  A. avromanicus Rech.f. [1009].
  Not climbing; Herb; Perennial.
  West Asia: Iran(N), Iraq(N).
  Description: 1009; Illustration: 1009.

A. gueldenstaedtiae Bunge [1233]
  Not climbing; Herb; Perennial.
  West Asia: Iran(N).
  Description: 1015.

A. heteromorphus Boriss. [1233]
  Not climbing; Herb; Perennial.
  West Asia: Iran(N); U.S.S.R.
  Description: 1150, 1234.

A. jolderensis B.Fedtsch. [1233]
  A. subalpinus Freyn [1233]; *A. tetanocarpus* Bornm. & Gauba [1233]; *A. trottianus* Parsa [1233].
  Not climbing; Herb; Perennial.
  West Asia: Iran(N); U.S.S.R.
  Description: 1015, 1150.

A. latifolius Lam. [1233]
  A. choicus Bunge [1102]; *A. latifolius* var. *choicus* Boiss. [1102].
  Not climbing; Herb; Perennial.
  West Asia: Iran(N); Middle East: Turkey(N); U.S.S.R.
  Description: 1015, 1102, 1150.
  The type of *A. choicus* is from Iran [1149].

A. leucophanus Bornm. [1233]
  Not climbing; Herb; Perennial.
  West Asia: Iran(N).
  Description: 1015.

A. longicuspis Bunge [1233]
  A. longicuspidatus St.Lager [1233].
  Not climbing; Herb; Perennial.
  West Asia: Iran(N); U.S.S.R.
  Description: 1015, 1150.
  The type is from Iran [1149].

A. lovensis Rech.f. [1233]
Not climbing; Herb; Perennial.
West Asia: Iran(N).
Description: 1233.

A. mangeri Bornm. [1233]
A. panjaoensis Sirj. & Rech.f. [1233].
Not climbing; Herb; Perennial.
West Asia: Afghanistan(N).
Description: 1052, 1218; Illustration: 1052.

A. mercklinii Boiss. & Buhse [1233]
A. holdichianus Aitch. & Baker [1233]; *A. maculatus* Bunge [1233]; *A. submaculatus* Boriss. [1233]; *A. superfluus* Rech.f. & Koie [1233]; *A. sykesiae* N.Simpson [1233].
Not climbing; Herb; Perennial.
West Asia: Afghanistan(N), Iran(N); South Asia: Pakistan(N); U.S.S.R.
Description: 1015, 1150, 1187; Illustration: 1011, 1052.
Fibre 1159.

A. micrancistrus Boiss. & Hausskn. [1009]
Not climbing; Herb; Perennial.
West Asia: Iran(N), Iraq(N).
Description: 1009, 1015; Illustration: 1009.
May occur in Turkey, according to Flora of Iraq [1009].

A. monozyx Bornm. [1233]
Not climbing; Herb; Perennial.
West Asia: Iran(N).
Description: 1015.

A. oproselius Rech.f. [1233]
Not climbing; Herb; Perennial.
West Asia: Iran(N).
Description: 1157.

A. platysematus Bunge [1233]
Not climbing; Herb; Perennial.
West Asia: Iran(N).
Description: 1015.
The type is from Iran[1149]

A. procerus Boiss. & Hausskn. [1009]
Not climbing; Herb; Insufficiently known; Perennial.
West Asia: Iran(N), Iraq(N).
Description: 1009, 1015.
Known only from the type, collected more than 100 years ago. The collecting locality is uncertain, and could be in either Iran or Iraq [1009].

A. pseudocyclophyllus Rech.f. [1233]
Not climbing; Herb; Perennial.
West Asia: Iran(N).
Description: 1235.

A. punctatus Bunge [1233]
Not climbing; Herb/Shrub; Perennial.
West Asia: Iran(N).
Description: 1015, 1150.
The type is from Iran [1149].

A. quinquefoliolatus Bunge [1233]
A. quinquefoliatus Bunge [1233].
Not climbing; Herb; Perennial.
West Asia: Iran(N).
Description: 1015.
The type is from Iran [1149].

A. refractus C.A. Meyer [1233]
Not climbing; Herb; Perennial.
West Asia: Iran(N).
Description: 1150; Illustration: 1150.

A. riouxii Rech.f. [1233]
Not climbing; Herb; Perennial.
West Asia: Iran(N).
Description: 1233.

A. robustus Bunge [1233]
A. subrobustus Boriss. [1233].
Not climbing; Herb; Perennial.
West Asia: Iran(N); U.S.S.R.
Description: 1015, 1150; Illustration: 1150.

A. rollovii Grossh. [1233]
Not climbing; Herb; Perennial.
West Asia: Iran(N).
Description: 1015, 1150.

A. rostratus C.Meyer [1150]
Not climbing; Herb; Perennial.
West Asia: Iran(N); U.S.S.R.
Description: 1150.

A. sangimashensis Rech.f. [1233]
Not climbing; Herb; Perennial.
West Asia: Afghanistan(N).
Description: 1233.

A. schirkuhicus Bornm. [1233]
A. shirkuhicus Bornm. [1015].
Not climbing; Herb; Perennial.
West Asia: Iran(N).
Description: 1015, 1188.

A. spruneri Boiss. [1233]
A. sprunerianus Bunge [1233]; *A. tempskyanus* Freyn & Bornm. [1233]; *A. thessalus* Boiss. [1233].
Not climbing; Herb; Perennial.
West Asia: Iran(N); Europe; Middle East: Turkey(N).
Description: 1015.
Parsa records this taxon from Iran, but this seems most unlikely [1015]; there is no record for Iran in Rechinger & Dulfer (1969) [1233].
Parsa 523 (K) was determined as this by him, but the pods appear too short and broad [1153].
Parsa mentions a var. *thessalus* Boiss. [1015].

A. subalpinus Boiss. & Buhse [1233]
Not climbing; Herb; Perennial.
West Asia: Iran(N).
Description: 1015.

A. supervisus (O.Kuntze) E.Sheldon [1190]
A. candolleanus Boiss. [1190].
Not climbing; Herb; Perennial.
West Asia: Iran(N); Middle East: Turkey(N); U.S.S.R.
Description: 1015, 1150; Illustration: 1150.
The taxon may be a depauperate form of *A. latifolius* [1102].
Parsa mentions a var. *acuminatus* G.Beck.
Notes on specimens at Kew (e.g. Cowan & Darlington 1088) state that the flower colour varies from yellow to red [1153].

A. taschkendicus Bunge [1098]
Not climbing; Herb; Perennial.
West Asia: Afghanistan(N).
Description: 1150; Illustration: 1144, 1145, 1150.

A. tenellus Bunge [1233]
    A. kuntzei E.Sheldon [1233].
    Not climbing; Herb; Perennial.
    West Asia: Iran(N).
    Description: 1015.

A. thionanthus Bornm. [1233]
    Not climbing; Herb; Perennial.
    West Asia: Iran(N).
    Description: 1015.

A. tigridis Boiss. [1009]
    Not climbing; Herb; Perennial.
    West Asia: Iran(N), Iraq(N); Middle East: Turkey(N).
    Description: 1009, 1015, 1102.
    The record from Iraq is very doubtful; the collecting locality is uncertain, but is probably in Turkey [1009].

A. ulothrix G.Beck [1015]
    Not climbing; Herb; Perennial.
    West Asia: Iran(N).
    Description: 1015.
    Not mentioned by Rechinger & Dulfer (1969) [1233].

A. xiphidiopsis Bornm. [1233]
    Not climbing; Herb; Perennial.
    West Asia: Iran(N).
    Description: 1015, 1188.

ASTRAGALUS Section **Irinaea**
Many-stemmed woody-based perennials, some prostrate. Hairs medifixed. Inflorescences loosely few-flowered. Pods short-stipitate, swollen at the base and tapering to a point.

A. saxifractor Rech.f. & Gilli [1052]
    Not climbing; Herb; Perennial.
    West Asia: Afghanistan(N).
    Description: 1052; Illustration: 1052.

ASTRAGALUS Section **Komaroviella**
Herbaceous perennials, caulescent (sometimes shortly). Hairs simple. Inflorescences lax, axillary. Pods long-stipitate, bilocular, inflated.

A. jabbor-khailii Kitam. [1011]
    A. bagramiensis Sirj. & Rech.f. [1011].
    Not climbing; Herb/Shrub; Perennial.
    West Asia: Afghanistan(N).
    Description: 1011, 1052; Illustration: 1011, 1052.

A. kuschakevitschii O.Fedtsch. [1185]
    Not climbing; Herb; Perennial.
    West Asia: Afghanistan(N); U.S.S.R.
    Description: 1150; Illustration: 1145, 1150.

A. truncato-alatus Sirj. & Rech.f. [1011]
    Not climbing; Herb; Perennial.
    West Asia: Afghanistan(N).
    Description: 1052; Illustration: 1052.

ASTRAGALUS Section **Laguropsis**
Herbaceous caulescent perennials with medifixed hairs. Leaves imparipinnate. Inflorescences long-pedunculate, many-flowered. Calyx becoming inflated. Pods remaining within the calyx, few-seeded, bilocular.

A. subsecundus Boiss. & Hohen. [1015]
Not climbing; Herb; Perennial.
West Asia: Iran(N); Middle East: Turkey(N).
Description: 1015, 1102.

A. urgutinus Lipsky [1015]
Not climbing; Herb/Shrub; Perennial.
West Asia: Iran(N); U.S.S.R.
Description: 1015, 1150; Illustration: 1144, 1145.
Parsa lists this taxon, but without any definite Iran record [1015].

ASTRAGALUS Section **Laxiflori**
Perennials with long stems and basifixed hairs. Leaves imparipinnate. Similar to section *Astragalus*, but inflorescences more than 15-flowered. Pod included in the calyx. See Agerer-Kirchhoff & Agerer (1977 — ref. 1223).

A. bracteosus Boiss. & Noe [1223]
A. butleri Dinsm. [1223].
Not climbing; Herb; Perennial.
West Asia: Iran(N); Middle East: Syria(N), Turkey(N).
Description: 1015, 1223.
Agerer-Kirchhoff & Agerer do not record this taxon from Iran [1223].

A. chlorostegius Boiss. & Hausskn. [1223]
Not climbing; Herb; Perennial.
West Asia: Iraq(N); Middle East: Syria(N).
Description: 1009, 1223; Distribution Map: 1223;
Illustration: 1223.
Iraq record must remain somewhat doubtful as the specimen is of an immature plant [1223].

A. dictyolobus Bunge [1223]
Not climbing; Herb; Perennial.
West Asia: Iran(N).
Description: 1015, 1223; Distribution Map: 1223; Illustration: 1223.

A. phlomoides Boiss. [1223]
A. baijiensis C.Towns. [1223].
Not climbing; Herb; Perennial.
West Asia: Iraq(N).
Description: 1009, 1132, 1223; Distribution Map: 1223; Illustration: 1009, 1132, 1223.

A. tawilicus C.Towns. [1223]
A. dictyolobus sensu auctt. [1009].
Not climbing; Herb; Perennial.
West Asia: Iran(N), Iraq(N).
Description: 1009, 1132, 1223; Distribution Map: 1223; Illustration: 1009, 1132, 1223.

ASTRAGALUS Section **Leucocercis**
Short-stemmed shrubby perennials. Leaves paripinnate; rachis thick, persistent, spinose. Hairs bifurcate. Inflorescences short-pedunculate, axillary. Pods unilocular, exceeding the calyx.

A. cornu-caprae Sirj. & Rech.f. [1224]
Not climbing; Herb/Shrub; Perennial.
West Asia: Iran(N).
Description: 1225.

A. crassispinus Bunge (provisional) [1224].
> Not climbing; Shrub; Perennial.
> West Asia: Afghanistan(N).
> Description: 1149.
> The type, which is from Afghanistan at Kew, is sterile, but Bunge regarded the taxon as very distinct [1224].

A. curviflorus Boiss. [1224]
> Not climbing; Herb, Herb/Shrub; Perennial.
> West Asia: Iran(N).
> Description: 1015.

A. mucronifolius Boiss. [1224]
> *A. taftanicus* Parsa [1157].
> Not climbing; Herb; Perennial.
> West Asia: Iran(N).
> Description: 1157, 1187.
> There are two subspecies, *brevidentatus* and *robustus* Sirj. & Rech.f. [1224].

subsp. **robustus** Sirj. & Rech.f. [1156]
> Not climbing; Herb; Perennial.
> West Asia: Iran(N).
> Description: 1156.

A. ovoideus Sirj. & Rech.f. [1224]
> Not climbing; Herb/Shrub; Perennial.
> West Asia: Iran(N).
> Description: 1156.

A. phyllokentrus Hausskn. & Bornm. [1224]
> Not climbing; Herb/Shrub; Perennial.
> West Asia: Iran(N).
> Description: 1015, 1158.

A. semnanensis Bornm. & Rech.f. [1224]
> Not climbing; Herb/Shrub; Perennial.
> West Asia: Iran(N).
> Description: 1015, 1168; Illustration: 1168.

A. talimansurensis Sirj. & Rech.f. [1224]
> Not climbing; Herb/Shrub; Perennial.
> West Asia: Iran(N).
> Description: 1226.

ASTRAGALUS Section **Macrocystis**
Perennials, some woody-based. Inflorescences axillary, long-pedunculate. Calyx soon inflated. Pod included within the calyx, bilocular.

A. neurophyllus Franchet [1181]
> *A. megalomerus* sensu Podl. [1181, 1185]; *A. neurocentron* Franchet [1181].
> West Asia: Afghanistan(N).

A. xanthomeloides Korovin & M.Popov [1181]
> Not climbing; Herb; Perennial.
> West Asia: Afghanistan(N); U.S.S.R.
> Description: 1150; Illustration: 1144, 1150.

ASTRAGALUS Section **Macrosemium**
Perennials, subacaulescent; hairs simple. Inflorescences very short. Calyx long-cylindric. Pods sessile, small, subunilocular, few-seeded. Monotypic.

A. paradoxus Bunge [1102]
Not climbing; Herb; Perennial.
West Asia: Iran(N); Middle East: Turkey(N).
Description: 1015, 1102, 1150; Illustration: 1151.

ASTRAGALUS Section **Malacothrix**

Caulescent or subacaulescent perennials, hairs simple, basifixed. Leaves imparipinnate. Inflorescence pedunculate, usually elongating later. Calyx slightly inflated, ruptured by the sessile or stipitate, bilocular or sub-bilocular pod. A very difficult section.

A. ahouicus Parsa [1103]
Not climbing; Herb; Perennial.
West Asia: Iran(N).
Description: 1103; Illustration: 1103.
Parsa distinguished a var. *pumilus*; the type (Hb.Parsa 20012) is at Kew [1103].

A. anserinifolius Boiss. [1179]
A. anserinaefolius Boiss. [1015].
Not climbing; Herb; Perennial.
West Asia: Iran(N).
Description: 1015.
The type is from Iran [1149].

A. babakhanloui A.A.R.Maassoumi & Podl. [1200]
Not climbing; Herb; Perennial.
West Asia: Iran(N).
Description: 1200; Illustration: 1200.
Known only from the type, collected in 1984 [1200].

A. beckii Bornm. [1179]
A. becki Bornm. [1015]; *A. viciaefolius* G.Beck [1158].
Not climbing; Herb; Perennial.
West Asia: Iran(N).
Description: 1015, 1158; Illustration: 1201.
Parsa mentions this without giving a full description. However, the type (Bornmuller 6770) is from Iran, and is at Kew [1153].

A. birjandicus Parsa [1015]
A. birjandieus Parsa [1015].
Not climbing; Herb; Perennial.
West Asia: Iran(N).
Description: 1015.
Not mentioned by Rechinger et al. [1179].

A. chahartaghensis A.A.R.Maassoumi & Podl. [1257]
Not climbing; Herb; Perennial.
West Asia: Iran(N).
Description: 1257.

A. comosus Bunge [1179]
Not climbing; Herb; Perennial.
West Asia: Iran(N).
Description: 1015.
The type is from Iran [1149].

A. elegans Bunge [1015]
Not climbing; Herb; Perennial.
West Asia: Iran(N).
Description: 1015, 1150.
The type is from Iran [1149].

A. eriopodus Boiss. [1179]
   A. stenostachys G.Beck [1179].
   Not climbing; Herb; Perennial.
   West Asia: Iran(N); Middle East: Turkey(N).
   Description: 1015, 1102.

A. eugenii Grossh. [1150]
   Not climbing; Herb; Perennial.
   West Asia: Iran(N); U.S.S.R.
   Description: 1150.
   There is also an *A.eugeniae* B.Fedtsch. [1190].

A. expectatus A.A.R.Maassoumi [1258]
   Not climbing; Herb; Perennial.
   West Asia: Iran(N).
   Description: 1258.

A. galbineus A.A.R.Maassoumi [1196]
   Not climbing; Herb; Perennial.
   West Asia: Iran(N).
   Description: 1196; Illustration: 1172, 1196.

A. gharemanii A.A.R.Maassoumi & Podl. [1200]
   Not climbing; Herb; Perennial.
   West Asia: Iran(N).
   Description: 1200.

A. ghoortapacensis A.A.R.Maassoumi [1258]
   Not climbing; Herb; Perennial.
   West Asia: Iran(N).
   Description: 1258.

A. griseus Boiss. [1179]
   A. deluensis Bunge [1179].
   Not climbing; Herb; Perennial.
   West Asia: Iran(N); Middle East: Turkey(N).
   Description: 1015.

A. heterodoxus Bunge [1172]
   A. heteroxus Bunge [1015].
   Not climbing; Herb; Perennial.
   West Asia: Iran(N).
   Description: 1015.

A. holopsilus Bunge [1179]
   Not climbing; Herb; Perennial.
   West Asia: Iran(N).
   Description: 1015.
   The type is from Iran [1149].

A. holosemius Bunge [1179]
   Not climbing; Herb; Perennial.
   West Asia: Iran(N).
   Description: 1015.
   The type is from Iran [1149].

A. idoneus A.A.R.Maassoumi & Podl. [1257]
   Not climbing; Herb; Perennial.
   West Asia: Iran(N).
   Description: 1257.

A. kabristanicus Grossh. [1150]
   Not climbing; Herb; Perennial.
   West Asia: Iran(N); U.S.S.R.
   Description: 1150.

A. kendewanensis Gilli [1179]
Not climbing; Herb; Perennial.
West Asia: Iran(N).
Description: 1015, 1227.

A. kermanicus Parsa [1103]
Not climbing; Herb; Perennial.
West Asia: Iran(N).
Description: 1103; Illustration: 1103.
Parsa also mentions his unpublished name *A. ahmad-farhadica* for this taxon [1103].
The type of *A. kermanicus* (Hb.Parsa 20008) is at Kew [1153].

A. khajiboulaghensis A.A.R.Maassoumi [1258]
Not climbing; Herb; Perennial.
West Asia: Iran(N).
Description: 1258.

A. laristanicus Bornm. & Gauba [1179]
A. kourosianus Parsa [1157].
Not climbing; Herb; Perennial.
West Asia: Iran(N).
Description: 1015, 1188.
The names *A. laristanicus* & *A. kourosianus* are typified by the same collection.
Rechinger et al. do not mention *A. kourosianus*.
Sirjaev placed *A. laristanicus* in a new section, *Bornmulleriana* [1167], which is not upheld by
   Rechinger et al. [1179]. However, Becht placed *A. kourosianus* in the section *Bornmulleriana*
   [1174].

A. lippertii A.A.R.Maassoumi & Podl. [1257]
Not climbing; Herb; Perennial.
West Asia: Iran(N).
Description: 1257.

A. longirostratus Pau [1173]
Not climbing; Herb; Perennial.
West Asia: Iran(N).
Description: 1015.
Not mentioned by Sirjaev [1189].
The type (at MA) has been re-examined by Podlech; he has transferred the species from
   Oxyglottis to *Malacothrix* [1173].

A. macrostachys DC. [1179]
A. hedysaroides Willd. [1179].
Not climbing; Herb; Perennial.
West Asia: Iran(N), Middle East: Turkey(N).
Description: 1015, 1102, 1150.

A. macrourus Fischer & Meyer [1179]
A. cylindraceus C.Meyer [1179]; *A. macrorus* Fischer & Meyer [1015].
Not climbing; Herb; Perennial.
West Asia: Iran(N); Middle East: Turkey(N); U.S.S.R.
Description: 1015, 1102, 1150; Illustration: 1150.

A. melanodon Boiss. [1179]
Not climbing; Herb; Perennial.
West Asia: Iran(N).
Description: 1015.

A. memnonius A.A.R.Maassoumi & Podl. [1200]
Not climbing; Herb; Perennial.
West Asia: Iran(N).
Description: 1172, 1200; Illustration: 1200.

A. mollis M.Bieb. [1009]
> *A. chrysotrichus* Boiss. [1009]; *A. entomophyllus* Boiss. & Hausskn. [1009]; *A. eriocarpus* M.Bieb. [1179]; *A. iranicus* Bunge [1009]; *A. longibracteatus* Sommier & Levier [1102]; *A. mollis* var. *iranicus* (Bunge)Boiss. [1009]; *A. sulukliensis* Freyn & Sint. [1102].
> Not climbing; Herb; Perennial.
> West Asia: Iran(N), Iraq(N); Middle East: Lebanon(N), Syria(N), Turkey(N); U.S.S.R.
> Description: 1009, 1015, 1102.
> Maassoumi regards both *A. eriocarpus* & *A. sulukliensis* as good species, and does not list *A. mollis* for Iran [1172].

A. nahavandicus A.A.R.Maassoumi [1258]
> Not climbing; Herb; Perennial.
> West Asia: Iran(N).
> Description: 1258.

A. patrius A.A.R.Maassoumi [1196]
> Not climbing; Herb; Perennial.
> West Asia: Iran(N).
> Description: 1196; Illustration: 1172, 1196.

A. pauperiflorus Bornm. [1179]
> Not climbing; Herb; Perennial.
> West Asia: Iran(N).
> Description: 1015, 1158.

A. plebeius Boiss. [1179]
> Not climbing; Herb; Perennial.
> West Asia: Iran(N).
> Description: 1015.

A. podocarpus C.Meyer [1179]
> Not climbing; Herb; Perennial.
> West Asia: Iran(N); U.S.S.R.
> Description: 1015, 1150.

A. protractus Boriss. [1179]
> Not climbing; Herb; Perennial.
> West Asia: Iran(N); U.S.S.R.
> Description: 1150.

A. pseudo-beckii Sirj. & Rech.f. [1179]
> Not climbing; Herb; Perennial.
> West Asia: Afghanistan(N), Iran(N); South Asia: Pakistan(N).
> Description: 1052; Illustration: 1052.

A. rawlinsianus Aitch. & Baker [1179]
> Not climbing; Herb/Shrub; Perennial.
> West Asia: Afghanistan(N), Iran(N).
> Description: 1015, 1150, 1159; Illustration: 1228.
> Forage 1159.

A. rivashensis A.A.R.Maassoumi [1196]
> Not climbing; Herb; Perennial.
> West Asia: Iran(N).
> Description: 1196; Illustration: 1172, 1196.

A. rusticus A.A.R.Maassoumi [1258]
> Not climbing; Herb; Perennial.
> West Asia: Iran(N).
> Description: 1258.

A. sachanewii Sirj. [1179]
Not climbing; Herb; Perennial.
West Asia: Iran(N); Middle East: Turkey(N).
Description: 1015, 1102.
Not definitely recorded from Iran [1179].
Flora of Turkey treats this, doubtfully, as a member of Section *Stereothrix* [1102].

A. sarae Eig [1009]
Not climbing; Herb; Perennial.
West Asia: Iraq(N).
Description: 1009; Illustration: 1009.
Forage 1153.

A. shahpouricus Parsa [1103]
Not climbing; Herb; Perennial.
West Asia: Iran(N).
Description: 1103.
The type (Hb.Parsa 20034) is at Kew [1103].

A. spachianus Boiss. [1179]
Not climbing; Herb; Perennial.
West Asia: Iran(N).
Description: 1015.

A. takhtadzjanii Grossh. [1150]
A. takhtadjanii Grossh. [1196].
Not climbing; Herb; Perennial.
West Asia: Iran(N); U.S.S.R.
Description: 1150.

A. tauricolus Boiss. [1179]
Not climbing; Herb; Perennial.
West Asia: Iran(N); Middle East: Turkey(N).
Description: 1015, 1102.
This species is not recorded from Iran by Rechinger et al., or elsewhere; the record (Parsa 1948) is surely wrong [1179].

A. tenuiscapus Freyn & Bornm. [1179]
Not climbing; Herb; Perennial.
West Asia: Iran(N).
Description: 1015.

A. termeanus A.A.R.Maassoumi & Podl. [1200]
Not climbing; Herb; Perennial.
West Asia: Iran(N).
Description: 1200.

A. typhaeformis A.A.R.Maassoumi [1196]
Not climbing; Herb; Perennial.
West Asia: Iran(N).
Description: 1196; Illustration: 1172, 1196.

A. vessalae A.A.R.Maassoumi & Podl. [1200]
Not climbing; Herb; Perennial.
West Asia: Iran(N).
Description: 1200; Illustration: 1200.

ASTRAGALUS Section **Megalocystis**
Perennial herbs, stemless or nearly so. Hairs simple, basifixed. Leaves paripinnate, the rachis ending in a spine, persistent. Inflorescence long-pedunculate. Calyx soon much inflated, membranous. Pod small, unilocular, enclosed within the calyx.

A. ardakensis Sirj. & Rech.f. [1229]
   Not climbing; Herb; Perennial.
   West Asia: Iran(N).
   Description: 1184.

A. bodeanus Fischer [1229]
   Not climbing; Shrub; Perennial.
   West Asia: Iran(N).
   Description: 1015.

A. coluteopsis Parsa [1015]
   Not climbing; Shrub; Perennial.
   West Asia: Iran(N).
   Description: 1015, 1124.
   Originally placed in Section *Platonychium*; Tietz moves to *Megalocystis* [1195].
   The type (Stapf 2765) is at Kew [1153].

A. distans Fischer [1229]
   Not climbing; Herb; Perennial.
   West Asia: Iran(N).
   Description: 1015.

A. eriostomus Bornm. [1229]
   Not climbing; Herb; Perennial.
   West Asia: Iran(N).
   Description: 1015, 1104; Illustration: 1154.
   Bornmuller suggested that this might represent a new section [1104].

A. flexilipes Bornm. [1229]
   Not climbing; Herb; Perennial.
   West Asia: Iran(N).
   Description: 1015, 1158; Illustration: 1154.

A. irmingardis Podl. [1185]
   Not climbing; Herb, Herb/Shrub; Perennial.
   West Asia: Afghanistan(N).
   Description: 1185; Illustration: 1185.

A. keratensis Bunge [1229]
   A. karetensis Bunge [1015].
   Not climbing; Herb; Perennial.
   West Asia: Afghanistan(N), Iran(N).
   Description: 1015.
   The type, which is poor, is from the Iran–Afghanistan border [1149].

A. khoshjailensis Sirj. & Rech.f. [1229]
   Not climbing; Herb; Perennial.
   West Asia: Iran(N).
   Description: 1225.

A. lalesarensis Bornm. [1229]
   A. wiesneri Bornm. [1229].
   Not climbing; Herb/Shrub; Perennial.
   West Asia: Iran(N).
   Description: 1015.

A. lumsdenianus Aitch. & Baker [1229]
   Not climbing; Herb; Perennial.
   West Asia: Afghanistan(N).
   Description: 1015, 1159; Illustration: 1159.
   Listed by Parsa without any definite Iran record [1015].

A. megalocystis Bunge [1229]
Not climbing; Herb; Perennial.
West Asia: Iran(N).
Description: 1015.
The type is from Iran [1149].

A. merkianus Aitch. & Baker [1229]
Not climbing; Herb; Perennial.
West Asia: Afghanistan(N).
Description: 1015, 1159.
Listed by Parsa without a definite Iran record [1015].

A. murinus Boiss. [1229]
A. malanogramma Bornm. [1104]; *A. melanogramma* Bornm. [1229].
Not climbing; Shrub; Perennial.
West Asia: Iran(N).
Description: 1015.
Rechinger et al. recognise a var. *melanogramma* [1229].

A. nishapurensis Sirj. & Rech.f. [1229]
Not climbing; Herb/Shrub; Perennial.
West Asia: Iran(N).
Description: 1225.

A. noziensis Sirj. & Rech.f. [1229]
Not climbing; Herb; Perennial.
West Asia: Afghanistan(N).
Description: 1052, 1225; Illustration: 1052.

A. pseudoszovitsii Sirj. & Rech.f. [1229]
Not climbing; Herb; Perennial.
West Asia: Iran(N).
Description: 1015, 1168; Illustration: 1168.

A. remotiflorus Boiss. [1229]
Not climbing; Herb; Perennial.
West Asia: Iran(N).
Description: 1015.

A. rubrolineatus Sirj. & Rech.f. [1229]
Not climbing; Herb/Shrub; Perennial.
West Asia: Iran(N).
Description: 1156.

A. submitis Boiss. & Hohen. [1229]
A. raswendicus Hausskn. & Bornm. [1229].
Not climbing; Herb/Shrub; Perennial.
West Asia: Iran(N).
Description: 1015.

A. szovitsii Fischer & C.Meyer [1229]
Not climbing; Herb; Perennial.
West Asia: Iran(N); Middle East: Turkey(N); U.S.S.R.
Description: 1015, 1102, 1150.
Not recorded for Iran by Rechinger et al. (1958) [1229]; according to Bunge [1149] the type locality is: "ad fines boreali-occidentalis Persiae". Boissier, however, gives the type locality as: "Armenia ad fines Persiae." [1121].

A. tortuosus DC. [1229]
A. micracme Boiss. [1009].
Not climbing; Herb/Shrub; Perennial.
West Asia: Iran(N), Iraq(N).
Description: 1009, 1015, 1150; Illustration: 1009.
Rechinger et al. treat *A. micracme* as a good species [1229], as does the Flora of U.S.S.R. [1150].

ASTRAGALUS Section **Mesocarpon**

Herbaceous perennial. Stems long. Hairs simple. Inflorescences pedunculate, many-flowered. Calyx tubular. Pod sessile, rigidly coriaceous, inflated, incompletely bilocular. Monotypic.

A. quisqualis Bunge [1015]
Not climbing; Herb; Perennial.
West Asia: Iran(N); U.S.S.R.
Description: 1015, 1150; Illustration: 1145, 1150.
Not definitely recorded from Iran [1015].
Parsa placed this species in Section *Christiania*; Agerer-Kirchhoff(1976) places it in *Mesocarpon* [1015].

ASTRAGALUS Section **Microphysa**

Much-branched perennial shrublets. Hairs simple. Leaves paripinnate. Rachis persisting as a spine. Inflorescence pedunculate, globose or ovate. Calyx scarcely accrescent. Pod unilocular, included in the calyx, 1-2 seeded. Confined to Iran. See Tietz (1988 — ref.1195).

A. ardahalicus Parsa [1256]
Not climbing; Herb/Shrub; Perennial.
West Asia: Iran(N).
Description: 1103, 1256; Illustration: 1103.

A. callistachys Buhse [1195]
Not climbing; Herb; Perennial.
West Asia: Iran(N).
Description: 1015, 1195; Distribution Map: 1195; Illustration: 1195.
2 subspecies, *callistachys* & *porphyrobaphis* [1195].

subsp. **callistachys** [1195]
A. callystachys Buhse [1015]; *A. yazdekhast* Parsa [1195].
Not climbing; Herb; Perennial.
West Asia: Iran(N).
Description: 1195; Distribution Map: 1195; Illustration: 1195.

subsp. **porphyrobaphis** (Fischer)S.Tietz [1195]
A. porphyrobaphis Fischer [1195].
Not climbing; Herb; Perennial.
West Asia: Iran(N).
Description: 1015, 1195; Distribution Map: 1195; Illustration: 1195.

A. carmanicus Bornm. [1195]
A. caramanicus Bornm. [1224].
Not climbing; Herb; Perennial.
West Asia: Iran(N).
Description: 1015, 1195; Distribution Map: 1195; Illustration: 1195.

A. cemerinus G.Beck [1195]
Not climbing; Herb; Perennial.
West Asia: Iran(N).
Description: 1015, 1195; Distribution Map: 1195; Illustration: 1195.

A. cephalanthus DC. [1195]
A. bulla Fischer [1195]; *A. cephalanthus* var. *schirasicus* (Fischer)Bornm. [1195]; *A. djiroftensis* Sirj. & Rech.f. [1195]; *A. schirasicus* Fischer [1195]; *A. schirazicus* Fischer [1015].
Not climbing; Herb; Perennial.
West Asia: Iran(N).
Description: 1195; Distribution Map: 1195; Illustration: 1195.
Intermediates occur with *A. ptychophyllus* [1195].

A. demavendicolus Bornm. & Gauba [1195]
Not climbing; Herb; Perennial.
West Asia: Iran(N).
Description: 1015, 1195; Distribution Map: 1195; Illustration: 1195.
2 subspecies, *demavendicolus* & *microphysopsis* [1195].

subsp. demavendicolus [1195]
Not climbing; Herb; Perennial.
West Asia: Iran(N).
Description: 1195; Distribution Map[1195; Illustration: 1195.

subsp. microphysopsis S.Tietz [1195]
Not climbing; Herb; Perennial.
West Asia: Iran(N).
Description: 1195; Distribution Map: 1195; Illustration: 1195.

A. fragiferus Bunge [1195]
A. porphyrocystis Bornm. [1195].
Not climbing; Herb; Perennial.
West Asia: Iran(N).
Description: 1195; Distribution Map: 1195; Illustration: 1195.
Occasional intermediates with *A. cephalanthus* occur[1195].

A. lurorum Bornm. [1195]
A. durudensis Sirj. & Rech.f. [1195]; *A. luristanicus* Bornm. [1195]; *A. lurorum* var. *chamchidensis* (Sirj. & Rech.f.)Parsa [1195]; *A. microphysa* var. *durudensis* (Sirj. & Rech.f.)Parsa [1195].
Not climbing; Herb; Perennial.
West Asia: Iran(N).
Description: 1015, 1195; Distribution Map: 1195; Illustration: 1195.

A. microphysa Boiss. [1195]
A. kellalensis Boiss. & Hausskn. [1195]; *A. microphysa* var. *paucijugus* Sirj. & Rech.f. [1195].
Not climbing; Herb; Perennial.
West Asia: Iran(N).
Description: 1195; Distribution Map: 1195; Illustration: 1195.

A. pseudofragiferus S.Tietz [1195]
Not climbing; Herb/Shrub; Perennial.
West Asia: Iran(N).
Description: 1195; Distribution Map: 1195; Illustration: 1195.
Known only from the type, collected in 1961 [1195].

A. ptychophyllus Boiss. [1195]
A. gandjehicus Parsa [1195]; *A. ptychophyllus* var. *longepedunculatus* Sirj. & Rech.f. [1195]; *A. ptychophyllus* var. *xerxis* Sirj. & Rech.f. [1195]; *A. sivandi* Parsa [1195].
Not climbing; Herb, Herb/Shrub; Perennial.
West Asia: Iran(N).
Description: 1195; Distribution Map: 1195; Illustration: 1195.

A. reuterianus Boiss. [1195]
A. khayamicus Parsa [1195].
Not climbing; Herb; Perennial.
West Asia: Iran(N).
Description: 1187, 1195; Distribution Map: 1195; llustration: 1103, 1195.

A. sanandajianus S.Tietz [1195]
Not climbing; Herb/Shrub; Perennial.
West Asia: Iran(N).
Description: 1195; Distribution Map: 1195; Illustration: 1195.
Known only from the type, collected in 1965 [1195].

ASTRAGALUS Section Mirae
Small annuals. Leaves imparipinnate with few leaflets and long simple hairs. Pod suborbicular, unilocular, one seeded. Monotypic.

A. migpo Kamelin [1230]
>A. mirus Sirj. & Rech.f. [1230]; *Dorycnium calycinum* Stocks [1214].
>Not climbing; Herb; Annual.
>West Asia: Afghanistan(N), Iran(N); South Asia: Pakistan(N).
>Description: 1052; Illustration: 1052.

ASTRAGALUS Section **Onobrychiopsis**
Acaulescent perennials from a woody base. Hairs medifixed. Inflorescence capitate on a long peduncle.

A. fursei Podl. [1181]
>Not climbing; Herb; Perennial.
>West Asia: Afghanistan(N).
>Description: 1181; Illustration: 1181.

ASTRAGALUS Section **Onobrychoidei**
Formerly Section *Onobrychium*. Caulescent perennial herbs, usually erect, with medifixed hairs. Leaves imparipinnate. Inflorescences dense, pedunculate. Calyx not inflated in fruit. Pod bilocular or nearly so, longer than the calyx. A difficult section, much in need of revision.

A. aduncus Willd. [1231]
>A. *expansus* Boiss. [1102]; *A. kotschyanus* Boiss. [1102]; *A. medicagineus* Boiss. [1102].
>Not climbing; Herb; Perennial.
>West Asia: Iran(N); Middle East: Lebanon(N), Syria(N), Turkey(N); U.S.S.R.
>Description: 1015, 1102, 1150.

A. brevidens Freyn & Sint. [1231]
>Not climbing; Herb/Shrub; Perennial.
>West Asia: Iran(N).
>Description: 1015, 1150.
>Rechinger describes a var. *brevifoliolatus* Sirj. & Rech.f [1157].

A. brevidentatiformis Sirj. & Rech.f. [1231]
>Not climbing; Herb/Shrub; Perennial.
>West Asia: Iran(N).
>Description: 1157.

A. bungeanus Boiss. [1231]
>A. *aduncus* M.Bieb. [1231]; *A. onobrychioides* M.Bieb. [1231]; *A. ustiurtensis* Grossh. [1231].
>Not climbing; Herb, Herb/Shrub; Perennial.
>West Asia: Iran(N); U.S.S.R.
>Description: 1015, 1150.

A. cancellatus Bunge [1231]
>A. *pseudocancellatus* Grossh. [1102].
>Not climbing; Herb; Perennial.
>West Asia: Iran(N); Middle East: Turkey(N).
>Description: 1015, 1102, 1150.

A. canus Bunge [1231]
>Not climbing; Herb; Perennial.
>West Asia: Iran(N).
>Description: 1015.
>The type is from Iran [1149].

A. chaborasicus Boiss. & Hausskn. [1231]
>A. *cephalotes* sensu auctt. [1009]; *A. xylobasis* sensu Rawi [1009, 1160].
>Not climbing; Herb; Perennial.
>West Asia: Iraq(N); Middle East: Syria(N), Turkey(N).
>Description: 1009, 1015; Illustration: 1009.
>Flora of Turkey regards this as synonymous with *A. aduncus* [1102].

A. darii Sirj. & Rech.f. [1231]
Not climbing; Herb; Perennial.
West Asia: Iran(N).
Description: 1170.

A. effusus Bunge [1231]
Not climbing; Herb; Perennial.
West Asia: Iran(N).
Description: 1015.
The type is from Iran [1149].

A. kadschorensis Bunge [1231]
A. kodzhorensis Grossh. [1231].
Not climbing; Herb; Perennial.
West Asia: Iran(N); U.S.S.R.
Description: 1015, 1102, 1150.
The type is from U.S.S.R. (Tblisi, Georgia) [1150]. Doubtfully recorded from Iran in the Flora of U.S.S.R [1150].

A. karputanus Boiss. & Noe [1231]
Not climbing; Herb; Perennial.
West Asia: Iran(N); Middle East: Turkey(N).
Description: 1015, 1102.
Parsa mentions a var. *huetianus* Boiss. [1015], and states that it is the same as *A. subacaulis* Bunge [1015]. Flora of Turkey places this species in Section *Hololeuce*, and regards it as possibly only a depauperate form of *A. bicolor* [1102]. Not recorded for Iran in Flora of Turkey [1102], but recorded for 'Kurdistan' by Rechinger et al [1231].

A. khorramabadensis Bornm. [1231]
Not climbing; Herb; Perennial.
West Asia: Iran(N).
Description: 1015, 1188.

A. lilacinus Boiss. [1231]
Not climbing; Herb; Perennial.
West Asia: Iran(N).
Description: 1015.

A. lunatus Pallas [1231]
Not climbing; Herb; Perennial.
West Asia: Iran(N); U.S.S.R.
Description: 1015, 1150.
"Species male nota floribus ignotis" [1121].
The type locality may not be in present-day Iran [1152].

A. mossulensis Bunge [1009]
A. mossulens var. *complicatus* Eig [1009]; *A. mossulensis* var. *multiflorus* Eig [1009]; *A. olginii* Bunge [1231]; *A. pseudomossulensis* Nab. [1009].
Not climbing; Herb; Perennial.
West Asia: Iran(N), Iraq(N); Middle East: Syria(N).
Description: 1009, 1015; Illustration: 1009.

A. onobrychis L. [1102]
Not climbing; Herb; Perennial.
West Asia: Iran(N); Europe; Middle East: Turkey(N); U.S.S.R.
Description: 1102, 1150.
The use of the Linnaean name for Iranian material is questionable, pending a sectional revision [1143].

A. strictipes Bornm. [1231]
Not climbing; Herb; Perennial.
West Asia: Iran(N).
Description: 1015, 1201.
Bornmuller was uncertain whether to place this species in Section *Onobrychium* or Section *Proselius* [1201].

A. teheranicus Boiss. [1231]
Not climbing; Herb; Perennial.
West Asia: Iran(N).
Description: 1015, 1150.

A. vegetus Bunge [1231]
Not climbing; Herb; Perennial.
West Asia: Iran(N).
Description: 1015.
The type is from Iran [1149].

A. xylobasis Freyn [1231]
A. *xylorrhizus* Freyn & Sint. [1231].
Not climbing; Herb/Shrub; Perennial.
West Asia: Iran(N); Middle East: Turkey(N).
Description: 1015.
Not recorded for Iran by Rechinger et al., nor by Flora of Turkey [1231].

ASTRAGALUS Section **Ophiocarpus**
Annuals with simple hairs. Leaves imparipinnate. Inflorescences axillary, sessile or very shortly pedunculate, few-flowered. Pod elongate, unilocular.

A. glaberrimus Sirj. & Rech.f. [1009]
Not climbing; Herb; Annual.
West Asia: Iran(N), Iraq(N).
Description: 1009; Illustration: 1009.

A. ophiocarpus Bunge [1009]
A. *aitchisonii* Baker [1163]; A. *ophiocarpus* var. *plurijugus* Sirj. [1009]; A. *ophiocarpus* var. *vermicularis* (Bornm.)Sirj. [1009]; A. *paulsenii* Freyn [1163]; A. *vermicularis* Bornm. [1009]; *Ophiocarpus paulsenii* (Freyn)Ikonn. [1190]; *Trigonella komarovii* Lipsky [1163].
Not climbing; Herb; Annual.
West Asia: Afghanistan(N), Iran(N), Iraq(N); East Asia: China(N) (Xizang Zizhiqu); South Asia: India(N);Pakistan(N); U.S.S.R.
Description: 1009, 1015, 1150, 1187; Illustration: 1009, 1144, 1145, 1150, 1187.

ASTRAGALUS Section **Ornithopodium**
Caulescent perennial herbs. Leaves imparipinnate, with medifixed hairs. Inflorescence pedunculate, many-flowered. Calyx not inflated. Pods narrowly cylindrical, elongated, bilocular.

A. brachyodontus Boiss. [1015]
Not climbing; Herb; Perennial.
West Asia: Iran(N).
Description: 1015.
The type is from Iran [1121].

A. brevipes Bunge [1015]
Not climbing; Herb; Perennial.
West Asia: Iran(N).
Description: 1015.

A. glochideus Boriss. [1150]
Not climbing; Herb; Perennial.
West Asia: Iran(N); U.S.S.R.
Description: 1150.

A. jodostachys Boiss. & Buhse [1015]
    Not climbing; Herb; Perennial.
    West Asia: Iran(N); Middle East: Turkey(N).
    Description: 1015, 1102, 1150.
    The type is from Iran [1121].

A. ornithopodioides Lam. [1015]
    A. horwoodii Eig [1102]; *A. multijugus* Grossh. [1102].
    Not climbing; Herb; Perennial.
    West Asia: Iran(N); Middle East: Turkey(N); U.S.S.R.
    Description: 1015, 1102.

A. schistosus Boiss. & Hohen. [1015]
    Not climbing; Herb; Perennial.
    West Asia: Iran(N).
    Description: 1015.
    The type is from Iran [1121].

A. stevenianus DC. [1015]
    A. achtalensis Conrath & Freyn [1102]; *A. applicatus* Boriss. [1102]; *A. conrathii* Freyn [1102]; *A. dahuricus* Koch [1102]; *A. ispirensis* Boiss. [1102]; *A. kochianus* Sosn. [1102].
    Not climbing; Herb; Perennial.
    West Asia: Irang(N); Middle East: Turkey(N); U.S.S.R.
    Description: 1015, 1102, 1150; Illustration: 1150.

A. trigonelloides Boiss. [1015]
    Not climbing; Herb; Perennial.
    West Asia: Iran(N).
    Description: 1015, 1150.
    Not recorded from Iran in Flora of U.S.S.R. The type is from Armenia [1150].

A. variistipula Turrill [1164]
    Not climbing; Herb; Perennial.
    West Asia: Iran(N).
    Description: 1015, 1164.

ASTRAGALUS Section Oroboidei
Formerly Section *Orobella*. Caulescent perennials with simple hairs. Inflorescences long-pedunculate. Pods shortly stipitate, membranous, sub-bilocular or sub-unilocular.

A. pseudobrachytropis Gontsch. [1185]
    Not climbing; Herb; Perennial.
    West Asia: Afghanistan(N); U.S.S.R.
    Description: 1150.

ASTRAGALUS Section Oxyglottis
Annuals with simple hairs. Leaves imparipinnate. Inflorescences sessile or pedunculate. Pod sessile, often falcate, bi- or semibilocular.

A. abbas-riazi Parsa (provisional) [1143]
    Not climbing; Herb; Annual.
    West Asia: Iran(N).
    Description: 1103; Illustration: 1103.
    The type (Hb.Parsa 20015) is at Kew.

A. asterias Ledeb. [1009]
    A. cruciatus sensu auctt. [1009].
    Not climbing; Herb; Annual.
    West Asia: Iran(N), Iraq(N), Saudi Arabia(N); Africa; Europe; Middle East: Israel(N), Jordan(N), Lebanon(N), Sinai(N), Syria(N), Turkey(N); U.S.S.R.
    Description: 1009, 1150; Illustration: 1009.

A. biovulatus Bunge [1189]
Not climbing; Herb; Annual.
West Asia: Afghanistan(N), Iran(N).
Description: 1015.

A. coronilla Bunge [1189]
Not climbing; Herb; Annual.
West Asia: Iran(N).
Description: 1015.

A. filicaulis Fischer & C.Meyer [1189]
A. agrestis Freyn [1015,1205]; *A. leptodermus* Bunge [1205].
Not climbing; Herb; Annual.
West Asia: Afghanistan(N), Iran(N); South Asia: Pakistan(N); U.S.S.R.
Description: 1015; Illustration: 1145.

A. kerkukiensis Bornm. [1009]
A. ammophilus var. *obtusatus* Popov [1009]; *A. kerkuensis* Bornm. [1189].
Not climbing; Herb; Annual.
West Asia: Iran(N), Iraq(N).
Description: 1009, 1015; Illustration: 1009.
2 varieties, *kerkukiensis* & *leiocarpus*. The latter occurs only in Iran [1009].

A. orthorhynchus Bornm. [1188]
Not climbing; Herb; Annual.
West Asia: Iran(N).
Description: 1015, 1188.
Close to *A. oxyglottis* & *A. biserrula* [1015].
Not mentioned by Sirjaev [1189].

A. persepolitanus Boiss. [1187]
A. ammophilus Karelin & Kir. [1187]; *A. ammophilus* var. *persepolitanus* (Boiss.)Ali [1011].
Not climbing; Herb; Annual.
West Asia: Afghanistan(N), Iran(N); South Asia: Pakistan(N); U.S.S.R.
Description: 1015, 1150, 1187; Illustration: 1145, 1150, 1187.

A. rytilobus Bunge [1011]
A. filicaulis subsp. *rytilobus* (Bunge)Popov [1189]; *A. nawabianus* Aitch. & Baker [1011].
Not climbing; Herb; Annual.
West Asia: Afghanistan(N), Iran(N); U.S.S.R.
Description: 1015, 1145, 1159; Illustration: 1144.
Forage 1159.

A. schimperi Boiss. [1009]
A. schimperi subsp. *magnus* Eig [1009].
Not climbing; Herb; Annual.
West Asia: Bahrain(N), Iran(N), Iraq(N), Kuwait(N), Oman(N), Qatar(N), Saudi Arabia(N), United Arab Emirates(N); Middle East: Israel(N), Jordan(N), Sinai(N), Syria(N).
Description: 1009; Illustration: 1001, 1009.

A. sesamoides Boiss. [1189]
Not climbing; Herb; Annual.
West Asia: Afghanistan(N), Iran(N); U.S.S.R.
Description: 1150; Illustration: 1144, 1145.

A. sinaicus Boiss. [1155]
A. sesameus Sibth. & Sm. [1102].
Not climbing; Herb; Annual.
West Asia: Saudi Arabia(N); Africa; Europe; Middle East: Turkey(N); U.S.S.R.
Description: 1102.

A. stella Gouan [1007]
Not climbing; Herb; Annual.
West Asia: Saudi Arabia(N).
Description: 1007.
Almost certainly based on a misidentification of *A. tribuloides* [1152].

A. tribuloides Del. [1009]

A. erpocaulos Boiss. [1009]; *A. kirghizicus* Stschegl. [1009]; *A. minutus* Boiss. [1009]; *A. perpusillus* Bertol. [1009]; *A. tribuloides* var. *kirghizicus* (Stschegl.)Sirj. [1009].
Not climbing; Herb; Annual.
West Asia: Afghanistan(N), Bahrain(N), Iran(N), Iraq(N), Kuwait(N), North Yemen(N), Oman(N), Qatar(N), Saudi Arabia(N); Africa; Middle East: Israel(N), Jordan(N), Lebanon(N), Syria(N), Turkey(N); South Asia: Pakistan(N); U.S.S.R.
Description: 1009, 1015, 1150; Illustration: 1001, 1009, 1145, 1150.
Forage 1014; Medicine 1163.
4 varieties, *tribuloides*, *minutus*, *mareoticus*, and *elarishensis* [1009].
The distribution of the varieties within Iraq is mapped in Fl.Iraq [1009].

A. triradiatus Bunge [1009]

Not climbing; Herb; Annual.
West Asia: Iraq(N); Middle East: Lebanon(N), Syria(N), Turkey(N).
Description: 1009, 1102.
The Iraq record is doubtful as the precise locality is uncertain [1009].

ASTRAGALUS Section Pelta

Spiny perennial dwarf shrubs. Hairs simple. Leaves imparipinnate, rachis spinescent. Leaflets more or less peltate, suborbicular. Inflorescences few-flowered, in the axils of the upper leaves. Pods firm, membranous.

A. distinctissimus Rech.f. & Edelb. [1052]

Not climbing; Shrub; Perennial.
West Asia: Afghanistan(N).
Description: 1052; Illustration: 1052.
Deml assigns this species to Section *Pelta*, not Section *Aegacantha* as in Kitamura (1960) [1171].

A. peltatus Podl. & Deml [1232]

Not climbing; Herb, Herb/Shrub; Perennial.
West Asia: Afghanistan(N).
Description: 1232; Illustration: 1145, 1232.

A. wilhelminae Kress-Deml [1259]

Not climbing; Shrub; Perennial.
West Asia: Afghanistan(N).
Description: 1259; Illustration: 1259.

ASTRAGALUS Section Pendulina

Herbaceous perennials, acaulescent or shortly caulescent. Hairs simple. Leaves imparipinnate. Inflorescences pedunculate, axillary. Pod stipitate, thinly coriaceous, elongated, bilocular or nearly so.

A. andersianus Podl. [1185]

Not climbing; Herb; Perennial.
West Asia: Afghanistan(N).
Description: 1185; Illustration: 1185.
Originally placed in Section *Caprini*; transferred to Section *Pendulina* in Podlech (1988) [1175].

A. dieterlei Podl. [1185]

Not climbing; Herb; Perennial.
West Asia: Afghanistan(N).
Description: 1185; Illustration: 1185.
Originally of uncertain sectional position; now placed in Section *Pendulina* by Podlech [1143].

A. merxmuelleri Podl. [1098]

Not climbing; Herb; Perennial.
West Asia: Afghanistan(N).
Description: 1098; Illustration: 1098.

ASTRAGALUS Section **Plagiophacos**

Perennial herbs; stems short. Hairs simple. Leaves imparipinnate. Inflorescences dense, pedunculate. Pods unilocular.

A. plagiophacos A.A.R.Maassoumi & Podl. [1200]
Not climbing; Herb; Perennial.
West Asia: Iran(N).
Description: 1200; Illustration: 1172, 1200.

ASTRAGALUS Section **Platyglottis**

Annuals or short-lived perennials. Hairs simple. Leaves imparipinnate. Inflorescences pedunculate. Pods sessile, straight or curved, semi- or sub-bilocular.

A. bombycinus Boiss. [1009]
A. bombycinus subsp. *sultanensis* (Bunge)Sirj. [1260]; *A. musilii* Velen. [1260]; *A. palaestinus* Eig ssp. *heteranthesmus* Eig [1260]; *A. palmyrensis* Post [1260]; *A. sultanensis* Bunge [1260].
Not climbing; Herb; Annual.
West Asia: Iran(N), Iraq(N), Kuwait(N), Saudi Arabia(N); Africa; Middle East: Israel(N), Jordan(N), Sinai(N), Syria(N).
Description: 1009, 1015, 1260; Illustration: 1001, 1009.
Forage 1009.

A. camptoceras Bunge [1260]
A. bungei Winkler & B.Fedtsch. [1260]; *A. hispidissimus* Grossh. [1260]; *A. spirorrhynchus* Bornm. [1260].
Not climbing; Herb; Annual.
West Asia: Iran(N); Middle East: Turkey(N); U.S.S.R.
Description: 1102, 1154, 1260; Illustration: 1145, 1150.

A. peregrinus Vahl [1260]
Not climbing; Herb; Annual.
West Asia: Saudi Arabia(N); Africa.
Description: 1007, 1260.
Migahid's record is not confirmed by Podlech (1990 — ref. 1260).

A. suberosus Banks & Sol. [1009]
Not climbing; Herb; Annual.
West Asia: Iran(N), Iraq(N); Europe; Middle East: Israel(N), Jordan(N), Lebanon(N), Syria(N), Turkey(N).
Description: 1009, 1015; Illustration: 1009.
2 subspecies, *suberosus* and *haarbachii*; only the former in West Asia.

subsp. **suberosus** [1009]
A. ancyleus Boiss. [1260]; *A. kirkukensis* Eig [1260]; *A. kotschyanus* Fischer [1260]; *A. pamphylicus* Boiss. [1260]; *A. suberosus* subsp. *ancyleus* (Boiss.)V.Matthews [1260]; *A. suberosus* subsp. *mersinensis* (Sirj.)V.Matthews [1260]; *A. tuberculosus* DC. [1260]; *A. tuberculosus* subsp. *mersinensis* Sirj. [1260].
Not climbing; Herb; Annual.
West Asia: Iran(N); Europe; Middle East: Israel(N), Jordan(N), Lebanon(N), Syria(N), Turkey(N).
Description: 1009, 1260.

ASTRAGALUS Section **Poliothrix**

Perennial caulescent herbs. Hairs simple. Leaves imparipinnate. Inflorescences capitate, long-pedunculate. Pods sessile, unilocular, about equalling the calyx. Monotypic.

A. leucocephalus Bunge [1011]
Not climbing; Herb; Perennial.
West Asia: Afghanistan(N); South Asia: India(N), Pakistan(N).
Illustration: 1088.

ASTRAGALUS Section **Poterion**

Perennial shrubs. Hairs basifixed. Leaves paripinnate, often dimorphic, some with persistent spinescent rachides. Flowers solitary, axillary, or in few-flowered axillary inflorescences. Calyx soon greatly inflated and bladder-like. Pod small, enclosed within the calyx. See Tietz (1988 — ref. 1195).

A. anisacanthus Boiss. [1195]
Not climbing; Herb/Shrub; Perennial.
West Asia: Afghanistan(N), Iran(N); South Asia: Pakistan(N).
Description: 1015, 1187, 1195; Distribution Map: 1195; Illustration: 1052, 1187, 1195.
2 subspecies; *anisacanthus* and *schurabicus* [1195].

subsp. **anisacanthus** [1195]
A. bludhistanicus Bunge [1052]; *A. bludhistanus* Bunge [1015]; *A. bludshistanus* Bunge [1195]; *A. jubatus* Bunge [1195]; *Caragana brachyantha* Rech.f. [1195].
Not climbing; Shrub; Perennial.
West Asia: Afghanistan(N); South Asia: Pakistan(N).
Description: 1195; Distribution Map: 1195; Illustration: 1052, 1195.

subsp. **schurabicus** (Bunge)S.Tietz [1195]
A. schurabicus Bunge [1195].
Not climbing; Herb,Shrub; Perennial.
West Asia: Afghanistan(N), Iran(N).
Description: 1195; Distribution Map: 1195; Illustration: 1195.

A. baba-alliar Parsa [1195]
Not climbing; Shrub; Perennial.
West Asia: Iran(N), Iraq(N).
Description: 1195; Distribution Map: 1195; Illustration: 1195.
2 subspecies, *baba-alliar* & *nudicarpus* [1195].

subsp. **baba-alliar** [1195]
Not climbing; Shrub; Perennial.
West Asia: Iran(N).
Description: 1195; Distribution Map: 1195; Illustration: 1195.

subsp. **nudicarpus** (Sirj. & Rech.f.)S.Tietz [1195]
A. bruguieri var. *nudicarpus* Sirj. & Rech.f. [1195]; *A. cornutus* var. *glaber* Parsa [1195]; *A. fasciculifolius* sensu C.Towns. [1009,1195]; *A. nudicarpus* Rech.f. [1195].
Not climbing; Shrub; Perennial.
West Asia: Iran(N), Iraq(N).
Description: 1195; Distribution Map: 1195; Illustration: 1195.
Intermediates with *A. bruguieri* occur [1195].

A. bruguieri Boiss. [1195]
A. brugieri Boiss. [1104]; *A. bruguieri* var. *leiocladus* Bornm. [1195]; *A. chlamydophorus* Bornm. [1195].
Not climbing; Herb/Shrub; Perennial.
West Asia: Iran(N), Iraq(N); Middle East: Syria(N), Turkey(N).
Description: 1009, 1195; Distribution Map: 1195; Illustration: 1195.

A. calliphysa Bunge [1195]
Not climbing; Herb/Shrub; Perennial.
West Asia: Iran(N).
Description: 1015, 1195; Distribution Map: 1195; Illustration: 1195.
2 subspecies, *calliphysa* & *angustifolius* [1195].

subsp. **angustifolius** S.Tietz [1195]
Not climbing; Herb/Shrub; Perennial.
West Asia: Iran(N).
Description: 1195; DiMap: 1195; Illustration: 1195.

subsp. **calliphysa** [1195]
A. heteracanthus Bornm. [1195]; *A. myriocystis* Bornm. [1195].
Not climbing; Herb/Shrub; Perennial.
West Asia: Iran(N).
Description: 1195; Distribution Map: 1195; Illustration: 1195.

A. **fasciculifolius** Boiss.[1195]
Not climbing; Shrub; Perennial.
West Asia: Iran(N), Oman(N).
Description: 1195; Distribution Map: 1195; Illustration: 1195.
2 subspecies, *fasciculifolius* & *arbusculinus* [1195].

subsp. **arbusculinus** (Bornm. & Gauba)S.Tietz [1195]
A. afghano-persicus Kitam. [1195]; *A. arbusculinus* Bornm. & Gauba [1195]; *A. cornutus* Bunge [1195].
Not climbing; Herb/Shrub,Shrub; Perennial.
West Asia: Iran(N), Oman(N).
Description: 1015, 1187, 1195; Distribution Map: 1195; Illustration: 1187, 1195.

subsp. **fasciculifolius** [1195]
Not climbing; Shrub; Perennial.
West Asia: Iran(N).
Description: 1195; Distribution Map: 1195; Illustration: 1195.

A. **glaucacanthos** Fischer [1195]
A. brachycladus Boiss. [1195]; *A. jubatus* var. *semiglaber* Bornm. [1195].
Not climbing; Herb/Shrub; Perennial.
West Asia: Iran(N).
Description: 1015, 1195; Distribution Map: 1195; Illustration: 1195.

A. **pachyrachis** Sirj. & Rech.f. [1195].
Not climbing; Herb/Shrub; Perennial.
West Asia: Afghanistan(N).
Description: 1195; Distribution Map: 1195; Illustration: 1195.

A. **porphyrophysa** Bornm. & Gauba [1195]
A. myriocystis var. *albiflorus* Parsa [1195].
Not climbing; Herb/Shrub; Perennial.
West Asia: Iran(N).
Description: 1195; Distribution Map: 1195; Illustration: 1195.

A. **russellii** Banks & Sol. [1195]
A. behen Bertol. [1195]; *A. rauwolfii* Pallas [1195]; *A. russelii* Banks & Sol. [1104]; *A. russellii* var. *hirsutus* Bornm. [1195]; *A. tumidus* Willd.
Not climbing; Herb/Shrub; Perennial
West Asia: Iran(N), Iraq(N); Middle East: Syria(N), Turkey(N).
Description: 1009, 1102, 1195; Distribution Map: 1195; Illustration: 1009, 1195.

A. **spinosus** (Forsskal) Muschler [1195]
A. armatus var. *libycus* Pampan. [1195]; *A. forskahlei* Boiss. [1195]; *A. kneuckeri* Freyn [1195]; *A. microthamnus* Boiss. & Hausskn. [1195]; *A. rauwolfii* Vahl [1195]; *A. scorpius* Boiss. [1195]; *A. spinosus* var. *hamrinensis* Eig [1195]; *A. spinosus* var. *kneuckeri* (Freyn)Tackh. & Boulos [1195].
Not climbing; Shrub; Perennial.
West Asia: Iran(N), Iraq(N), Kuwait(N), Saudi Arabia(N); Africa; Middle East: Israel(N), Jordan(N), Lebanon(N), Sinai(N), Syria(N).
Description: 1009, 1195; Distribution Map: 1195; Illustration: 1001, 1009, 1195.
Forage 1009; Weed 1009.
A variable species; the variation is discussed by Tietz, pp.302–303 [1195].

ASTRAGALUS Section **Rhabdotus**
Perennial subacaulescent herbs. Hairs simple. Leaves imparipinnate. Inflorescences few-flowered, pedunculate. Calyx large, somewhat accrescent, eventually ruptured by the sessile unilocular pod.

A. venulosus Boiss. [1009]
Not climbing; Herb; Perennial.
West Asia: Iraq(N); Middle East: Turkey(N).
Description: 1009, 1102; Illustration: 1009.

ASTRAGALUS Section **Scheremeteviana**

Perennial acaulescent herbs. Hairs simple. Leaves imparipinnate. Inflorescences long-pedunculate, many-flowered. Pods stipitate, inflated, unilocular.

A. longistipitatus Boriss. [1181]
Not climbing; Herb; Perennial.
West Asia: Afghanistan(N); U.S.S.R.
Description: 1150.

A. mundulus Podl. [1181]
Not climbing; Herb; Perennial.
West Asia: Afghanistan(N).
Description: 1181; Illustration: 1181.

A. ovczinnikovii Boriss. [1185]
Not climbing; Herb; Perennial.
West Asia: Afghanistan(N).
Description: 1150; Illustration: 1144, 1145.

A. scheremetevianus B.Fedtsch. [1126]
Not climbing; Herb; Perennial.
West Asia: Afghanistan(N); U.S.S.R.
Description: 1150; Illustration: 1145, 1150.

ASTRAGALUS Section **Sesbanella**

Herbaceous perennials with long stems. Hairs simple. Leaves imparipinnate. Pods long-stipitate, inflated, semibilocular.

A. otiporensis Boiss. [1011]
Not climbing; Herb; Perennial.
West Asia: Afghanistan(N).

A. paktiensis Podl. [1185]
Not climbing; Herb; Perennial.
West Asia: Afghanistan(N).
Description: 1185; Illustration: 1185.

A. pseudohofmeisteri Sirj. & Rech.f. [1011]
Not climbing; Herb; Perennial.
West Asia: Afghanistan(N).
Description: 1052; Illustration: 1052.

ASTRAGALUS Section **Sewerzowia**

Annuals. Hairs simple. Inflorescences few-flowered, loose, pedunculate. Pods sessile, ovoid–oblong, each valve ridged along the middle, the ridges toothed. The peculiar pod has led to this section sometimes being treated as a genus.

A. biserrula Bunge [1015]
A. oxyglottoides Rech.f. & Gauba [1015].
Not climbing; Herb; Annual.
West Asia: Iran(N).
Description: 1015.
Sirjaev places this taxon in Section *Sewerzowia*, but Parsa (1948) puts it in *Oxyglottis* [1189].

A. crispocarpus Nab. [1009]
A. mardabadensis Bornm. & Gauba [1009].
Not climbing; Herb; Annual.
West Asia: Iran(N), Iraq(N).
Description: 1009, 1015; Illustration: 1009.

A. oxyglottis M.Bieb. [1009]
A. psiloglottis DC. [1214].
Not climbing; Herb; Annual.
West Asia: Afghanistan(N), Iran(N), Iraq(N); Middle East: Syria(N), Turkey(N); South Asia: Pakistan(N); U.S.S.R.
Description: 1009, 1102, 1150; Illustration: 1145, 1150, 1187.
2 varieties, *oxyglottis* & *psiloglottis* [1009].

A. schmalhausenii Bunge [1190]
Sewerzowia turkestanica Regel & Schmalh. [1190].
Not climbing; Herb; Annual.
West Asia: Afghanistan(N), Iran(N); U.S.S.R.
Description: 1150; Illustration: 1145, 1150.
Forage 1150.

A. vicarius Lipsky [1185]
Not climbing; Herb; Annual.
West Asia: Afghanistan(N), Iran(N); U.S.S.R.
Description: 1015, 1150; Illustration: 1144, 1145, 1150.

ASTRAGALUS Section **Stereothrix**
Herbaceous perennials with simple hairs. Leaves imparipinnate. Inflorescences capitate, dense, shortly pedunculate. Pods small, sessile, shorter than the calyx, bilocular.

A. barbatus Lam. [1102]
A. hispidus Labill. [1102].
Not climbing; Herb; Perennial.
West Asia: Iran(N); Middle East: Lebanon(N), Turkey(N).
Description: 1102.

A. capito Boiss. [1179]
Not climbing; Herb; Perennial.
West Asia: Iran(N); Middle East: Turkey(N).
Description: 1015, 1102.
Rechinger et al. mention a var. *violaceus* Bornm. & Gauba from Iran [1179].

A. doshman-ziariensis A.A.R.Maassoumi & Podl. [1200]
Not climbing; Herb; Perennial.
West Asia: Iran(N).
Description: 1200; Illustration: 1200.

A. hirtus Bunge [1179]
Not climbing; Herb; Perennial.
West Asia: Iran(N).
Description: 1015.
The type is from Iran [1149]].

A. koelzii Barneby [1236]
A. unifoliolatus Sirj. & Rech.f. [1236].
Not climbing; Herb; Perennial.
West Asia: Iran(N).
Description: 1213, 1236; Illustration: 1236.
Originally placed in the monotypic Section *Koelziana*, but is linked to Section *Stereothrix* by *A. ledinghamii* (q.v.) [1236].

A. ledinghamii Barneby [1236]
Not climbing; Herb; Perennial.
West Asia: Iran(N).
Description: 1236; Illustration: 1236.

A. podosphaerus Boiss. & Hausskn. [1179]
Not climbing; Herb; Perennial.
West Asia: Iran(N).
Description: 1015.
Close to *A. capito* (q.v.) and possibly conspecific [1102].

A. saganlugensis Trautv. [1179]
A. sapozhnikovii Schischkin [1102].
Not climbing; Herb; Perennial.
West Asia: Iran(N); Middle East; Turkey(N); U.S.S.R.
Description: 1015, 1102, 1150; Illustration: 1151.

A. setulosus Gontsch. [1150]
Not climbing; Herb; Perennial.
West Asia: Iran(N); U.S.S.R.
Description: 1150.

A. sphaeranthus Boiss. [1009]
Not climbing; Herb; Perennial.
West Asia: Iran(N), Iraq(N); Middle East: Turkey(N).
Description: 1009, 1015, 1102; Illustration: 1009.

A. ulodjensis Sirj. & Rech.f. [1179]
Not climbing; Herb; Perennial.
West Asia: Iran(N).
Description: 1170.

ASTRAGALUS Section **Stipitella**
Perennial shrubs. Hairs simple. Leaves imparipinnate, the rachides persistent, strongly spinescent. Inflorescences axillary, pedunculate. Pods long-stipitate, bilocular. See Podlech (1975 — ref. 1237).

A. cuneifolius Bunge [1237]
Not climbing; Shrub; Perennial.
West Asia: Afghanistan(N).
Description: 1237; Distribution Map: 1237; Illustration: 1237.

A. stipitatus Bunge [1237]
Not climbing; Herb/Shrub; Perennial.
West Asia: Afghanistan(N), U.S.S.R.
Description: 1237; Distribution Map: 1237.
3 subspecies, *stipitatus, angustifructus* & *shatuensis* [1237].

subsp. **angustifructus** Podl. [1237]
A. eustrophacanthus Rech.f. & Edelb. [1237].
Not climbing; Herb/Shrub; Perennial.
West Asia: Afghanistan(N).
Description: 1237; Distribution Map: 1237; Illustration: 1052.

subsp. **shatuensis** Podl. [1237]
Not climbing; Herb/Shrub; Perennial.
West Asia: Afghanistan(N).
Description: 1237; Distribution Map: 1237.

subsp. **stipitatus** [1237]
A. babensis Sirj. & Rech.f. [1237]; *A. massagetovii* B.Fedtsch. [1237]; *A. neostipitatus* Kitam. [1237].
Not climbing; Herb/Shrub; Perennial.
West Asia: Afghanistan(N); U.S.S.R.
Description: 1052, 1150, 1237; Distribution Map: 1237; Illustration: 1052, 1145, 1150.

ASTRAGALUS Section **Tamias**

Perennial caulescent herbs, with both simple and bifurcate hairs. Leaves imparipinnate. Imflorescences shortly pedunculate, loose. Pods short-stipitate, bilocular.

A. turczaninovii Karelin & Kir. [1185]
Not climbing; Herb; Perennial.
West Asia: Afghanistan(N); U.S.S.R.
Description: 1150; Illustration: 1145.

ASTRAGALUS Section **Thaumasiophaca**

Perennial herbs, acaulescent or nearly so. Hairs simple. Inflorescences 2–4 flowered, in the axils of the upper leaves. Pods more or less unilocular. Monotypic.

A. thaumasios Podl. [1181]
Not climbing; Herb; Perennial.
West Asia: Afghanistan(N).
Description: 1181; Illustration: 1181.

ASTRAGALUS Section **Theiochrus**

Herbaceous perennials with long stems. Hairs simple. Leaves imparipinnate. Inflorescences axillary, lax, long-pedunculate. Pods shortly stipitate, bilocular. See Podlech & Kosik (1983 — ref. 1238).

A. remotifolius Boiss. & Hausskn. [1015]
Not climbing; Herb; Perennial.
West Asia: Iran(N).
Description: 1015.

A. shelkovnikovii Grossh. [1015]
Not climbing; Herb; Perennial.
West Asia: Iran(N); Middle East: Turkey(N); U.S.S.R.
Description: 1015, 1102.

A. siliquosus Boiss. [1238]
Not climbing; Herb; Perennial.
West Asia: Afghanistan(N), Iran(N); Middle East: Syria(N), Turkey(N); U.S.S.R.
Description: 1015, 1102, 1238; Distribution Map: 1238; Illustration: 1098, 1238.
2 subspecies, *siliquosus* and *stramineus*; only the former occurs in West Asia [1238].

subsp. siliquosus [1238]
A. conjecturalis Schischkin [1238]; *A. ispahanicus* Boiss. [1238]; *A. laxus* Boiss. & Hausskn. [1238]; *A. sulfureus* Bunge [1238]; *A. tetragonocarpus* Boiss. [1238].
Not climbing; Herb; Perennial.
West Asia: Afghanistan(N), Iran(N); Middle East: Turkey(N); U.S.S.R.
Description: 1015, 1150, 1238; Distribution Map: 1238; Illustration: 1098, 1145, 1150.

subsp. stramineus (Boiss.)E.Kozik & Podl. [1238]
A. stramineus Boiss. & Kotschy [1238].
Not climbing; Herb; Perennial.
Middle East: Syria(N).
Description: 1238; Distribution Map: 1238.

ASTRAGALUS Section **Thlaspidium**

Annuals with simple hairs. Inflorescences few-flowered, axillary. Pods sessile, bilocular, strongly compressed at right angles to the septum, with a flat marginal wing-like flange. The peculiar pods have led to this section being treated as a separate genus.

A. thlaspi Lipsky [1185]
>Thlaspidium thlaspi (Lipsky)Rassul. [1190].
>Not climbing; Herb; Annual.
>West Asia: Afghanistan(N); U.S.S.R.
>Description: 1150; Illustration: 1144, 1145, 1150.

ASTRAGALUS Section Trachycercis

Low perennials, usually acaulescent. Hairs medifixed. Flowers axillary, 1-4 together, or in very short-pedunculate inflorescences. Pods very small, bilocular or rarely subunilocular.

A. humilis M.Bieb. [1009]
>A. vanensis Sosn. [1102].
>Not climbing; Herb; Perennial.
>West Asia: Iran(N), Iraq(N); Middle East: Turkey(N); U.S.S.R.
>Description: 1009, 1015,1102,1150; Illustration: 1009.

A. mikrophyton Sirj. & Rech.f. [1052]
>Not climbing; Herb; Perennial.
>West Asia: Afghanistan(N).
>Description: 1052; Illustration: 1052.
>Grows on exposed ridges in coarse sand with only the leaf-tips visible [1143].

A. poliotrichus Bornm. [1154]
>Not climbing; Herb; Perennial.
>West Asia: Iran(N).
>Description: 1154.

ASTRAGALUS Section Tricholobus

Perennial subshrubs. Leaves paripinnate, the rachis lignifying and becoming spinose. Inflorescences axillary, long-pedunculate, compactly capitate. Pods sub-bilocular.

A. hohenackeri Boiss. [1015]
>Not climbing; Shrub; Perennial.
>West Asia: Iran(N); U.S.S.R.
>Description: 1015, 1150; Illustration: 1150.

A. kashgakius Parsa [1103]
>Not climbing; Herb; Perennial.
>West Asia: Iran(N).
>Description: 1103; Illustration: 1103.
>The type (Hb.Parsa 20031) is at Kew [1153]].

A. leptorhaphis Bornm. & Gauba [1015]
>Not climbing; Herb/Shrub; Perennial.
>West Asia: Iran(N).
>Description: 1015, 1215.

A. magistratus A.A.R.Maassoumi et al. [1239]
>Not climbing; Herb; Perennial.
>West Asia: Iran(N).
>Description: 1239.

A. tricholobus DC. [1015]
>A. aciphyllus Freyn [1195]; A. triocholobus DC. [1015].
>Not climbing; Shrub; Perennial.
>West Asia: Iran(N).
>Description: 1015.

ASTRAGALUS Section **Uliginosi**

Formerly Section *Euodmus*. Herbaceous perennials with long stems. Hairs medifixed. Leaves imparipinnate. Pods exceeding the calyx, bilocular or nearly so.

A. odoratus Lam. [1015]
A. vidaeus Parsa [1157].
Not climbing; Herb; Perennial.
West Asia: Iran(N); Middle East: Turkey(N).
Description: 1015, 1102, 1150.

A. peduncularis Royle [1126]
Not climbing; Herb; Perennial.
West Asia: Afghanistan(N).
Description: 1150; Illustration: 1145, 1150.

ASTRAGALUS Section **Vesicarii**

Formerly Section *Cystodes*.

A. alamliensis Rech.f. [1015]
Not climbing; Herb; Perennial.
West Asia: Iran(N).
Description: 1015, 1211.
The type is from Iran [1211].

ASTRAGALUS Section **Xiphidium**

Perennials, mostly herbaceous. Hairs bifurcate. Leaves imparipinnate. Inflorescences lax or sometimes compact, long-pedunculate. Pods many-seeded, bilocular, coriaceous. A large and difficult section.

A. afghanomontanus Sirj. & Rech.f. [1011]
Not climbing; Herb; Perennial.
West Asia: Afghanistan(N).
Description: 1052; Illustration: 1052.

A. al-hamedensis Rech.f. (provisional) [1182]
A. sanctus var. *hamadensis* Velen. [1182].
Not climbing; Shrub; Perennial.
West Asia: Saudi Arabia(N).
Description: 1182; Illustration: 1182.
The locality is uncertain and may not be in Saudi Arabia [1182]. Not mentioned by Hedge & Podlech (1987) [1155].

A. angustatus Boiss. [1015]
Not climbing; Shrub; Perennial.
West Asia: Iran(N).
Description: 1015.
The type is from Iran [1121].

A. argyroides G.Beck [1102]
A. angustatus Bunge [1150]; *A. novus* Grossh. [1102].
Not climbing; Herb; Perennial.
West Asia: Iran(N); Middle East: Turkey(N); U.S.S.R.
Description: 1015, 1102, 1150.

A. aucheri Boiss. [1015]
Not climbing; Herb,Herb/Shrub; Perennial.
West Asia: Iran(N); Middle East: Turkey(N).
Description: 1015, 1102.
Not recorded from Iran in Flora of Turkey [1102].

A. bifoliolatus Sirj. & Rech.f. [1011]
Not climbing; Herb; Perennial.
West Asia: Afghanistan(N).
Description: 1052; Illustration: 1052.

A. bijugus Sirj. & Rech.f. [1011]
Not climbing; Herb; Perennial.
West Asia: Afghanistan(N).
Description: 1052; Illustration: 1052.

A. brachylobus Fischer [1015]
Not climbing; Herb,Shrub; Perennial.
West Asia: Iran(N); U.S.S.R.
Description: 1015, 1150.
Not recorded for Iran in Flora of the U.S.S.R., or by Bunge [1150].

A. cornutus Pallas [1102]
A. fruticosus Pallas [1102]; *A. vimineus* Pallas [1102].
Not climbing; Shrub; Perennial.
West Asia: Iran(N); Middle East: Turkey(N); U.S.S.R.
Description: 1015, 1102, 1150; Illustration: 1151.
Parsa records *A. vimineus* from Iran [1015], but it is not recorded for Iran in Flora of Turkey or in the Flora of the U.S.S.R. [1102].

A. dendroproselius Rech.f. [1009]
A. quercetorum Rech.f. [1009].
Not climbing; Shrub; Perennial.
West Asia: Iraq(N).
Description: 1009; Illustration: 1009.

A. eburneus Bornm. & Gauba [1211]
Not climbing; Herb/Shrub; Perennial.
West Asia: Iran(N).
Description: 1015.

A. juratzkanus Freyn & Sint. [1190]
A. macrotropis sensu auctt. [1190]; *A. neilreichianus* Freyn & Sint. [1190]; *A. variifolius* Freyn & Sint. [1190]; *A. xanthoxiphidium* Freyn & Sint. [1190].
Not climbing; Herb; Perennial.
West Asia: Iran(N); U.S.S.R.
Description: 1015.
A. macrotropis Bunge is recorded from Iran by Parsa [1015], but in this strict sense, it is not recorded for Iran in the Flora of the U.S.S.R. or in Flora of Tadjikistan [1150].
In this strict sense, *A. macrotropis* is described & illustrated in the Flora of the U.S.S.R. & Flora of Tadjikistan [1150].

A. kabadianus Lipsky [1190]
A. cisdarwasicus Gontsch. [1190].
Not climbing; Herb; Perennial.
West Asia: Afghanistan(N); U.S.S.R.
Description: 1150; Illustration: 1145, 1150.

A. kandaharensis Sirj. & Rech.f. [1011]
Not climbing; Herb; Perennial.
West Asia: Afghanistan(N).
Description: 1052; Illustration: 1052.

A. khadjouicus Parsa [1103]
Not climbing; Herb; Perennial.
West Asia: Iran(N).
Description: 1103; Illustration: 1103.
The type (Hb.Parsa 20005) is at Kew [1153].

A. lancifolius Gontsch. [1085]
>Not climbing; Herb; Perennial.
>West Asia: Afghanistan(N); U.S.S.R.
>Description: 1150.

A. maverranagri Popov [1085]
>Not climbing; Herb; Perennial.
>West Asia: Afghanistan(N); U.S.S.R.
>Description: 1150; Illustration: 1145.

A. melanocalyx Boiss. & Buhse [1015]
>Not climbing; Herb/Shrub; Perennial.
>West Asia: Iran(N).
>Description: 1015.

A. microcalycinus Sirj. & Rech.f. [1011]
>Not climbing; Herb; Perennial.
>West Asia: Afghanistan(N).
>Description: 1052; Illustration: 1052.

A. nigrolineatus Sirj. & Rech.f. [1052]
>Not climbing; Herb; Perennial.
>West Asia: Afghanistan(N).
>Description: 1052; Illustration: 1052.

A. nitens Boiss. & Heldr. [1009]
>Not climbing; Herb; Perennial.
>West Asia: Iran(N);Iraq(N); Middle East: Turkey(N).
>Description: 1009, 1102.
>Fl. Turkey [1102] treats as synonym of *A. viridis* Bunge (q.v.).

A. patulepilosus Sirj. & Rech.f. [1052]
>Not climbing; Herb; Perennial.
>West Asia: Afghanistan(N).
>Description: 1052; Illustration: 1052.

A. pseudopendulinus Sirj. & Rech.f. [1213]
>Not climbing; Herb; Perennial.
>West Asia: Iran(N).
>Description: 1213.

A. puligumrensis Rech.f. [1052]
>*A. puligumuriensis* Rech.f. [1011].
>Not climbing; Herb; Perennial.
>West Asia: Afghanistan(N).
>Description: 1052; Illustration: 1052.

A. ruscifolius Boiss. [1015]
>Not climbing; Shrub; Perennial.
>West Asia: Iran(N).
>Description: 1015.
>The type is from Iran [1121].

A. sitiens Bunge [1015]
>Not climbing; Herb/Shrub; Perennial.
>West Asia: Iran(N).
>Description: 1015.
>The type is from Iran [1149].

A. steineranus Podl. [1185]
>Not climbing; Herb; Perennial.
>West Asia: Afghanistan(N).
>Description: 1185; Illustration: 1185.

A. subuliformis DC. [1205]
  A. anfractuosus Bunge [1205]; *A. subulatus* M.Bieb. [1205].
  Not climbing; Shrub; Perennial.
  West Asia: Afghanistan(N); South Asia: Pakistan(N); U.S.S.R.
  Description: 1015, 1150.
  Parsa lists this taxon, but without a definite Iran record [1015].

A. suffalcatus Bunge [1011]
  A. macrosyrinx Rech.f. [1011].
  Not climbing; Herb; Perennial.
  West Asia: Afghanistan(N).
  Description: 1052; Illustration: 1052.
  Ali(1958) treats this as a synonym of *A. subuliformis* (q.v.) [1205].

A. tolgorensis Sirj. & Rech.f. [1170]
  Not climbing; Herb; Perennial.
  West Asia: Iran(N).
  Description: 1170.

A. trachoniticus Post [1009]
  Not climbing; Herb; Not Threatened
  Perennial.
  West Asia: Iraq(N);Saudi Arabia(N); Middle East: Jordan(N);Syria(N).
  Description: 1009; Illustration: 1001, 1009.

A. triqueter Bornm. & Gauba [1015]
  Not climbing; Herb/Shrub; Perennial.
  West Asia: Iran(N).
  Description: 1015, 1180; Illustration: 1180.

A. variegatus Franchet [1185]
  A. sarypulensis B.Fedtsch. [1190].
  Not climbing; Herb; Perennial.
  West Asia: Afghanistan(N); U.S.S.R.
  Description: 1150; Illustration: 1150.

A. versipilus Rech.f. & Koie [1052]
  A. vespilus Rech.f. & Koie [1011].
  Not climbing; Herb; Perennial.
  West Asia: Afghanistan(N).
  Description: 1052; Illustration: 1052.

A. viridiformis Sirj. [1102]
  Not climbing; Herb/Shrub; Perennial.
  West Asia: Iran(N); Middle East: Turkey(N).
  Description: 1015, 1102.
  Doubtfully recorded from Iran in Flora of Turkey [1102].

A. viridis Bunge [1102]
  Not climbing; Shrub; Perennial.
  West Asia: Iran(N); Middle East: Turkey(N); U.S.S.R.
  Description: 1015, 1102, 1150.
  The type is from Iran [1149].

A. xanthoxiphidiopsis Rech.f. [1052]
  Not climbing; Herb; Perennial.
  West Asia: Afghanistan(N).
  Description: 1052; Illustration: 1052.

ASTRAGALUS: Section Uncertain
  Listed here are several species whose sectional position remains uncertain pending further study. The material of some is unsatisfactory; some, on the other hand, may prove to be the basis of new sections.

A. congestus Baker (provisional) [1011].
Not climbing; Herb/Shrub; Perennial.
West Asia: Afghanistan(N); South Asia: Pakistan(N).
Description: 1240.
Kitamura(1960) placed this in Section *Brachycarpus* [1011]; but Rechinger et al. (1959) placed it in Section *Campylanthus* [1220].
In her revision of the section, Tietz (1988) excluded it from *Campylanthus*, but did not assign it to a section; she also suggested that it is probably synonymous with *A. diopogon*, also of uncertain sectional position [1195].

A. curvidens Freyn & Bornm. [1222]
A. courvidens Freyn & Bornm. [1015].
Not climbing; Herb; Perennial.
West Asia: Iran(N).
Description: 1015.
This species may belong to Section *Hypoglottoidei* [1222].

A. diopogon Bunge [1011]
West Asia: Afghanistan(N); South Asia: Pakistan(N).
Description: 1015.
Placed in Section *Acanthophace* by Kitamura (1960) [1011], and transferred to Section *Acidodes* by Deml (1971). Tietz (1988) suggested that it could form the basis of a new section [1195].

A. erythrolepis Boiss. [1015]
Not climbing; Shrub; Perennial.
West Asia: Iran(N).
Description: 1015.
Parsa placed this in Section *Acanthace* [1015]; Deml transferred it to Section Campylanthus [1171]. Tietz excluded it from *Campylanthus* in her revision of the section, but did not reassign it [1195].

A. hafez-shirazi Parsa (provisional) [1015].
Not climbing; Shrub; Perennial.
West Asia: Iran(N).
Description: 1015.
The type is Stapf 2567 (K) [1015]. Parsa placed this in Section *Campylanthus*, from which Tietz excluded it in her revision of Campylanthus, but without assigning it to another section [1195].
The type has been determined by Rechinger as *A. albispinus* (= *Astracantha albispina*, q.v.) [1153].

CARAGANA Lam.

A genus of shrubs or small trees, with perhaps as many as 80 species in the colder parts of Eurasia.

C. afghanica Kitam. (provisional) [1011]
Not climbing; Shrub; Perennial.
West Asia: Afghanistan(N).
Description: 1011; Illustration: 1011.
Kitamura places *C. nuristanica* as a synonym of this [1011].
Rechinger (1984) does not mention *C. afghanica* and makes *C. nuristanica* a synonym of *C. gerrardiana* [1058].

C. aurantiaca Koehne [1058]
C. grandiflora Bunge [1058].
Not climbing; Shrub; Perennial.
West Asia: Afghanistan(N); South Asia: Pakistan(N); U.S.S.R.
Description: 1058; Illustration: 1058.

C. decorticans Hemsley [1058]
C. prainii C.Schneider [1058].
Not climbing; Shrub; Perennial.
West Asia: Afghanistan(N); South Asia: Pakistan(N).
Description: 1058; Illustration: 1058.

C. gerrardiana Benth. [1058]
C. nuristanica Rech.f. & Edelb. [1058].
Not climbing; Shrub; Perennial.
West Asia: Afghanistan(N); South Asia: India(N), Pakistan(N).
Description: 1015, 1058; Illustration: 1058.

C. grandiflora (M.Bieb.) DC. [1058]
Not climbing; Shrub; Perennial.
West Asia: Iran(N); Middle East: Turkey(N); U.S.S.R.
Description: 1058; Illustration: 1151.

C. maimanensis Rech.f. [1058]
Not climbing; Shrub; Perennial.
West Asia: Afghanistan(N).
Description: 1058; Illustration: 1058.
Endemic to Afghanistan [1058].

C. ulicina Stocks [1058]
Not climbing; Shrub; Perennial.
West Asia: Afghanistan(N); South Asia: Pakistan(N).
Description: 1015, 1058; Illustration: 1058.

C. versicolor Benth. [1058]
C. pygmaea sensu auctt. [1058].
Not climbing; Shrub; Perennial.
West Asia: Afghanistan(N); East Asia: China(N) (Xizang Zizhiqu); South Asia: India(N).
Description: 1058; Illustration: 1088.

CHESNEYA Endl.

Perennial herbs, sometimes woody at the base; about 20 species in West Asia, Himalaya and south-central Asia. Very close to *Astragalus*.

C. acaulis (Baker) Popov [1058]
Caragana acaulis Baker [1058].
Not climbing; Herb; Perennial.
West Asia: Afghanistan(N); South Asia: Pakistan(N).
Description: 1058; Illustration: 1058.

C. afghanica Rech.f. & Koie [1058]
Not climbing; Herb; Perennial.
West Asia: Afghanistan(N).
Description: 1052, 1058; Illustration: 1052, 1058.
Endemic to Afghanistan.

C. astragalina Jaub. & Spach [1058]
Not climbing; Herb; Perennial.
West Asia: Iran(N); U.S.S.R.
Description: 1015, 1058; Illustration: 1058.

C. crassipes Boriss [1058]
Not climbing; Herb; Perennial.
West Asia: Afghanistan(N); South Asia: Pakistan(N); U.S.S.R.
Description: 1058; Illustration: 1058, 1144.

C. cuneata (Benth.)Ali [1058]
Calophaca cuneata (Benth.)Komarov [1058]; *Caragana cuneata* (Benth.)Ali [1058] *Gueldenstaedtia cuneata* Benth. [1058].
Not climbing; Herb; Perennial.
West Asia: Afghanistan(N); East Asia: China(N) (Xizang Zizhiqu); South Asia: India(N), Pakistan(N); U.S.S.R.
Description: 1058; Illustration: 1058.

C. gaubaeana Bornm. [1058]
C. gaubeana Bornm. [1015].
Not climbing; Herb; Perennial.
West Asia: Afghanistan(N), Iran(N).
Description: 1015, 1058; Illustration: 1058.

C. kopetdaghensis Boriss [1058]
Not climbing; Herb; Perennial.
West Asia: Iran(N); U.S.S.R.
Description: 1058; Illustration: 1058.

C. kotschyi Boiss. [1058]
Not climbing; Herb; Perennial.
West Asia: Iran(N).
Description: 1015, 1058; Illustration: 1058.
Endemic to Iran [1058].

C. parviflora Jaub. & Spach [1058]
C. macranica Rech.f. & Esfand. [1058]; C. makranica Rech.f. & Esfand. [1058].
Not climbing; Herb; Perennial.
West Asia: Iran(N); South Asia: Pakistan(N).
Description: 1015, 1058; Illustration: 1058.

C. rytidosperma Jaub. & Spach [1009]
C. microphylla Vahl [1058]; C. oliverii Jaub. & Spach [1009]; C. olivierii Jaub. & Spach [1058]; C. vaginalis sensu Rawi [1009]; C. velutina Jaub. & Spach [1058].
Not climbing; Herb; Perennial.
West Asia: Iran(N), Iraq(N); Middle East: Syria(N), Turkey(N).
Description: 1009, 1058; Illustration: 1009, 1058.

C. tadzhikistana Boriss [1058]
C. vesiculifera C.Towns. [1058].
Not climbing; Herb; Perennial.
West Asia: Afghanistan(N); U.S.S.R.
Description: 1058, 1087; Illustration: 1058, 1087, 1145.

C. ternata (Korsh.) Popov [1058]
Not climbing; Herb; Perennial.
West Asia: Afghanistan(N).
Description: 1058; Illustration: 1058, 1145.

C. vaginalis Jaub. & Spach [1009]
Not climbing; Herb; Perennial.
West Asia: Iran(N).
Description: 1009.
Very similar to C. rytidosperma and probably synonymous [1009].
No collection apart from the type (Aucher-Eloy, 1835), which is almost certainly from Iran, but locality unclear [1009].

C. volkii Rech.f. [1058]
Not climbing; Shrub; Perennial.
West Asia: Afghanistan(N).
Description: 1052, 1058; Illustration: 1052, 1058.
Endemic to Afghanistan [1058].

COLUTEA L.

Shrubs, subshrubs or small trees. 25-30 species in southern Europe, the Mediterranean region and central Asia.

C. afghanica Browicz [1058]
Not climbing; Shrub; Perennial.
West Asia: Afghanistan(N), Iran(N).
Description: 1058, 1084; Illustration: 1058.
Endemic to Afghanistan [1058].

C. buhsei (Boiss.) Shap. [1058]
C. buhsei var. densiflora Browicz [1058]; *C. persica* sensu auctt. [1058]; *C. persica var. buhsei* Boiss. [1058].
Not climbing; Shrub; Perennial.
West Asia: Iran(N); U.S.S.R.
Description: 1058, 1084; Illustration: 1058, 1084.

C. cilicica Boiss. & Bal. [1009]
Not climbing; Shrub; Perennial.
West Asia: Iran(N), Iraq(N); Europe; Middle East: Israel(N), Lebanon(N), Syria(N), Turkey(N); U.S.S.R.
Description: 1009, 1058, 1084; Illustration: 1009, 1058.

C. gifana Parsa [1058]
Not climbing; Shrub; Perennial.
West Asia: Iran(N).
Description: 1015, 1058, 1084; Illustration: 1084.
Endemic to Iran [1058].

C. gracilis Freyn & Sint. [1058]
C. persica var. gracilis (Freyn) Parsa [1058].
Not climbing; Shrub; Perennial.
West Asia: Afghanistan(N), Iran(N).
Description: 1058, 1084.
Forms 'transitional to *C. persica*' occur [1058].

C. istria Miller [1001]
C. halepica Lam. [1084]; *C. haleppica* Lam. [1064].
Not climbing; Shrub; Perennial.
West Asia: Saudi Arabia(N); Middle East: Israel(N), Jordan(N), Lebanon(N), Syria(N), Turkey(N).
Description: 1001, 1084; Illustration: 1001.

C. komarovii Takht. [1058]
Not climbing; Shrub; Perennial.
West Asia: Iran(N); U.S.S.R.
Description: 1084.

C. nepalensis Sims [1058]
C. persica var. buhsei sensu Rech.f. [1084]; *C. rostrata* Gilli, p.p. [1058].
Not climbing; Shrub; Perennial.
West Asia: Afghanistan(N); East Asia: China(N) (Xizang Zizhiqu); South Asia: India(N), Nepal(N), Pakistan(N).
Description: 1058, 1084; Illustration: 1058.

C. paulsenii Freyn [1058]
C. rostrata Gilli p.p. [1058].
Not climbing; Shrub; Perennial.
West Asia: Afghanistan(N); South Asia: Pakistan(N).
Description: 1052, 1058, 1084; Illustration: 1058, 1084.
2 subspp., paulsenii & mesantha.

subsp. mesantha (Ali) Ali [1058]
C. karakoramensis Kitam. [1058]; *C. mesantha* Ali [1058]; *C. paulsenii var. menantha* (Ali) Browicz [1058].
Not climbing; Shrub; Perennial.
West Asia: Afghanistan(N).
Description: 1058; Illustration: 1084.

subsp. paulseni [1058]
Not climbing; Shrub; Perennial.
West Asia: Afghanistan(N).
Description: 1058; Illustration: 1058.

C. persica Boiss. [1058]
    Not climbing; Shrub; Perennial.
    West Asia: Afghanistan(N), Iran(N).
    Description: 1058, 1084; Illustration: 1058.

C. porphyrogramma Rech.f. [1058]
    Not climbing; Shrub; Perennial.
    West Asia: Iran(N).
    Description: 1058, 1084; Illustration: 1058.
    Endemic to Iran [1058].

C. uniflora G.Beck [1058]
    Not climbing; Shrub; Perennial.
    West Asia: Iran(N).
    Description: 1015, 1058, 1084; Illustration: 1084.
    Endemic to Iran [1058].

GALEGA L.

A genus of about six species of perennial herbs, some cultivated as fodder or as ornamentals. Mainly in the Mediterranean region.

G. officinalis L. [1058]
    G. persica (Boiss.) E.Small [1052].
    Not climbing; Herb/Shrub; Perennial.
    West Asia: Afghanistan(N), Iran(N); Europe; South Asia: Pakistan(N).
    Description: 1015, 1058; Illustration: 1058.
    Koie & Rechinger (1957) recognise a var. albiflora Boiss. [1052].

GLYCYRRHIZA L.

A genus of about 20 species, mostly in Europe and Asia, but some elsewhere. The roots of several species yield liquorice, and one (*G. glabra*) is cultivated for this purpose.

G. aspera Pallas [1058]
    Astragalus glandulosus G.Beck [1058]; *Glycyrrhiza asperrima* L.f. [1058].
    Not climbing; Herb; Perennial.
    West Asia: Afghanistan(N), Iran(N); Europe; U.S.S.R.
    Description: 1015, 1052, 1058; Illustration: 1058, 1145.

G. bucharica Regel [1058]
    Not climbing; Herb; Perennial.
    West Asia: Afghanistan(N); U.S.S.R.
    Description: 1058; Illustration: 1058.

G. echinata L. [1058]
    Not climbing; Herb; Perennial.
    West Asia: Iran(N); U.S.S.R.
    Description: 1015, 1058; Illustration: 1058.

G. erythrocarpa (Vassilcz.)Abdull. [1190]
    G. triphylla sensu Parsa [1190]; *Meristotropis erythrocarpa* Vassilcz. [1190]; *Meristotropis xanthioides* Vassilcz.
    Not climbing; Herb; Perennial.
    West Asia: Afghanistan(N), Iran(N).
    Description: 1015, 1052, 1058; Illustration: 1058.
    Gunn (1983) does not recognise *Meristotropis*; Rechinger (1984) does so [1058,1090].

G. glabra L. [1009]
> *G. glabra* var. *violacea* (Boiss.)Boiss. [1009]; *G. glandulifera* Waldst. & Kit. [1009].
> Not climbing; Herb; Perennial.
> West Asia: Afghanistan(N), Iran(N), Iraq(N); Africa; Europe; Middle East: Israel(N), Jordan(N), Lebanon(N), Syria(N), Turkey(N); South Asia: India(N); U.S.S.R.
> Description: 1009, 1052, 1058; Illustration: 1009, 1058, 1145, 1151.
> Domestic [1009]; Food or drink [1009]; Medicine [1009]; Wood [1009].
> Common Liquorice [1009].
> 2 varieties, *glabra* & *glandulifera*.[1009,1058].

G. uralensis Fischer [1058]
> Not climbing; Herb; Perennial.
> West Asia: Afghanistan(N); East Asia: China(N); South Asia: Pakistan(N).
> Description: 1058; Illustration: 1058.

HALIMODENDRON DC.

A monospecific genus. A halophytic shrub with persistent spiny leaf rachides, occurring in the arid saline regions of central Asia.

H. halodendron (Pallas) Voss [1058]
> *H. argenteum* (Lam.)DC. [1058].
> Not climbing; Shrub; Perennial.
> West Asia: Afghanistan(N), Iran(N); Middle East: Turkey(N); South Asia: Pakistan(N); U.S.S.R.
> Description: 1015, 1058; Illustration: 1058.

OREOPHYSA (Boiss.)Bornm.

A monospecific genus; a herbaceous perennial of the mountains of Iran.

O. microphylla (Jaub. & Spach) Browicz [1058]
> *Colutea triphylla* Boiss. [1058]; *O. triphylla* Boiss. [1058]; *Sphaerophysa microphylla* Jaub. & Spach [1058].
> Not climbing; Herb; Perennial.
> West Asia: Iran(N).
> Description: 1015, 1058; Illustration: 1058.
> Endemic to Iran; Elburz Mts [1058].

OXYTROPIS DC.

A large genus of small herbs, many of high pastures and rocky places. Kept separate from *Astragalus* largely by tradition and for convenience; the main difference is in the shape of the keel. The differences between many of the species appear small and there is a need for a critical revision.

O. admiranda Rech.f. [1058]
> Not climbing; Herb; Perennial.
> West Asia: Afghanistan(N).
> Description: 1058; Illustration: 1058.
> Endemic to Afghanistan [1058].

O. aellenii Vassilcz. [1058]
> Not climbing; Herb; Perennial.
> West Asia: Iran(N).
> Description: 1058; Illustration: 1058.
> Endemic to Iran; known only from the type [1058].

O. afghanica Rech.f. & Koie [1058]
   Not climbing; Herb; Perennial.
   West Asia: Afghanistan(N).
   Description: 1052, 1058; Illustration: 1052, 1058.
   Endemic to Afghanistan.

O. alavae Rech.f. [1058]
   Not climbing; Herb; Perennial.
   West Asia: Iran(N).
   Description: 1058; Illustration: 1058.
   Endemic to Iran; known only from the type [1058].

O. andersii Vassilcz. [1249]
   Not climbing; Herb; Perennial.
   West Asia: Afghanistan(N).
   Description: 1249.
   Type (Anders 8067) coll. 1971; no other material cited [1249]. This number cited as *O. cobresietorum* in Podlech & Anders 1977 [1126] and as *O. trichosphaera* in Flora Iranica [1058].

O. assadliensis Vassilcz. [1058]
   Not climbing; Herb; Perennial.
   West Asia: Iran(N).
   Description: 1058; Illustration: 1058.
   Endemic to Iran; known only from the type [1058].

O. asterocarpa Vassilcz. [1058]
   Oxytropis spec.I Podl. & Anders [1058,1126]; *Oxytropis spec.III* Podl. & Anders [1058,1126].
   Not climbing; Herb; Perennial.
   West Asia: Afghanistan(N).
   Description: 1058; Illustration: 1058.
   Endemic to Afghanistan [1058].

O. astragaloides Boriss. [1058]
   Not climbing; Herb; Perennial.
   West Asia: Afghanistan(N); U.S.S.R.
   Description: 1058; Illustration: 1145.
   "Doubtless to be found...in Afghanistan" [1058].

O. aucheri Boiss. [1058]
   Not climbing; Herb; Perennial.
   West Asia: Afghanistan(N), Iran(N).
   Description: 1015, 1058; Illustration: 1058.

O. baburi Vassilcz. [1058]
   Not climbing; Herb; Perennial.
   West Asia: Afghanistan(N).
   Description: 1058; Illustration: 1058.
   Endemic to Afghanistan [1058].

O. bella O. Fedtsch. [1058]
   Not climbing; Herb; Perennial.
   West Asia: Afghanistan(N); U.S.S.R.
   Description: 1058; Illustration: 1058.

O. bicornis Vassilcz. [1058]
   Not climbing; Herb; Perennial.
   West Asia: Iran(N).
   Description: 1058; Illustration: 1058.
   Endemic to Iran [1058].

O. binaludensis Vassilcz. [1058]
   Not climbing; Herb; Perennial.
   West Asia: Iran(N).
   Description: 1058; Illustration: 1058.
   Endemic to Iran; known only from the type [1058].

O. boguschi B.Fedtsch. [1058]
   Oxytropis darvasica B.Fedtsch. [1058].
   Not climbing; Herb; Perennial.
   West Asia: Afghanistan(N); U.S.S.R.
   Description: 1058; Illustration: 1145.
   Record for Afghanistan based on a single sterile specimen [1058].

O. cabulica (Boiss.)Boiss. [1058]
   Astragalus cabulicus Boiss. [1058].
   Not climbing; Herb; Perennial.
   West Asia: Afghanistan(N).
   Description: 1052, 1058; Illustration: 1058.
   Endemic to Afghanistan [1058].

O. callophylla Vassilcz. [1249]
   Not climbing; Herb; Perennial.
   West Asia: Afghanistan(N).
   Description: 1249.
   Type (Breckle 2678) coll. 1969; no other material cited [1249].

O. caraganetorum Vassilcz. [1058]
   Not climbing; Herb; Perennial.
   West Asia: Afghanistan(N), Iran(N).
   Description: 1058; Illustration: 1058.

O. carduchorum Hedge (provisional) [1058]
   Perennial.
   West Asia: Afghanistan(N), Iraq(N); Middle East: Turkey(N).
   Description: 1058.
   Townsend (1974) regards as synonym of *O. savellanica* [1009]; Vassilczenko keeps distinct [1058].

O. chiliophylla Benth. [1058]
   Not climbing; Herb; Perennial.
   West Asia: Afghanistan(N); East Asia: China(N) (Xizang Zizhiqu); South Asia: Pakistan(N); U.S.S.R.
   Description: 1058; Illustration: 1058, 1144.

O. chionophylla Schrenk [1058]
   Not climbing; Herb; Perennial.
   West Asia: Afghanistan(N); U.S.S.R.
   Description: 1058.
   No definite W. Asian record in Rechinger (1984).

O. chitralensis Ali [1058]
   Not climbing; Herb; Perennial.
   West Asia: Afghanistan(N); South Asia: Pakistan(N).
   Description: 1058.

O. chrysocarpa Boiss. [1058]
   Not climbing; Herb; Perennial.
   West Asia: Iran(N).
   Description: 1015, 1058; Illustration: 1058.
   Endemic to Iran [1058].

O. cinerea Vassilcz. [1058]
   Not climbing; Herb; Perennial.
   West Asia: Iran(N).
   Description: 1058; Illustration: 1058.
   Endemic to Iran; known only from the type [1058].

O. crassiuscula Boriss. [1058]
   Not climbing; Herb; Perennial.
   West Asia: Afghanistan(N); South Asia: Pakistan(N).
   Description: 1058; Illustration: 1058.

O. czapan-daghi B.Fedtsch. [1058]
Not climbing; Herb; Perennial.
West Asia: Iran(N); U.S.S.R.
Description: 1058.

O. danorum Rech.f. [1058]
Not climbing; Herb; Perennial.
West Asia: Afghanistan(N).
Description: 1058; Illustration: 1058.
Endemic to Afghanistan; known only from the type [1058].

O. dashtinavarensis Vassilcz. [1058]
Not climbing; Herb; Perennial.
West Asia: Afghanistan(N).
Description: 1058; Illustration: 1058.
Endemic to Afghanistan [1058].

O. farsi Vassilcz. [1058]
Not climbing; Herb; Perennial.
West Asia: Afghanistan(N).
Description: 1058, 1089.
Endemic to Afghanistan; known only from the type [1058].

O. fohlenensis Vassilcz. [1058]
Not climbing; Herb; Perennial.
West Asia: Afghanistan(N).
Description: 1058.
Endemic to Afghanistan [1058].

O. fuscescens Vassilcz. [1058]
Not climbing; Herb; Perennial.
West Asia: Afghanistan(N).
Description: 1058; Illustration: 1058.
Endemic to Afghanistan; known only from the type [1058].

O. glacialis Benth. [1058]
Not climbing; Herb; Perennial.
West Asia: Afghanistan(N); East Asia; China(N) (Xizang Zizhiqu).
Description: 1058.
Likely to occur in Pakistan; not otherwise recorded [1058].

O. gorbunovii Boriss. [1058]
Not climbing; Herb; Perennial.
West Asia: Afghanistan(N); U.S.S.R.
Description: 1058; Illustration: 1058.

O. gracillima Vassilcz. [1058]
Not climbing; Herb; Perennial.
West Asia: Iran(N).
Description: 1058.
Endemic to Iran; known only from the type [1058].

O. griffithii Boiss. [1058]
Not climbing; Herb; Perennial.
West Asia: Afghanistan(N).
Description: 1058; Illustration: 1058.

O. gubanovii Vassilcz. [1058]
Not climbing; Herb; Perennial.
West Asia: Afghanistan(N).
Description: 1058, 1089.
Endemic to Afghanistan; only known from the type [1058].

O. guntensii B.Fedtsch. [1058]
Not climbing; Herb; Perennial.
West Asia: Afghanistan(N).
Description: 1058.
No definite Afghanistan record in Rechinger (1984) [1058].

O. hedgeii Vassilcz. [1249]
Not climbing; Herb; Perennial.
West Asia: Afghanistan(N).
Description: 1249.
Type (Hedge et al.9125) coll.1969. No other material cited [1249].

O. heratensis Bunge [1058]
Not climbing; Herb; Perennial.
West Asia: Afghanistan(N), Iran(N).
Description: 1015, 1058; Illustration: 1052, 1058.

O. hindukushensis Vassilcz. [1058]
Not climbing; Herb; Perennial.
West Asia: Afghanistan(N).
Description: 1058; Illustration: 1058.
Endemic to Afghanistan [1058].

O. hirsutiuscula Freyn [1058].
Not climbing; Herb; Perennial.
West Asia: Afghanistan(N), Iran(N); South Asia: Pakistan(N); U.S.S.R.
Description: 1058; Illustration: 1058.

O. hypsophila Bunge [1058]
Oxytropis gypsophila sensu Parsa.
Not climbing; Herb; Perennial.
West Asia: Iran(N).
Description: 1015, 1058.
Endemic to Iran [1058].

O. immersa (Baker) B.Fedtsch. [1058]
Astragalus immersus Baker [1058]; *Oxytropis incanescens* Freyn [1058]; *Oxytropis pamirica* Danguy [1058].
Not climbing; Herb; Perennial.
West Asia: Afghanistan(N), Iran(N), Iraq(N); South Asia: Pakistan(N); U.S.S.R.
Description: 1058; Illustration: 1058, 1145.
A var. *jinaliensis* Ali has been recognised [1058].

O. iranica Vassilcz. [1058]
Not climbing; Herb; Perennial.
West Asia: Iran(N).
Description: 1058; Illustration: 1058.
Endemic to Iran [1058].

O. karjaginii Grossh. [1058]
Not climbing; Herb; Perennial.
West Asia: Iran(N); Middle East: Turkey(N); U.S.S.R.
Description: 1058; Illustration: 1058, 1151.

O. kazidanica Vassilcz. [1058]
Not climbing; Herb; Perennial.
West Asia: Afghanistan(N).
Description: 1058.
Endemic to Afghanistan; known only from the type [1058].

O. kermanica Freyn & Bornm. [1058]
Not climbing; Herb; Perennial.
West Asia: Iran(N).
Description: 1015, 1058; Illustration: 1058.
Endemic to Iran [1058].

O. khinjahi Vassilcz. [1249]
    Not climbing; Herb; Perennial.
    West Asia: Afghanistan(N).
    Description: 1249.
    Type (Hedge et al. W 8558) coll.1969; no other material cited [1249].

O. kopetdaghensis Gontch. [1058]
    Not climbing; Herb; Perennial.
    West Asia: Iran(N); U.S.S.R.
    Description: 1058; Illustration: 1058.

O. kotschyana Boiss. & Hohen. [1058]
    Astragalus sorkhabadensis Sirj. & Rech.f. [1142].
    Not climbing; Herb; Perennial.
    West Asia: Iran(N).
    Description: 1015, 1058; Illustration: 1058.
    Endemic to Iran [1058].

O. kuchanensis Vassilcz. [1058]
    Not climbing; Herb; Perennial.
    West Asia: Iran(N).
    Description: 1058; Illustration: 1058.
    Endemic to Iran [1058].

O. kukkonenii Vassilcz. [1058]
    Not climbing; Herb; Perennial.
    West Asia: Afghanistan(N).
    Description: 1058.
    Endemic to Afghanistan; known only from the type [1058].

O. kunarensis Vassilcz. [1249]
    Not climbing; Herb; Perennial.
    West Asia: Afghanistan(N).
    Description: 1249.
    Type (Hedge & Wendelbo 9389) coll.1969. 2 other collections cited [1249].

O. lapponica (Wahlenb.)Gay [1058]
    Not climbing; Herb; Perennial.
    West Asia: Afghanistan(N); East Asia; China(N) (Xizang Zizhiqu); Europe; South Asia: Pakistan(N); U.S.S.R.
    Description: 1058; Illustration: 1058.

O. laxiracemosa Vassilcz. [1249]
    Not climbing; Herb; Perennial.
    West Asia: Afghanistan(N).
    Description: 1249.
    Type (Breckle 2218) coll. 1969. No other collections cited [1249]. This number cited as *O. viae-amicitiae* in Flora Iranica [1249]. The diagnosis gives distinctions between the two taxa [1249].

O. liliputa Vassilcz. [1058]
    Not climbing; Herb; Perennial.
    West Asia: Afghanistan(N).
    Description: 1058; Illustration: 1058.
    Endemic to Afghanistan; known only from the type [1058].

O. linczevskii Gontch. [1058]
    Not climbing; Herb; Perennial.
    West Asia: Afghanistan(N); U.S.S.R.
    Description: 1058; Illustration: 1145.

O. lupinoides Grossh. [1058]
    Not climbing; Herb; Perennial.
    West Asia: Iran(N); U.S.S.R.
    Description: 1058.
    No definite record from Iran [1058].

O. luteo-coerulea (Baker) Ali [1058]
   Astragalus luteo-coerulea Baker [1058].
   Not climbing; Herb; Perennial.
   West Asia: Afghanistan(N); South Asia: Pakistan(N).
   Description: 1058.

O. lydiae Vassilcz. [1058]
   Not climbing; Herb; Perennial.
   West Asia: Afghanistan(N).
   Description: 1058; Illustration: 1058.
   Endemic to Afghanistan; known only from the type [1058].

O. marco-poloi Vassilcz. [1058]
   Not climbing; Herb; Perennial.
   West Asia: Afghanistan(N).
   Description: 1058; Illustration: 1058.

O. margacea Vassilcz. [1058]
   Not climbing; Herb; Perennial.
   West Asia: Afghanistan(N).
   Description: 1058; Illustration: 1058.
   Endemic to Afghanistan; known only from the type [1058].

O. masanderanensis Vassilcz. [1058]
   Not climbing; Herb; Perennial.
   West Asia: Iran(N).
   Description: 1058; Illustration: 1058.
   Endemic to Iran [1058].

O. microsphaera Bunge [1058]
   Not climbing; Herb; Perennial.
   West Asia: Afghanistan(N); East Asia: China(N); U.S.S.R.
   Description: 1058; Illustration: 1058, 1145.

O. minjanensis Rech.f. [1058]
   Not climbing; Herb; Perennial.
   West Asia: Afghanistan(N).
   Description: 1052, 1058; Illustration: 1052, 1058.
   Endemic to Afghanistan [1058].

O. neo-rechingeriana Vassilcz. [1058]
   Not climbing; Herb; Perennial.
   West Asia: Iran(N).
   Description: 1058; Illustration: 1058.
   Endemic to Iran [1058].

O. nuristanica Vassilcz. [1058]
   Not climbing; Herb; Perennial.
   West Asia: Afghanistan(N).
   Description: 1058; Illustration: 1058.
   Endemic to Afghanistan; known only from the type [1058].

O. oreophila Vassilcz. [1058]
   Oxytropis spec.II Podl. & Anders [1058,1126].
   Not climbing; Herb; Perennial.
   West Asia: Afghanistan(N).
   Description: 1058.
   Endemic to Afghanistan [1058].

O. oroboides Vassilcz. (provisional) [1058]
   Not climbing; Herb; Perennial.
   West Asia: Afghanistan(N).
   Description: 1058.
   Endemic to Afghansitan; known only from the type [1058].
   Probably not a validly published name.

O. pagobia Bunge [1058]
Not climbing; Herb; Perennial.
West Asia: Afghanistan(N); U.S.S.R.
Description: 1058; Illustration: 1058, 1145.

O. pakistanica Vassilcz. [1058]
Not climbing; Herb; Perennial.
West Asia: Afghanistan(N); South Asia: Pakistan(N).
Description: 1058; Illustration: 1058.

O. panjshirica Podl. & Deml [1058]
Not climbing; Herb; Perennial.
West Asia: Afghanistan(N).
Description: 1058, 1085; Illustration: 1058, 1085.
Endemic to Afghanistan [1058].

O. parvanensis Vassilcz. [1058]
Not climbing; Herb; Perennial.
West Asia: Afghanistan(N).
Description: 1058; Illustration: 1058.
Endemic to Afghanistan [1058].

O. persica Boiss. [1058]
Not climbing; Herb; Perennial.
West Asia: Iran(N), Iraq(N).
Description: 1015, 1058; Illustration: 1058.

O. platonychia Bunge [1058]
Not climbing; Herb; Perennial.
West Asia: Afghanistan(N); East Asia: China(N); South Asia: Pakistan(N); U.S.S.R.
Description: 1058; Illustration: 1058.

O. podlechii Vassilcz. [1058]
Not climbing; Herb; Perennial.
West Asia: Afghanistan(N).
Description: 1058; Illustration: 1058.
Endemic to Afghanistan; known only from the type [1058].

O. poncinsii Franchet [1139]
O. poincinsii sensu Rech.f.
Not climbing; Herb; Perennial.
West Asia: Afghanistan(N); East Asia: China(N); U.S.S.R.
Description: 1058, 1139; Illustration: 1058.

O. proxima Boriss. [1058]
Astragalus sikaramensis (Sirj. & Rech.f.) Ali [1058]; *O. sikaramensis* (Sirj. & Rech.f.) Ali [1058].
Not climbing; Herb; Perennial.
West Asia: Afghanistan(N); South Asia: Pakistan(N); U.S.S.R.
Description: 1052, 1058; Illustration: 1052, 1145.
Record is from the Afghanistan/Pakistan border.

O. pseudohirsutiuscula Vassilcz. [1058]
Not climbing; Herb; Perennial.
West Asia: Afghanistan(N).
Description: 1058.
Endemic to Afghanistan; known only from the type [1058].

O. puberula Boriss. [1058]
Not climbing; Herb; Perennial.
West Asia: Afghanistan(N); East Asia: China(N); U.S.S.R.
Description: 1058.
Without doubt grows in North-east Afghanistan [1058].

O. pusilloides Vassilcz. [1058]
   Not climbing; Herb; Perennial.
   West Asia: Afghanistan(N), Iran(N).
   Description: 1058.

O. rechingeri Vassilcz. [1058]
   Not climbing; Herb; Perennial.
   West Asia: Iran(N).
   Description: 1058; Illustration: 1058.
   Endemic to Iran; known only from the type [1058].

O. rhodontha Vassilcz. [1249]
   Not climbing; Herb; Perennial.
   West Asia: Iran(N).
   Description: 1249.
   Type (Foroughi s.n.) coll. 1972. No other material cited [1249].

O. riparia Litv. [1058]
   Not climbing; Herb; Perennial.
   West Asia: Afghanistan(N); U.S.S.R.
   Description: 1058; Illustration: 1058.

O. rudbariensis Vassilcz. [1249]
   Not climbing; Herb; Perennial.
   West Asia: Iran(N).
   Description: 1249.
   Type (Pichler s.n.) coll. 1882. Lacks pods; systematic position uncertain [1249].

O. salangensis Podl. & Deml [1058]
   Not climbing; Herb; Perennial.
   West Asia: Afghanistan(N).
   Description: 1058, 1085; Illustration: 1058.
   Endemic to Afghanistan [1058].

O. salicetorum Vassilcz. [1249]
   Not climbing; Herb; Perennial.
   West Asia: Afghanistan(N).
   Description: 1249.
   Type (Grey-Wilson & Hewer 1614) coll.1971; no other material cited [1249].
   This no. cited as *O. hirtiuscula* by Podlech & Anders 1977 [1126]; distinctions from this species are given in the diagnosis [1249].

O. saperlebulensis Vassilcz. [1058]
   Not climbing; Herb; Perennial.
   West Asia: Afghanistan(N).
   Description: 1058; Illustration: 1058.
   Endemic to Afghanistan [1058].

O. sata-kandaonensis Vassilcz. [1058]
   Not climbing; Herb; Perennial.
   West Asia: Afghanistan(N).
   Description: 1058; Illustration: 1058.
   Endemic to Afghanistan [1058].

O. savellanica Boiss. [1009]
   Not climbing; Herb; Perennial.
   West Asia: Afghanistan(N), Iran(N), Iraq(N); Middle East: Turkey(N); South Asia: Pakistan(N); U.S.S.R.
   Description: 1009, 1015, 1058; Illustration: 1009, 1058, 1145, 1151.
   Environmental 1009.
   Townsend (1974) includes *O. carduchorum* and *O. persica* (q.v.) here [1009].

O. shirkuhi Vassilcz. [1249]
   Not climbing; Herb; Perennial.
   West Asia: Iran(N).
   Description: 1249.
   Type (Foroughi & Assadi 17989) coll. 1975. No other specimens [1249].

O. siah-sangi Vassilcz. [1058]
   Not climbing; Herb; Perennial.
   West Asia: Afghanistan(N).
   Description: 1058; Illustration: 1058.
   Endemic to Afghanistan; known only from the type [1058].

O. sojakii Vassilcz. [1058]
   Not climbing; Herb; Perennial.
   West Asia: Afghanistan(N), Iran(N).
   Description: 1058.

O. straussii Bornm. [1058]
   Not climbing; Herb; Perennial.
   West Asia: Iran(N); Middle East: Turkey(N).
   Description: 1015, 1058.
   Parsa (1948) mentions a var. *glabrescens* Bornm [1015].

O. suavis Boriss. [1058]
   Not climbing; Herb; Perennial.
   West Asia: Iran(N); U.S.S.R.
   Description: 1058; Illustration: 1058.

O. surculosa Rech.f. [1058]
   Not climbing; Herb; Perennial.
   West Asia: Afghanistan(N).
   Description: 1052, 1058; Illustration: 1052, 1058.
   Endemic to Afghanistan; known only from the type [1058].

O. surmandehi Vassilcz. [1249]
   Not climbing; Herb; Perennial.
   West Asia: Iran(N).
   Description: 1249.
   Type (Rechinger 47570) coll. 1974; no other material cited [1249].

O. szovitsii Boiss. & Buhse [1058]
   Not climbing; Herb; Perennial.
   West Asia: Iran(N).
   Description: 1058; Illustration: 1058.
   Endemic to Iran [1058].

O. tachtensis Franchet [1058]
   Not climbing; Herb; Perennial.
   West Asia: Afghanistan(N); East Asia: China(N); U.S.S.R.
   Description: 1058; Illustration: 1058, 1145.

O. takhti-soleimanii Vassilcz. [1058]
   Not climbing; Herb; Perennial.
   West Asia: Iran(N).
   Description: 1058.
   Endemic to Iran [1058].

O. tatarica Baker [1058]
   Not climbing; Herb; Perennial.
   West Asia: Afghanistan(N); East Asia: China(N) (Xizang Zizhiqu); South Asia: India(N), Pakistan(N).
   Description: 1058; Illustration: 1058.

O. thaumasiomorpha Rech.f. [1058]
  Not climbing; Herb; Perennial.
  West Asia: Iran(N).
  Description: 1058; Illustration: 1058.
  Endemic to Iran; known only from the type [1058].

O. tianschanica Bunge [1058]
  Oxytropis spec.V Podl. & Anders [1058,1126].
  Not climbing; Herb; Perennial.
  West Asia: Afghanistan(N); East Asia: China(N).
  Description: 1058.

O. trichosphaera Freyn [1058]
  Not climbing; Herb; Perennial.
  West Asia: Afghanistan(N); U.S.S.R.
  Description: 1058; Illustration: 1058.

O. tunnellii Vassilcz. [1249]
  Not climbing; Herb; Perennial.
  West Asia: Afghanistan(N).
  Description: 1249.
  Type (W-960) coll.1969; collector not stated. No other material cited [1249].

O. vadimii Vassilcz. [1058]
  Not climbing; Herb; Perennial.
  West Asia: Afghanistan(N).
  Description: 1058, 1089.
  Endemic to Afghanistan; known only from the type [1058].

O. vakhdzhiri Vassilcz. [1249]
  Not climbing; Herb; Perennial.
  West Asia: Afghanistan(N).
  Description: 1249.
  Type is Anders 7623, in Hb.Podlech [1249].
  This number cited as *O. pagobia* (q.v.) in Podlech & Anders 1977 [1126] & in Fl.Iranica [1249].

O. vavilovii Vassilcz. [1058]
  Not climbing; Herb; Perennial.
  West Asia: Afghanistan(N).
  Description: 1058; Illustration: 1058.
  Endemic to Afghanistan [1058].

O. viae-amicitiae Vassilcz. [1058]
  Not climbing; Herb; Perennial.
  West Asia: Afghanistan(N).
  Description: 1058; Illustration: 1058.
  Endemic to Afghanistan [1058].

O. volkii Rech.f. [1058]
  Not climbing; Herb; Perennial.
  West Asia: Afghanistan(N).
  Description: 1052, 1058; Illustration: 1052, 1058.
  Endemic to Afghanistan [1058].

O. vositensis Vassilcz. [1249]
  Not climbing; Herb; Perennial.
  West Asia: Afghanistan(N).
  Description: 1249.
  Type (Breckle 1369) coll.1968. No other material cited.

O. wendelboi Vassilcz. [1058]
  Not climbing; Herb; Perennial.
  West Asia: Afghanistan(N), Iran(N).
  Description: 1058; Illustration: 1058.
  Determination of Iran material doubtful [1058].

O. yazdi Vassilcz. [1249]
   Not climbing; Herb; Perennial.
   West Asia: Iran(N).
   Description: 1249.
   Type (Foroughi & Assadi 18019) coll. 1974 [1249].

O. zangolehensis Vassilcz. [1249]
   Not climbing; Herb; Perennial.
   West Asia: Iran(N).
   Description: 1249.
   Type at Teheran; collector unknown 15239, in 1973 [1249].

SMIRNOWIA Bunge

A monospecific genus; perennial desert shrubs.

S. turkestana Bunge [1058,1090]
   S. iranica Vassilcz. [1058].
   Not climbing; Shrub; Perennial.
   West Asia: Afghanistan(N), Iran(N); U.S.S.R.
   Description: 1058; Illustration: 1058.

SPHAEROPHYSA DC.

Perennial herbs or subshrubs; two species, both halophytes, mainly in Central Asia. Very similar to the Australian *Swainsonia*.

S. salsula (Pallas) DC. [1009,1129]
   Phaca salsula Pallas [1058]; *Swainsonia salsula* (Pallas) Taubert [1058].
   Not climbing; Herb; Perennial.
   West Asia: Afghanistan(N), Iran(N), Iraq(I); East Asia: China(N); U.S.S.R.
   Description: 1009, 1058; Illustration: 1058, 1144, 1145
   Medicine [1009].

GENISTEAE

ARGYROLOBIUM Ecklon & Zeyher

Herbs or small shrubs. About 70 species, mostly in South Africa, but a few extend to tropical Africa, the Mediterranean region and India. Much in need of taxonomic revision.

A. arabicum (Decne.)Jaub. & Spach [1001]
   A. abyssinicum Jaub. & Spach [1066]
   Not climbing; Herb; Annual.
   West Asia: Iran(N), North Yemen(N), Saudi Arabia(N), South Yemen(N); Africa.
   Description: 1001, 1007; Illustration: 1001.

A. biebersteinii P.Ball [1058]
    A. calycinum (M.Bieb.)Boiss.; *Cytisus calycinus* M.Bieb.
    Not climbing; Shrub; Perennial.
    Middle East: Turkey(N); U.S.S.R.
    Description: 1058; Illustration: 1058.
    No W. Asian material at Kew.

A. confertum Polh. [1001]
    Not climbing; Herb; Perennial.
    West Asia: North Yemen(N), Saudi Arabia(N).
    Description: 1001; Illustration: 1001.

A. crotalarioides Jaub. & Spach [1009]
    Not climbing; Shrub; Perennial.
    West Asia: Iraq(N); Middle East: Lebanon(N), Syria(N), Turkey(N).
    Description: 1009, 1058; Illustration: 1009, 1058.
    Saudi Arabian material (Collenette 4381 & 4795) has been determined at Kew as "sp.aff.crotalarioides".

A. lotoides Trautv. [1058]
    West Asia: Iran(U); U.S.S.R.
    Not definately recorded from W. Asia.[1058]

A. roseum (Cambess.)Jaub. & Spach [1009]
    A. kotschyi Boiss.; *A. prostratus* Boiss.
    Not climbing; Herb; Perennial.
    West Asia: Afghanistan(N), Iran(N), Iraq(N), North Yemen(N), Oman(N), United Arab Emirates(N); South Asia: India(N), Pakistan(N).
    Description: 1009, 1058; Illustration: 1009.

  subsp. **ornithopodioides** (Jaub. & Spach)S.M.H.Jafri & Ali [1058]
    Not climbing; Herb; Perennial.
    West Asia: Afghanistan(N), Iran(N), South Yemen(N); South Asia: Pakistan(N).
    Description: 1058; Illustration: 1058.

  subsp. **roseum** [1058]
    Not climbing; Herb; Perennial.
    West Asia: Afghanistan(N), Iran(N), Iraq(N); South Asia: Pakistan(N).
    Description: 1058; Illustration: 1058.

A. rupestre (E.Meyer)Walp. [1066]
    Not climbing; Herb; Perennial.
    West Asia: North Yemen(N).
    Description: 1064.

  subsp. **remotum** (A.Rich.)Polhill [1064]
    Argyrolobium virgatum Baker [1064].
    Not climbing; Herb; Perennial.
    West Asia: North Yemen(N).
    Description: 1064.
    The material from Yemen at Kew has been determined by Polhill.

A. stenophyllum Boiss. [1011]
    Not climbing; Herb; Perennial.
    West Asia: Afghanistan(N), Iran(N); South Asia: Pakistan(N).
    Description: 1058; Illustration: 1058.

A. trigonelloides Jaub. & Spach [1058]
    Not climbing; Herb; Annual.
    West Asia: Iran(N).
    Description: 1058; Illustration: 1058.
    Endemic to Iran [1058]; includes var. *subuniflorum* [1015].

A. uniflorum (Decne.)Jaub. & Spach [1007]
   Not climbing; Herb/Shrub; Perennial.
   West Asia: Saudi Arabia(N), South Yemen.
   Description: 1007.

GENISTA L.

A genus of about 80 species, mainly shrubs, often spiny; centred in the Mediterranean region. Several groups are treated either as genera or subgenera according to the taxonomy being used.

G. tinctoria L. [1009]
   Not climbing; Shrub; Perennial.
   West Asia: Afghanistan(I), Iran(I), Iraq(I); Europe; Middle East: Turkey(N); U.S.S.R.
   Description: 1009,1058.
   Domestic 1009; Environmental 1009; Medicine 1009.
   Indicator of the poverty of the land [1009].

RETAMA Raf.

A small genus of 4 shrubby species, mostly occurring in very dry places.

R. raetam (Forsskal)Webb [1001]
   Genista raetam Forsskal; *Lygos raetam* (Forsskal)Heyw.
   Not climbing; Shrub; Perennial.
   West Asia: Saudi Arabia(N), South Yemen.
   Description: 1001, 1007; Illustration: 1001.

SPARTIUM L.

A monospecific genus from the Mediterranean region, widely planted for its ornamental yellow flowers.

S. junceum L. [1009]
   Not climbing; Shrub; Perennial.
   West Asia: Afghanistan(I), Iraq(I); Africa; Europe; Middle East: Israel(N), Lebanon(N), Syria(N), Turkey(N).
   Description: 1009, 1058.
   Environmental 1009; Fibre 1009; Forage 1009; Medicine 1009.
   Spanish Broom 1009.

HEDYSAREAE

EBENUS L.

A genus of about 20 species, mostly herbs or sub-shrubs, with its centre of diversity in Turkey and surrounding countries.

E. lagopus (Jaub. & Spach)Boiss. [1015]
   Not climbing; Herb; Perennial.
   West Asia: Iran(N).
   Description: 1015.
   Not mentioned by Rechinger [1058]. Boissier records: "Hab. in Persia australi. (Aucher)" [1121].

E. stellata Boiss. [1058]
>*E. ferruginea* Jaub. & Spach [1058]; *E. horrida* Jaub. & Spach [1058]; *E. tragacanthoides* Jaub. & Spach [1058].
>Not climbing; Shrub; Perennial.
>West Asia: Afghanistan(N), Iran(N), Oman(N); South Asia: Pakistan(N).
>Description: 1015, 1058; Illustration: 1012, 1058.

EVERSMANNIA Bunge

A monospecific genus from the dry regions of west and central Asia.

E. subspinosa (Fischer)B.Fedtsch. [1058]
>*E. hedysaroides* Bunge [1058].
>Not climbing; Shrub; Perennial.
>West Asia: Afghanistan(N), Iran(N); Europe; U.S.S.R.
>Description: 1015, 1058; Illustration: 1058.

HEDYSARUM L.

A genus of about 100 species, herbs or subshrubs, occurring throughout the north temperate regions but most diverse in the eastern Mediterranean region and in west Asia. A few are cultivated as fodder or for their ornamental value.

H. astragaloides Benth. [1058]
>West Asia: Afghanistan(N); South Asia: Pakistan(N).
>Description: 1122.

H. atropatanum Boiss. [1058]
>Not climbing; Herb; Perennial.
>West Asia: Iran(N); U.S.S.R.
>Description: 1015, 1058; Illustration: 1058.
>A.J.Dobson 70 at Kew from Afghanistan is similar.

H. bellevii (Prain)Bornm. (provisional) [1058]
>West Asia: Afghanistan(N).
>Included by Rechinger as "species dubia" [1058].

H. brahuicum Boiss. [1058]
>Not climbing; Herb; Perennial.
>West Asia: Afghanistan(N).
>Description: 1058.
>Endemic to Afghanistan; known only from the type [1058].

H. bucharicum B.Fedtsch. [1058]
>*Hedysarum purpureo-pilosum* Kitam. [1058].
>Not climbing; Herb; Perennial.
>West Asia: Afghanistan(N).
>Description: 1052, 1058; Illustration: 1011, 1058.

H. callithrix Boiss. [1058]
>Not climbing; Herb; Perennial.
>West Asia: Iran(N).
>Description: 1015, 1058.
>Endemic to Iran [1058].

H. cappadocicum Boiss. [1102]
>*Hedysarum grandiflorum* subsp. *cappadocicum* Boiss.
>West Asia: Afghanistan(N).

H. criniferum Boiss. [1058]
Hedysarum ecbatanum Beck [1058].
Not climbing; Herb; Perennial.
West Asia: Iran(N).
Description: 1015, 1058; Illustration: 1058.
Endemic to Iran [1058].
The varieties *melanotrichum* Boiss. & *pallidiflora* Bornm. are not upheld by Rechinger 1984 [1058].

H. damghanicum Rech.f. [1058]
Not climbing; Herb; Perennial.
West Asia: Iran(N).
Description: 1058; Illustration: 1058.
Endemic to Iran; known only from the type material [1058].

H. elbursense Bornm. & Gauba [1058]
Not climbing; Herb; Perennial.
West Asia: Iran(N).
Description: 1015, 1058; Illustration: 1133.
Endemic to Iran; known only from the type material [1058].

H. elymaiticum Boiss. & Hausskn. [1058]
Not climbing; Herb; Perennial.
West Asia: Iran(N), Iraq(N).
Description: 1015, 1058; Illustration: 1058.

H. falconeri Baker [1058]
Not climbing; Herb; Perennial.
West Asia: Afghanistan(N); South Asia: Pakistan(N).
Description: 1058.
Rechinger does not record this taxon from Afghanistan [1058].

H. fallacinum Rech.f. & Aellen [1123]
Not climbing; Herb/Shrub; Perennial.
West Asia: Iran(N).
Description: 1123.
Not mentioned by Rechinger (1984) [1058].

H. farinosum Parsa [1015]
Not climbing; Herb; Perennial.
West Asia: Iran(N).
Description: 1015, 1124.
Not mentioned by Rechinger (1984) [1058].

H. formosum Basiner [1058]
Not climbing; Herb; Perennial.
West Asia: Afghanistan(N), Iran(N), Iraq(N); Europe; U.S.S.R.
Description: 1015, 1058; Illustration: 1058.

H. halophilum Bornm. & Gauba [1058]
Not climbing; Herb; Perennial.
West Asia: Iran(N).
Description: 1058.
Endemic to Iran; known only from the type [1058].

H. hyrcanum Bornm. & Gauba [1058]
Not climbing; Herb; Perennial.
West Asia: Iran(N).
Description: 1015, 1058.
Endemic to Iran [1058].
2 varieties, *hyrcanum* & *incanescens* [1058]; var. *viridissimum* is a synonym of var. *hyrcanum* [1058].
Both varieties are known only from the type material [1058].

H. ibericum M.Bieb. [1058]
Hedysarum glaucescens Ledeb. [1058].
Not climbing; Herb; Perennial.
West Asia: Iran(N); U.S.S.R.
Description: 1058; Illustration: 1058.

H. kopetdaghi Boriss [1058]
Not climbing; Herb; Perennial.
West Asia: Afghanistan(N), Iran(N); U.S.S.R.
Description: 1058; Illustration: 1058.

H. kotschyi Boiss. [1009]
Not climbing; Herb; Perennial.
West Asia: Iraq(N); Middle East: Syria(N), Turkey(N).
Description: 1009, 1058; Illustration: 1058.

H. lehmannianum Bunge [1058]
Not climbing; Herb; Perennial.
West Asia: Afghanistan(N).
Description: 1058; Illustration: 1058.

H. mahrense Rech.f. [1058]
Not climbing; Herb; Perennial.
West Asia: Afghanistan(N).
Description: 1052, 1058; Illustration: 1052, 1058.
Endemic to Afghanistan; known only from the type [1058].

H. maitlandianum Aitch. & Baker [1058]
Not climbing; Herb; Perennial.
West Asia: Afghanistan(N).
Description: 1015, 1058.
Endemic to Afghanistan [1058].

H. micropterum Boiss. [1058]
Not climbing; Herb; Perennial.
West Asia: Afghanistan(N), Iran(N); U.S.S.R.
Description: 1015, 1058.

H. minjanense Rech.f. [1058]
Hedysarum cephalotes Franchet [1058].
Not climbing; Herb; Perennial.
West Asia: Afghanistan(N); U.S.S.R.
Description: 1052, 1058; Illustration: 1052, 1058.

H. papillosum Boiss. [1058]
Not climbing; Herb; Perennial.
West Asia: Iran(N).
Description: 1015, 1058.
Endemic to Iran [1058].

H. plumosum Boiss. & Hausskn. [1058]
Not climbing; Herb; Perennial.
West Asia: Iran(N).
Description: 1015,1058.
Endemic to Iran; known only from the type, collected by Haussknecht [1058].

H. praticolum Rech.f. [1058]
Not climbing; Herb; Perennial.
West Asia: Afghanistan(N).
Description: 1052, 1058; Illustration: 1052, 1058.
Endemic to Afghanistan [1058].

H. renzii Rech.f. [1058]
Not climbing; Herb; Perennial.
West Asia: Iran(N).
Description: 1058, 1125.
Endemic to Iran; known only from the type [1058].

H. sauzakense Rech.f. [1058]
   Not climbing; Herb; Perennial.
   West Asia: Afghanistan(N).
   Description: 1052, 1058; Illustration: 1052, 1058.
   Endemic to Afghanistan [1058].

H. sericatum Kitam. [1011]
   Not climbing; Herb; Perennial.
   West Asia: Afghanistan(N).
   Description: 1011; Illustration: 1011.
   Not mentioned by Rechinger (1984) [1058].

H. sericeum M.Bieb. [1058]
   Not climbing; Herb; Perennial.
   West Asia: Iran(N); Middle East: Turkey(N); U.S.S.R.
   Description: 1058; Illustration: 1058.

H. singarense Boiss. & Hausskn. [1009]
   Hedysarum pannosum sensu auctt.
   Not climbing; Herb; Perennial.
   West Asia: Iraq(N); Middle East: Syria(N), Turkey(N).
   Description: 1009, 1058; Illustration: 1058.
   H. pannosum Boiss. forma *assyriacum* is a synonym.

H. varium Willd. [1009]
   Not climbing; Herb; Perennial.
   West Asia: Iran(N), Iraq(N); Europe; Middle East: Turkey(N); U.S.S.R.
   Description: 1009, 1015, 1058; Illustration: 1009.

   subsp. **syriacum** (Boiss.)C.Towns. [1009]
      Hedysarum syriacum Boiss. [1009]
      Not climbing; Herb; Perennial.
      West Asia: Iran(N), Iraq(N); Middle East: Turkey(N); U.S.S.R.
      Description: 1009, 1058; Illustration: 1009.

H. volkii Rech.f. [1058]
   Not climbing; Herb; Perennial.
   West Asia: Afghanistan(N).
   Description: 1052, 1058; Illustration: 1052, 1058.
   Endemic to Afghanistan [1058].

H. wakhanicum Podl. & Anders [1058]
   Not climbing; Herb; Perennial.
   West Asia: Afghanistan(N).
   Description: 1058, 1126; Illustration: 1058.
   Endemic to Afghanistan; known only from the type [1058].

H. wrightianum Aitch. & Baker [1058]
   Not climbing; Herb; Perennial.
   West Asia: Afghanistan(N), Iran(N); South Asia: Pakistan(N); U.S.S.R.
   Description: 1015, 1058; Illustration: 1058.

ONOBRYCHIS Miller

About 130 species in the temperate regions of the Old World; most diverse in the eastern Mediterranean and west and central Asia. Mainly herbs. One species is widely cultivated as fodder.

O. acaulis Bornm. [1009]

Onobrychis koeieana Sirj. & Rech.f. [1058]; *Onobrychis sirinae* Nab. [1009]; *Onobrychis sirinae* var. *behboudii* Sirj. & Rech.f. [1058]; *Onobrychis unicornis* Pau [1058].
Not climbing; Herb; Perennial.
West Asia: Iran(N), Iraq(N).
Description: 1009, 1015, 1058; Illustration: 1009, 1058.

O. aequidentata (Smith)Urv. [1058]

Not climbing; Herb; Annual.
West Asia: Iran(N), Iraq(N); Middle East: Syria(N), Turkey(N); U.S.S.R.
Description: 1009, 1058; Illustration: 1009, 1058.

O. afghanica Sirj. & Rech.f. [1058]

Not climbing; Shrub; Perennial.
West Asia: Afghanistan(N).
Description: 1052, 1058; Illustration: 1052.
3 subspecies, all endemic to Afghanistan [1058].

subsp. afghanica [1058]

Not climbing; Shrub; Perennial.
West Asia: Afghanistan(N).
Description: 1058; Illustration: 1058.

subsp. brachycalyx Rech.f. [1058]

Not climbing; Shrub; Perennial.
West Asia: Afghanistan(N).
Description: 1058; Illustration: 1058.

subsp. codringtonii (Sirj. & Rech.f.)Rech.f. [1058]

Not climbing; Shrub; Perennial.
West Asia: Afghanistan(N).
Description: 1058; Illustration: 1052, 1058.

O. altissima Grossh. [1058]

Onobrychis viciaefolia var. *persica* Sirj.
Not climbing; Herb; Perennial.
West Asia: Iran(N); Middle East: Turkey(N); U.S.S.R.
Description: 1058; Illustration: 1058.

O. alyassinicus Parsa [1103]

Not climbing; Herb; Perennial.
West Asia: Iran(N).
Description: 1103; Illustration: 1103.

O. amoena Popov & Vved. [1058]

Not climbing; Herb; Perennial.
West Asia: Afghanistan(N), Iran(N); U.S.S.R.
Description: 1015, 1058.

subsp. amoena [1058]

Not climbing; Herb; Perennial.
West Asia: Afghanistan(N), Iran(N).
Description: 1058; Illustration: 1058.

subsp. meshhedensis Sirj. & Rech.f. [1058]

Not climbing; Herb; Perennial.
West Asia: Iran(N).
Description: 1058; Illustration: 1058.
Endemic to Iran [1058].

O. andalanica Bornm. [1058]
   Not climbing; Herb; Perennial.
   West Asia: Iran(N).
   Description: 1058; Illustration: 1058.
   Endemic to Iran [1058].

O. arnacantha Boiss. [1058]
   Not climbing; Shrub; Perennial.
   West Asia: Afghanistan(N), Iran(N).
   Description: 1015, 1058; Illustration: 1058.

O. atropatana Boiss. [1058]
   Not climbing; Herb; Perennial.
   West Asia: Iran(N); U.S.S.R.
   Description: 1015, 1058; Illustration: 1058.

O. aucheri Boiss. [1058]
   Not climbing; Herb; Annual.
   West Asia: Afghanistan(N), Iran(N).
   Description: 1015, 1058.
   3 subspecies [1058].

   subsp. **aucheri** [1058]
      Not climbing; Herb; Annual.
      West Asia: Iran(N).
      Description: 1058.
      Endemic to Iran [1058]

   subsp. **psammophila** (Bornm.)Rech.f. [1058]
      Onobrychis psammophila Bornm. [1058].
      Not climbing; Herb; Annual.
      West Asia: Afghanistan(N), Iran(N).
      Description: 1058; Illustration: 1058.

   subsp. **teheranica** (Bornm.)Rech.f. [1058]
      Onobrychis spinescens Bornm. [1058]; *Onobrychis teheranica* Bornm. [1058].
      Not climbing; Herb; Annual.
      West Asia: Afghanistan(N), Iran(N).
      Description: 1058; Illustration: 1058.

O. bicornis Vassilcz. (provisional)
   Not climbing; Herb; Perennial.
   West Asia: Iran(U).
   The Kew specimen is marked 'TYPUS', but publication is not recorded in Index Kewensis.
   Not mentioned by Rechinger 1984.

O. buhseana Boiss. [1058]
   Not climbing; Herb; Perennial.
   West Asia: Iran(N); Middle East: Turkey(N); U.S.S.R.
   Description: 1015, 1058.

O. bungei Boiss. [1058]
   Not climbing; Herb; Perennial.
   West Asia: Iran(N); U.S.S.R.
   Description: 1015, 1058; Illustration: 1058.

O. cadmea Boiss. (provisional) [1015]
   Not climbing; Herb/Shrub; Perennial.
   West Asia: Iran(N).
   Description: 1015, 1101.
   In Flora of Turkey this taxon is treated as a subspecies of *O. montana* DC.

O. cana (Boiss.)Hand.-Mazz.
Perennial.
West Asia: Iran(N).
Description: 1101.
Not mentioned by Rechinger (1984).

O. caput-galli (L.)Lam. [1009]
Not climbing; Herb; Annual.
West Asia: Iran(N), Iraq(N); Middle East: Israel(N), Syria(N), Turkey(N); U.S.S.R.
Description: 1009, 1015, 1058; Illustration: 1009, 1058.

O. carduchorum C.Towns. [1009]
Onobrychis fallax sensu auctt. [1009]; *Onobrychis pindicola* sensu Rawi [1009].
Not climbing; Herb; Perennial.
West Asia: Iran(N), Iraq(N); Middle East: Syria(N), Turkey(N).
Description: 1009, 1058; Illustration: 1009, 1025.
Forage 1009.
A form (*nabelekii*) of *O. fallax* is a synonym [1009].

O. chorassanica Boiss. [1058]
Onobrychis megalobotrys Aitch. & Baker [1058]; *Onobrychis sykesiae* N.Simpson [1058].
Not climbing; Herb; Perennial.
West Asia: Afghanistan(N), Iran(N); U.S.S.R.
Description: 1015, 1058; Illustration: 1058.

O. cornuta (L.)Desv. [1009]
Not climbing; Herb/Shrub; Perennial.
West Asia: Afghanistan(N), Iran(N), Iraq(N); Middle East: Lebanon(N), Syria(N), Turkey(N); South Asia: Pakistan(N); U.S.S.R.
Description: 1009, 1015, 1058; Illustration: 1009, 1058.
Includes var. *elbursensis* Sirj. & Rech.f [1058].

subsp. cornuta [1058]
Not climbing; Herb/Shrub; Perennial.
West Asia: Afghanistan(N), Iran(N), Iraq(N).
Description: 1058; Illustration: 1058.

subsp. leptacantha Rech.f.
Not climbing; Shrub; Perennial.
West Asia: Afghanistan(N); South Asia: Pakistan(N).
Description: 1058; Illustration: 1058.

O. crista-galli (L.)Lam. [1009]
Onobrychis crista-galli subsp. *trilocarpa* Quezel & Santa [1009]; *Onobrychis squarrosa* Viv. [1009].
Not climbing; Herb; Annual.
West Asia: Iran(N), Iraq(N); Africa; Europe; Middle East: Israel(N), Jordan(N), Lebanon(N), Syria(N), Turkey(N).
Description: 1009, 1015, 1058; Illustration: 1009, 1058.
Forage 1009.
Townsend distinguishes a var. *trilophocarpa* (Sirj.)Townsend.

O. dealbata Stocks [1058]
Onobrychis dasycephala Baker [1058].
Not climbing; Herb; Perennial.
West Asia: Afghanistan(N); South Asia: Pakistan(N).
Description: 1015, 1058; Illustration: 1058.
A specimen at Kew from Iran (Merton 4076) may be this.

O. depauperata Boiss. [1058]
Not climbing; Herb; Perennial.
West Asia: Iran(N).
Description: 1015, 1058.
Endemic to Iran; known only from the fragmentary type collected in 1828 [1058].

O. echidna Lipsky [1058]
Astragalus yosiianus Kitam. [1058]
Not climbing; Shrub; Perennial.
West Asia: Afghanistan(N).
Description: 1058; Illustration: 1011, 1058.

O. elymaitiaca Boiss. & Hausskn. [1058]
Not climbing; Herb/Shrub; Perennial.
West Asia: Iran(N).
Description: 1015, 1058.
Endemic to Iran [1058].

O. eubrychidea Boiss. [1058]
Onobrychis charikarensis Podl.
Not climbing; Herb; Perennial.
West Asia: Afghanistan(N).
Description: 1052, 1058; Illustration: 1058.

O. fallax Freyn & Sint.
Not climbing; Herb; Perennial.
West Asia: Iran(N), Iraq(N); Middle East: Turkey(N).
Description: 1101.
Notes on sheets at Kew suggest it may be synonymous with *O. carduchorum*.
Not mentioned by Townsend or by Rechinger [1058].

O. freitagii Rech.f. [1058]
Not climbing; Herb; Perennial.
West Asia: Afghanistan(N).
Description: 1058; Illustration: 1058.
Endemic to Afghanistan; known only from the type [1058].

O. galegifolia Boiss. [1009]
Onobrychis aurantiaca Boiss. [1009]; *Onobrychis aurantiaca* var. *velutina* Post [1009]; *Onobrychis galegifolia* var. *luristanica* Sirj. & Rech.f. [1058].
Not climbing; Herb; Perennial.
West Asia: Iran(N), Iraq(N); Middle East: Syria(N), Turkey(N).
Description: 1009, 1058; Illustration: 1009, 1058.

O. gaubae Bornm. [1058]
Onobrychis porphyrogramma Rech.f. [1058]; *Onobrychis semnanensis* Sirj. & Rech.f. [1058].
Not climbing; Herb; Perennial.
West Asia: Iran(N).
Description: 1015, 1058; Illustration: 1058.
Endemic to Iran [1058].
Parsa mentions a forma *obscura* Sirj.

O. grandis Lipsky [1058]
Not climbing; Herb; Perennial.
West Asia: Afghanistan(N).
Description: 1058; Illustration: 1058.
Endemic to Iran [1058].
Includes var. *edelbergii* Sirj. & Rech.f. [1058].

O. gypsicola Rech.f. [1058]
Not climbing; Herb; Perennial.
West Asia: Iran(N).
Description: 1058; Illustration: 1058.
Endemic to Iran [1058]; most records are from gypsum substrates.

O. haussknechtii Boiss. [1009]
Onobrychis bicolor Bornm. [1009].
Not climbing; Herb; Perennial.
West Asia: Iran(N), Iraq(N); Middle East: Syria(N), Turkey(N).
Description: 1009, 1015, 1058; Illustration: 1009, 1058.

O. heliocarpa Boiss. [1058]
Not climbing; Herb; Annual.
West Asia: Iran(N).
Description: 1015, 1058; Illustration: 1058.
Endemic to Iran [1058].

O. heterophylla C.Meyer [1058]
Not climbing; Herb; Perennial.
West Asia: Iran(N).
Description: 1015, 1058; Illustration: 1058.
Endemic to Iran [1058].
Parsa (1948) places *O. spinescens* as a synonym of this taxon.
Rechinger (1984) places *O. spinescens* as a synonym of *O. aucheri* ssp. *teheranica*.

O. hohenackeriana C.Meyer [1058]
Onobrychis kachetica sensu auctt. [1058]; *Onobrychis michauxii* sensu Sirj.,p.p. [1058].
Not climbing; Herb; Perennial.
West Asia: Iran(N); U.S.S.R.
Description: 1058; Illustration: 1058.

O. hypargyrea Boiss. [1009]
Not climbing; Herb; Perennial.
West Asia: Iraq(N).
Description: 1101.
Townsend mentions the gathering as probably mixed.
The locality is only "Mesopotamia", so the precise place of collection is uncertain.

O. iberica Grossh. (provisional) [1015]
Not climbing; Herb; Perennial.
West Asia: Iran(N); U.S.S.R.
Description: 1015.
Not mentioned by Rechinger or by Townsend; notes on sheets at Kew suggest that it may be the same as *O. gracilis* Besser.

O. iranica Bornm. [1104]
Not climbing; Herb; Perennial.
West Asia: Iran(N).
Description: 1104.
Not mentioned by Rechinger, Townsend or Parsa.

O. iranshahrii Rech.f. [1058]
Not climbing; Herb/Shrub; Perennial.
West Asia: Iran(N).
Description: 1058; Illustration: 1058.
Endemic to Iran [1058].

O. kermanensis (Sirj. & Rech.f.)Rech.f. [1058]
Onobrychis melanotricha var. *kermanensis* Sirj. & Rech.f. [1058].
Not climbing; Herb; Perennial.
West Asia: Iran(N).
Description: 1058; Illustration: 1058.
Endemic to Iran [1058].

O. kotschyana Fenzl [1009]
Not climbing; Herb; Perennial.
West Asia: Iran(N), Iraq(N); Middle East: Israel(N), Jordan(N), Lebanon(N), Syria(N), Turkey(N).
Description: 1009, 1058; Illustration: 1009, 1058.

O. lahidjanicus Parsa [1103]
Not climbing; Herb; Perennial.
West Asia: Iran(N).
Description: 1103; Illustration: 1103.

O. laxiflora Baker [1058]
Onobrychis schugnanica B.Fedtsch. [1058].
Not climbing; Herb; Perennial.
West Asia: Afghanistan(N); South Asia: Pakistan(N).
Description: 1058; Illustration: 1058.
A variable species; Rechinger recognises 5 subspecies.
O. schugnanica variously treated as a forma or variety of *O. laxiflora*.
Rechinger (1984) saw no material af *O. schugnanica*.

subsp. kabulica Rech.f. [1058]
Not climbing; Herb; Perennial.
West Asia: Afghanistan(N).
Description: 1058; Illustration: 1058.

subsp. laxiflora [1058]
Not climbing; Herb; Perennial.
West Asia: Afghanistan(N); South Asia: Pakistan(N).
Description: 1058; Illustration: 1058.

subsp. macrodonta Rech.f. [1058]
Not climbing; Herb; Perennial.
West Asia: Afghanistan(N); South Asia: Pakistan(N).
Description: 1058; Illustration: 1058.

subsp. shahrestanica Rech.f. [1058]
Not climbing; Herb; Perennial.
West Asia: Afghanistan(N).
Description: 1058; Illustration: 1058.

subsp. taftanica Rech.f. [1058]
Not climbing; Herb; Perennial.
West Asia: Iran(N).
Description: 1058; Illustration: 1058.

O. longipes Bunge [1058]
Not climbing; Herb; Perennial.
West Asia: Iran(N).
Description: 1015, 1058.
Endemic to Iran; known only from the type, collected by Bunge.

O. lunata Boiss. [1009]
Onobrychis oxypteroides Sirj. & Rech.f. [1058].
Not climbing; Herb; Perennial.
West Asia: Iran(N).
Description: 1009, 1015, 1058; Illustration: 1009, 1058.
Doubtful records from Syria & Iraq; probably based on poorly-localised specimens.

O. luristanica Rech.f. [1058]
Not climbing; Herb; Perennial.
West Asia: Iran(N).
Description: 1058; Illustration: 1058.
Endemic to Iran [1058].

O. macrorrhiza Rech.f. [1058]
Not climbing; Herb; Perennial.
West Asia: Afghanistan(N).
Description: 1052, 1058; Illustration: 1052, 1058.
Endemic to Afghanistan [1058].

O. major (Boiss.)Hand.-Mazz. [1058]
Onobrychis major var. *angustifolia* Sirj. & Rech.f. [1058].
Not climbing; Herb; Perennial.
West Asia: Iran(N); Middle East: Turkey(N).
Description: 1058; Illustration: 1058.

O. mazanderanica Rech.f. [1058]
  Not climbing; Herb; Perennial.
  West Asia: Iran(N).
  Description: 1058; Illustration: 1058.
  Endemic to Iran [1058].

O. megataphros Boiss. [1009]
  Not climbing; Herb; Perennial.
  West Asia: Iran(N), Iraq(N); Middle East: Syria(N), Turkey(N).
  Description: 1009, 1015, 1058; Illustration: 1009, 1058.
  Forage 1009.

O. melanotricha Boiss. [1058]
  Onobrychis belangeri Boiss. & Buhse [1058]; *Onobrychis linearis* Pau & Vicioso [1058].
  Not climbing; Herb; Perennial.
  West Asia: Iran(N).
  Description: 1015, 1058; Illustration: 1058.
  Endemic to Iran [1058]. Rechinger (1984) distinguishes vars. *melanotricha* & *villosa*.

O. merxmuelleri Podl. [1058]
  Not climbing; Herb; Perennial.
  West Asia: Afghanistan(N).
  Description: 1058, 1098; Illustration: 1058.
  Endemic to Afghanistan [1058].

O. michauxii DC. [1058]
  Not climbing; Herb; Perennial.
  West Asia: Iran(N); U.S.S.R.
  Description: 1015, 1058; Illustration: 1058.

O. micrantha Schrenk [1058]
  Not climbing; Herb; Annual.
  West Asia: Afghanistan(N), Iran(N); South Asia: Pakistan(N); U.S.S.R.
  Description: 1015, 1058; Illustration: 1058.

O. microptera Baker [1058]
  Not climbing; Herb; Annual.
  West Asia: Afghanistan(N); South Asia: Pakistan(N).
  Description: 1058; Illustration: 1058.
  May be biennial rather than annual [1058].

O. nummularia Boiss. [1058]
  Not climbing; Herb; Annual.
  West Asia: Afghanistan(N), Iran(N); South Asia: Pakistan(N).
  Description: 1015, 1058; Illustration: 1058.

O. oxyptera Boiss. [1058]
  Not climbing; Herb; Perennial.
  West Asia: Iran(N).
  Description: 1015, 1058; Illustration: 1058.
  Endemic to Iran; known only from the type material collected by Kotschy in 1842 [1058].

O. persica Sirj. & Rech.f. [1058]
  Not climbing; Herb; Perennial.
  West Asia: Iran(N).
  Description: 1015, 1058; Illustration: 1058.
  Endemic to Iran [1058].

O. plantago Bornm. [1058]
  Not climbing; Herb; Perennial.
  West Asia: Iran(N).
  Description: 1015, 1058; Illustration: 1058.
  Endemic to Iran [1058].

O. poikilantha Rech.f. [1058]
Not climbing; Herb; Perennial.
West Asia: Afghanistan(N).
Description: 1058; Illustration: 1058.
Endemic to Afghanistan [1058].

O. psoraleifolia Boiss. [1058]
Not climbing; Herb; Perennial.
West Asia: Iran(N).
Description: 1015, 1058; Illustration: 1058.
Endemic to Iran [1058].
Rechinger (1984) distinguishes vars. *psoraleifolia* & *pleiophylla*.

O. ptolemaica (Del.)DC. [1058]
Onobrychis gaillardoti Boiss. [1058]; *Onobrychis lanata* Boiss. [1058]; *Onobrychis olivieri* Boiss. [1058]; *Onobrychis pinnata* (Bertol.)Hand.-Mazz. [1058]; *Onobrychis ptolemaica* subsp. *macroptera* C.Towns. [1058]; *Onobrychis sirjaevi* Nab. [1058].
Not climbing; Herb; Perennial.
West Asia: Iran(N), Iraq(N), Kuwait(N), North Yemen(N), Saudi Arabia(N); Africa; Middle East: Jordan(N), Sinai(N), Syria(N), Turkey(N).
Description: 1001, 1009, 1058; Illustration: 1001, 1009, 1058.
Forage 1009.

O. ptychophylla Sirj. & Rech.f. [1058]
Not climbing; Herb/Shrub; Perennial.
West Asia: Iran(N).
Description: 1015, 1058; Illustration: 1058.
Endemic to Iran; known only from the type [1058].

O. pulchella Schrenk [1058]
Onobrychis caloptera Aitch. & Baker [1058]; *Onobrychis elegans* Franchet [1058].
Not climbing; Herb; Annual.
West Asia: Afghanistan(N), Iran(N); U.S.S.R.
Description: 1015, 1058; Illustration: 1058.

O. radiata (Desf.)M.Bieb. [1102]
Not climbing; Herb; Perennial.
West Asia: Iran(N).
Description: 1101.
Recorded for Iran in Flora of Turkey.

O. rechingerorum Wendelbo [1058]
Not climbing; Herb; Annual.
West Asia: Iran(N).
Description: 1058, 1099; Illustration: 1058.
Endemic to Iran; known only from the type [1058].

O. samanganica Rech.f. [1058]
Not climbing; Shrub; Perennial.
West Asia: Afghanistan(N).
Description: 1058; Illustration: 1058.
Endemic to Afghanistan [1058].

O. saravschanica B.Fedtsch. [1100]
Onobrychis baldzuanica Sirj.; *Onobrychis seravschanica* B.Fedtsch.
Not climbing; Herb; Perennial.
West Asia: Afghanistan(N); U.S.S.R.
Description: 1058, 1100; Illustration: 1058.

O. sauzakensis Sirj. & Rech.f. [1058]
Not climbing; Herb; Perennial.
West Asia: Afghanistan(N).
Description: 1052, 1058; Illustration: 1052.
Endemic to Afghanistan [1058].

O. schahuensis Bornm. [1058]
Not climbing; Herb; Perennial.
West Asia: Iran(N), Iraq(N); Middle East: Syria(N), Turkey(N).
Description: 1009, 1058; Illustration: 1009.
Forage 1009.

O. scrobiculata Boiss. [1058]
Onobrychis marginata Beck [1058]
Not climbing; Herb; Perennial.
West Asia: Iran(N).
Description: 1015, 1058; Illustration: 1058.
Endemic to Iran [1058].

O. shahpurensis Rech.f. [1058]
Not climbing; Herb; Perennial.
West Asia: Iran(N); Middle East: Turkey(N).
Description: 1058; Illustration: 1058.

O. sintenisii Bornm. [1058]
Not climbing; Herb; Perennial.
West Asia: Iran(N); U.S.S.R.
Description: 1015, 1058; Illustration: 1058.
Furse 7681 from Afghanistan (K) is similar.

O. sirdjanicus Parsa [1103]
Not climbing; Herb; Annual.
West Asia: Iran(N).
Description: 1103; Illustration: 1103.

O. sojakii Rech.f. [1058]
Not climbing; Herb; Perennial.
West Asia: Iran(N).
Description: 1058; Illustration: 1058.
Endemic to Iran [1058].

O. spinosissima Baker [1058]
Not climbing; Shrub; Perennial.
West Asia: Afghanistan(N).
Description: 1058; Illustration: 1058.
Endemic to Afghanistan [1058].

O. splendida Rech.f. & Podl. [1058]
Not climbing; Herb; Perennial.
West Asia: Afghanistan(N).
Description: 1058; Illustration: 1058.
Endemic to Afghanistan; known only from the type [1058].

O. subacaulis Boiss. [1058]
Onobrychis picta Bornm. [1058].
Not climbing; Herb; Annual.
West Asia: Iran(N).; U.S.S.R.
Description: 1015, 1058; Illustration: 1058.

O. subnitens Bornm. [1058]
Not climbing; Herb; Perennial.
West Asia: Iran(N).
Description: 1015, 1058; Illustration: 1058.
Endemic to Iran [1058].

O. susiana Nab. [1058]
Not climbing; Herb; Perennial.
West Asia: Iran(N).
Description: 1009, 1058.
Could perhaps occur in Iraq [1009].

O. szovitsii Boiss. [1058]
  Not climbing; Herb; Perennial.
  West Asia: Iran(N).
  Description: 1015, 1058.
  Endemic to Iran; known only from the type [1058].

O. talagonica Rech.f. [1058]
  Not climbing; Herb; Perennial.
  West Asia: Iran(N).
  Description: 1058; Illustration: 1058.
  Endemic to Iran [1058].

O. tavernieraefolia Boiss. [1058]
  Not climbing; Herb; Annual.
  West Asia: Afghanistan(N), Iran(N); South Asia: Pakistan(N).
  Description: 1015, 1058; Illustration: 1058.

O. transcaspica V.V.Nikitin [1058]
  Not climbing; Herb; Perennial.
  West Asia: Afghanistan(N), Iran(N); U.S.S.R.
  Description: 1058; Illustration: 1058.

O. transcaucasica Grossh. [1058]
  Not climbing; Herb; Perennial.
  West Asia: Iran(N); Middle East: Turkey(N); U.S.S.R.
  Description: 1058.

O. verae Sirj. [1058]
  Onobrychis lipskyi Korovin [1058]; *Onobrychis verae* var. *rechingeri* Sirj. [1058].
  Not climbing; Herb/Shrub; Perennial.
  West Asia: Afghanistan(N), Iran(N); U.S.S.R.
  Description: 1015, 1058; Illustration: 1058.

O. viciifolia Scop. (accepted)
  Onobrychis sativa Lam. [1101]; *Onobrychis sativa* var. *subinermis* Boiss. [1101]; *Onobrychis viciaefolia* Scop.
  Not climbing; Herb; Perennial.
  West Asia: Iran(N).
  Description: 1015, 1101.
  Forage 1009.

O. wettsteinii Nab. (provisional) [1015]
  Not climbing; Herb; Perennial.
  West Asia: Iran(N).
  Description: 1015, 1101.
  Not in Rechinger or Townsend; according to Parsa, is said to occur in "Arabia".

TAVERNIERA DC.

Woody herbs or shrubs; about ten species, mainly in the countries bordering the Red Sea. See Thulin (1985 — ref.1097).

T. aegyptiaca Boiss. [1097]
  Not climbing; Herb/Shrub; Perennial.
  West Asia: North Yemen(N), Qatar(N), Saudi Arabia(N).
  Description: 1001, 1014, 1097; Distribution Map: 1097; Illustration: 1001, 1014, 1097.

T. albida Thulin [1097]
  Not climbing; Shrub; Perennial.
  West Asia: South Yemen(N).
  Description: 1097; Distribution Map: 1097; Illustration: 1097.

T. brevialata Thulin [1097]
   Not climbing; Herb/Shrub; Perennial.
   West Asia: Oman(N).
   Description: 1097; Distribution Map: 1097; Illustration: 1097.

T. cuneifolia (Roth)Arn. [1058]
   Taverniera glabra Boiss. [1097].
   Not climbing; Herb/Shrub; Perennial.
   West Asia: Iran(N), Oman(N), United Arab Emirates(N); South Asia: India(N), Pakistan(N).
   Description: 1015, 1058, 1097; Distribution Map: 1097; Illustration: 1058, 1097.

T. echinata V.Mozaffarian [1250]
   Not climbing; Shrub; Perennial.
   West Asia: Iran(N).
   Description: 1250; Illustration: 1250.

T. glauca Edgew. [1097]
   Not climbing; Herb/Shrub; Perennial.
   West Asia: North Yemen(N), South Yemen(N).
   Description: 1097; Distribution Map: 1097.

T. lappacea (Forsskal)DC. [1097]
   Taverniera stefaninii Chiov. [1097].
   Not climbing; Herb/Shrub; Perennial.
   West Asia: Oman(N), Saudi Arabia(N), South Yemen(N); Africa; South Asia: Pakistan(N).
   Description: 1001, 1058, 1097; Distribution Map: 1097; Illustration: 1001.

T. multinoda Thulin [1097]
   Not climbing; Herb/Shrub; Perennial.
   West Asia: Oman(N), South Yemen(N).
   Description: 1097; Distribution Map: 1097.
   South Yemen record doubtful; based on sterile material [1097].

T. nummularia DC. [1009]
   Taverniera persica Boiss. & Hausskn.
   Not climbing; Herb/Shrub; Perennial.
   West Asia: Iran(N), Iraq(N); South Asia: India(N), Pakistan(N).
   Description: 1009, 1058, 1097; Distribution Map: 1097; Illustration: 1009, 1058, 1097.

T. spartea (Burm.f.)DC. [1058]
   Not climbing; Herb/Shrub; Perennial.
   West Asia: Bahrain(N), Iran(N), Oman(N), Qatar(N), Saudi Arabia(N), United Arab Emirates(N);
      South Asia: Pakistan(N).
   Description: 1015, 1058, 1097; Distribution Map: 1097; Illustration: 1058, 1097.

INDIGOFEREAE

CYAMOPSIS DC.

Three species, native to the dry areas of Africa and Arabia. All are herbs; one is cultivated, particularly in India, as fodder and for the seed gum. See Gillett (1958 — ref. 1083).

C. senegalensis Guillemin & Perrottet [1001]
   Not climbing; Herb; Annual.
   West Asia: Saudi Arabia(N); Africa.
   Description: 1001; Illustration: 1001.

C. tetragonoloba (L.)Taubert [1058]
   C. tetragonolobus (L.)Taubert.
   Not climbing; Herb; Annual.
   West Asia: Afghanistan(I), North Yemen(I); South Asia: India(N), Pakistan(N).
   Description: 1058.
   Occurs in "Arabia" [1058].

INDIGOFERA L.

A pantropical genus of about 700 herbs, woody herbs, and shrubs. Some are cultivated as dye-plants. See Gillett (1958).

I. amorphoides Jaub. & Spach [1064]
   Not climbing; Herb/Shrub; Perennial.
   West Asia: North Yemen(N), Saudi Arabia(N), South Yemen(N); Africa.
   Description: 1064.

I. arabica Jaub. & Spach [1058]
   Not climbing; Herb; Perennial.
   West Asia: North Yemen(N), Oman(N), Saudi Arabia(N), South Yemen(N); Middle East: Sinai(N); South Asia: Pakistan(N).
   Description: 1001, 1058; Illustration: 1001.

I. argentea Burm.f. [1083]
   I. burmannii Boiss.
   Not climbing; Herb/Shrub; Perennial.
   West Asia: Iran(N), North Yemen(N), Oman(N), Qatar(N), South Yemen(N); United Arab Emirates(N); Africa; South Asia: Pakistan(N).
   Description: 1058; Illustration: 1058.

I. arrecta A.Rich. [1083]
   Not climbing; Herb/Shrub; Perennial.
   West Asia: North Yemen(N), Saudi Arabia(N); Africa.
   Description: 1064.
   Cover crop 1083; Domestic 1064, 1083.

I. articulata Gouan [1058]
   Not climbing; Herb/Shrub; Perennial.
   West Asia: Iran(N), North Yemen(N), Oman(N), Qatar(N), Saudi Arabia(N); South Yemen(N); Africa; Middle East: Sinai(N), Syria(N); South Asia: Pakistan(N).
   Description: 1001, 1014, 1058; Illustration: 1001, 1014, 1058.

I. atriceps Hook.f. [1064]
   Not climbing; Herb; Annual.
   West Asia: North Yemen(N); Africa.
   Description: 1064.
   Several subspecies, of which subsp. *kaessneri* occurs in West Asia.

   subsp. **kaessneri** (Baker f.)J.B.Gillett [1064].
   Not climbing; Herb; Annual.
   West Asia: North Yemen(N); Africa.
   Description: 1064.

I. atropurpurea Hornem. [1058]
   Not climbing; Shrub; Perennial.
   West Asia: Afghanistan(N); East Asia: China(N); South Asia: India(N), Nepal(N), Pakistan(N).
   Description: 1058; Illustration: 1058.
   Some doubt as to identity of Afghan plant [1058].

I. brevicalyx Baker f. [1064]
   Not climbing; Herb; Perennial.
   West Asia: North Yemen(N); Africa.
   Description: 1064.

I. coerulea Roxb. [1058]

I. argentea sensu auctt. [1083]; *I. articulata* Gouan [1064]; *I. caerulea* Roxb. [1083].
Not climbing; Herb/Shrub; Perennial.
West Asia: Iran(N), North Yemen(N), Oman(N), Saudi Arabia(N), South Yemen(N); Africa; South Asia: India(N), Pakistan(N).
Description: 1016, 1058; Illustration: 1016.
Domestic 1016, 1064; Medicine 1016.
2 varieties, *occidentalis* & *coerulea* occur in Oman [1016].

I. colutea (Burm.f.)Merr. [1058]

I. viscosa Lam. [1083].
Not climbing; Herb; Perennial.
West Asia: Afghanistan(N), North Yemen(N), Saudi Arabia(N); Africa; South Asia: India(N), Pakistan(N).
Description: 1001, 1058, 1064; Illustration: 1001.
Includes the variety *colutea*.

I. contorta J.B.Gillett [1083]

Not climbing; Herb; Perennial.
West Asia: South Yemen(N).
Description: 1083.

I. cordifolia Roth [1058]

Not climbing; Herb; Annual.
West Asia: Afghanistan(N), Oman(N); Africa; South Asia: India(N), Pakistan(N).
Description: 1058.
Food or drink 1083; Forage 1083.

I. costata Guillemin & Perrottet [1064]

Not climbing; Herb; Annual.
West Asia: North Yemen(N); Africa.
Description: 1064.
The West Asian plants belong to subspecies *gonioides*.

subsp. gonioides (Baker)J.B.Gillett [1064]

Not climbing; Herb; Annual.
West Asia: North Yemen(N); Africa.
Description: 1064.

I. deflersii Baker f. [1120]

Not climbing; Herb; Perennial.
West Asia: North Yemen(N)
Description: 1120.

I. disjuncta J.B.Gillett [1001]

Not climbing; Herb; Annual.
West Asia: Saudi Arabia(N); Africa.
Description: 1001; Illustration: 1001.
Forage.

I. heterantha Brandis [1058]

I. gerardiana Baker [1058]; *I. gerardiana* var. *heterantha* (Brandis)Baker [1058].
Not climbing; Herb/Shrub; Perennial.
West Asia: Afghanistan(N); South Asia: India(N), Nepal(N), Pakistan(N).
Description: 1052, 1058; Illustration: 1052, 1058.
2 varieties, *heterantha* & *gerardiana*, both of which occur in Afghanistan [1058].

I. himalayensis Ali [1058]

Not climbing; Herb; Perennial.
West Asia: Afghanistan(N); South Asia: India(N), Pakistan(N).
Description: 1058.

I. hochstetteri Baker [1058]

I. anabaptista Steudel [1083]
Not climbing; Herb; Annual.
West Asia: Afghanistan(N), North Yemen(N), Oman(N), Saudi Arabia(N), South Yemen(N); Africa; South Asia: India(N), Pakistan(N).
Description: 1001, 1058; Illustration: 1001.

I. insularis Chiov. [1064]

Not climbing; Herb; Perennial.
West Asia: North Yemen(N), Oman(N), Saudi Arabia(N), South Yemen(N).
Description: 1064.

I. intricata Boiss. [1058]

Not climbing; Herb; Perennial.
West Asia: Iran(N), Oman(N), South Yemen(N); South Asia: Pakistan(N).
Description: 1058; Illustration: 1058.

I. linifolia (L.f)Retz. [1058]

Not climbing; Herb; Annual.
West Asia: Afghanistan(N), North Yemen(N), Saudi Arabia(N); Africa; Australasia; South Asia: India(N), Pakistan(N).
Description: 1001, 1015, 1058; Illustration: 1001, 1058.
Food or drink 1083; Forage 1083; Toxins 1083.

I. nephrocarpa Balf.f. [1058]

Not climbing; Herb; Perennial.
West Asia: Oman(N); South Asia: Pakistan(N).
Description: 1058.

I. oblongifolia Forsskal [1058]

Not climbing; Shrub; Perennial.
West Asia: Bahrain(N), Iran(N), North Yemen(N), Oman(N), Saudi Arabia(N); South Yemen(N); Africa; South Asia: India(N), Pakistan(N).
Description: 1001, 1016; Illustration: 1001, 1016.
Forage 1083; Medicine 1083.

I. parviflora Wight & Arn. [1064]

Not climbing; Herb; Annual.
West Asia: North Yemen(N); Africa; Australasia; South Asia: India(N).
Description: 1064.

I. phillipsiae Baker f. [1058]

Not climbing; Herb; Perennial.
West Asia: North Yemen(N), Oman(N), Saudi Arabia(N); Africa.
Description: 1001; Illustration: 1001.

I. pseudointricata J.B.Gillett [1083]

Not climbing; Herb/Shrub; Perennial.
West Asia: Oman(N), Qatar(N), South Yemen(N), United Arab Emirates(N).
Description: 1083.

I. schimperi Jaub. & Spach [1083]

Not climbing; Herb/Shrub; Perennial.
West Asia: North Yemen(N), Saudi Arabia(N); Africa.
Description: 1001, 1064; Illustration: 1001.
Forage 1083
A very variable taxon; some forms are very difficult to separate from *I. oblongifolia* [1083].

I. semitrijuga Forsskal [1001]

Not climbing; Herb; Annual.
West Asia: North Yemen(N), Oman(N), Qatar(N), Saudi Arabia(N), South Yemen(N); Africa.
Description: 1001, 1015; Illustration: 1001.

I. sessiliflora DC. [1001]
I. tribuloides Boiss. [1083].
Not climbing; Herb; Annual.
West Asia: Afghanistan(N), North Yemen(N), Saudi Arabia(N); Africa; South Asia: Pakistan(N).
Description: 1001, 1007, 1058; Illustration: 1001, 1058.

I. spicata Forsskal [1083]
Not climbing; Herb; Perennial.
West Asia: North Yemen(N); Africa; South Asia: India(N); South-East Asia.
Cover crop 1083; Forage 1083; Toxins 1083; Weed 1083.

I. spiniflora Boiss. [1001]
Not climbing; Herb/Shrub; Perennial.
West Asia: North Yemen(N), Oman(N), Saudi Arabia(N), South Yemen(N); Africa.
Description: 1001; Illustration: 1001.

I. spinosa Forsskal [1001]
Not climbing; Herb/Shrub; Perennial.
West Asia: North Yemen(N), Saudi Arabia(N), South Yemen(N); Africa.
Description: 1001, 1007; Illustration: 1001, 1007.

I. tinctoria L. [1064]
Not climbing; Herb/Shrub; Perennial.
West Asia: North Yemen(U);Saudi Arabia(U); Africa.
Description: 1064.
Domestic 1064.

I. trita L.f. [1001]
Not climbing; Herb/Shrub; Perennial.
West Asia: North Yemen(I), Oman(I), Saudi Arabia(I); Africa; Australasia; South Asia: India(N).
Description: 1001, 1064; Illustration: 1001.
2 varieties, *maffei* (Chiov.)Ali & *scabra* (Roth.)Ali.

I. tritoides Baker [1001]
Not climbing; Herb/Shrub; Perennial.
West Asia: North Yemen(N), Saudi Arabia(N), South Yemen(N); Africa.
Description: 1001, 1007; Illustration: 1001.

I. sp.3234 S.Collenette (provisional) [1001]
Not climbing; Herb/Shrub; Perennial.
West Asia: Saudi Arabia(N)
Description: 1001; Illustration: 1001.

I. sp.aff.volkensii S.Collenette (provisional) [1001]
Not climbing; Herb; Perennial.
West Asia: Saudi Arabia(N)
Description: 1001; Illustration: 1001.

LOTEAE

ANTHYLLIS L.

A genus of about 20 herbaceous species, mainly in the Mediterranean area. Flowers ornamental. *A. vulneraria* is a very complex species with a large number of subspecies, variously treated. See Cullen (1976).

A. boissieri Sagorski [1058]
A. lachnophora Juz. [1058]; *A. vulneraria* subsp. *boissieri* (Sagorski) Bornm. (Sagorski)Bornm. [1058].
Not climbing; Herb; Perennial.
West Asia: Iran(N); Middle East: Turkey(N); U.S.S.R.
Description: 1058; Illustration: 1058.

A. vulneraria L. [1015]
   Not climbing; Herb; Perennial.
   West Asia: Iran(N).
   Description: 1015.
   Parsa's record is from the same locality as Rechinger's *A. boissieri* [1015,1058]; the records may well refer to the same taxon.

CORONILLA L.

A genus of about 20 herbaceous or sub-shrubby species, mainly in the Mediterranean area. Some are cultivated for ornament and erosion control.

C. coronata L. [1102]
   C. monatana Scop. [1102].
   Not climbing; Herb; Perennial.
   West Asia: Iran(N); Middle East: Syria(N), Turkey(N).

C. scorpioides (L.)Koch [1009]
   Not climbing; Herb; Annual.
   West Asia: Iran(N), Iraq(N); Africa; Europe; Middle East: Israel(N), Jordan(N), Lebanon(N), Syria(N); U.S.S.R.
   Description: 1009, 1015, 1058; Illustration: 1009, 1058.
   Environmental 1009; Forage 1009; Medicine 1009.

DORYCNIUM Miller

A small herbaceous genus of about 10 species, sometimes considered to be part of *Lotus*.

D. intermedium Ledeb. [1058]
   Not climbing; Herb; Perennial.
   West Asia: Iran(N); Middle East: Turkey(N); U.S.S.R.
   Description: 1058; Illustration: 1058.

D. pentaphyllum Scop. [1009]
   Not climbing; Herb/Shrub; Perennial.
   West Asia: Iraq(N); Africa; Europe; Middle East: Lebanon(N), Syria(N), Turkey(N); U.S.S.R.
   Description: 1009.
   2 subspecies, *pentaphyllum* & *haussknechtii*; only the latter occurs in W. Asia [1009].

   subsp. **haussknechtii** (Boiss.)Gams [1009]
      D. haussknechtii Boiss. [1009].
      Not climbing; Herb/Shrub; Perennial.
      West Asia: Iraq(N); Middle East: Syria(N), Turkey(N).
      Description: 1009, 1058; Illustration: 1009.

HIPPOCREPIS L.

Mostly herbs, but recent changes in the definition of the genus have led to the inclusion of some shrubby species. Mainly Mediterranean; about 20 species.

H. bicontorta Lois. [1009]
   H. cornigera Boiss. [1009].
   Not climbing; Herb; Annual.
   West Asia: Bahrain(N), Iran(N), Iraq(N), Kuwait(N), Qatar(N), Saudi Arabia(N), United Arab Emirates(N); Africa; Middle East: Israel(N), Sinai(N).
   Description: 1009, 1015, 1058; Illustration: 1001, 1009, 1058.
   Forage 1009, 1014.

H. bornmulleri Hausskn. [1015]
Not climbing; Herb; Annual.
West Asia: Iran(N).
Description: 1015.
Close to H. unisiliquosa [1015].

H. ciliata Willd. [1009]
Not climbing; Herb; Annual.
West Asia: Iraq(N), Saudi Arabia(N); Africa; Europe; Middle East: Jordan(N), Syria(N); U.S.S.R.
Description: 1001, 1009; Illustration: 1001.
Occurrence in Iraq very doubtful [1009].

H. constricta Kunze [1001]
Not climbing; Herb; Annual.
West Asia: Iran(N), Qatar(N), Saudi Arabia(N); Africa; South Asia: Pakistan(N).
Description: 1001, 1015, 1058; Illustration: 1001, 1007, 1014.

H. cyclocarpa Murb. [1007]
Not climbing; Herb; Annual.
West Asia: Saudi Arabia(N); Africa; Europe.
Description: 1007; Illustration: 1007.

H. multisiliquosa L. [1007]
Not climbing; Herb; Annual.
West Asia: Qatar(N), Saudi Arabia(N); Africa; Europe.
Description: 1007, 1014; Illustration: 1007, 1014.

H. unisiliquosa L. [1009]
Not climbing; Herb; Annual.
West Asia: Bahrain(N), Iran(N), Iraq(N), Kuwait(N), Qatar(N), Saudi Arabia(N); Africa; Europe;
    Middle East: Israel(N), Lebanon(N), Syria(N), Turkey(N); U.S.S.R.
Description: 1009, 1015, 1058; Illustration: 1001, 1007.
Forage 1009.

subsp. biflora (Sprengel)O.Bolos & Vigo [1009]
H. biflora Sprengel; *H. bisiliqua* Forsskal [1009]; *H. unisiliquosa* subsp. *bisiliqua* (Forsskal)Bornm.
Not climbing; Herb; Annual.
West Asia: Iran(N), Iraq(N), Oman(N).
Description: 1009, 1015, 1058; Illustration: 1009, 1058.
Forage 1009.
Sometimes regarded as a good species.

HYMENOCARPOS Savi

A monospecific genus confined to the Mediterranean region. Small annual herbs with discoid pods.

H. circinnatus (L.)Savi [1009]
Not climbing; Herb; Annual.
West Asia: Iran(N), Iraq(N), Kuwait(N), Qatar(N), Saudi Arabia(N); Africa; Europe; Middle East:
    Israel(N), Lebanon(N), Sinai(N), Syria(N), Turkey(N).
Description: 1009, 1015, 1058; Illustration: 1009.
Forage 1009.
Spanish Medick 1009.
Rechinger (1984) recognises 2 subspecies, *circinnatus* & *nummularius* [1058], but Townsend
    (1974) does not recognise any infraspecific taxa [1009].

subsp. circinnatus [1058]
Not climbing; Herb; Annual.
West Asia: Iran(N), Iraq(N); Europe; Middle East.
Description: 1058; Illustration: 1058.

subsp. **nummularius** (DC.)Chrtek & B.Slavik [1058]
>H. nummularius (DC.)Boiss. [1058].
>Not climbing; Herb; Annual.
>West Asia: Iran(N), Iraq(N), Saudi Arabia(N); Europe; Middle East.
>Description: 1015, 1058; Illustration: 1058.

LOTUS L.

A north temperate zone genus of about 90 species. The delimitation of species is controversial; some works (e.g. Townsend & Guest 1974) recognize fewer taxa at the species level than are listed here.

L. aegeus (Griseb.)Boiss. [1009]
>L. sulphureus Boiss. [1009].
>Not climbing; Herb; Perennial.
>West Asia: Iraq(N); Europe; Middle East: Turkey(N).
>Description: 1009.

L. angustissimus L. [1058]
>Not climbing; Herb; Perennial.
>West Asia: Iran(N); Europe; Middle East.
>Description: 1058; Illustration: 1058.

L. arabicus L. [1007]
>L. glinoides sensu Baker [1007]; L. mossamedensis Baker [1007]; L. roseus Forsskal [1007].
>Not climbing; Herb; Annual.
>West Asia: Saudi Arabia(N); Africa.
>Description: 1007; Illustration: 1007.
>Radcliffe-Smith 3694 ex Oman (K) is this or a related taxon.

L. corniculatus L. [1009]
>Not climbing; Herb/Shrub; Perennial.
>West Asia: Afghanistan(N), Iran(N), Iraq(N), North Yemen(N), Saudi Arabia(N); Africa; Europe; Middle East; U.S.S.R.
>Descriptrion [1009,1058].
>Cover crop 1009; Forage 1009.
>Birds-Foot Trefoil 1009.
>Townsend recognised two varieties, *corniculatus* & *tenuifolius*, but regarded the taxon as very variable, and most infraspecific taxa as of little worth [1009]. Chrtkova-Zertova in Rechinger (1984), however, treated var. *tenuifolius* as a good species (*L. tenuis*), and recognised 4 subspecies, *corniculatus, fruticosus, frondosus* & *afghanicus* [1058].

subsp. **afghanicus** Chrtkova-Zert. [1058]
>Not climbing; Herb; Perennial.
>West Asia: Afghanistan(N); South Asia: Pakistan(N).
>Description: 1058; Illustration: 1058.

subsp. **corniculatus** [1009]
>L. corniculatus var. alpinus DC. [1009].
>Not climbing; Herb; Perennial.
>West Asia: Afghanistan(N), Iran(N), Iraq(N).
>Description: 1009, 1058; Illustration: 1058.
>4 varieties, *corniculatus, kochii, hirsutus,* & *brachyodon* [1058].

subsp. **frondosus** Freyn [1058]
>L. frondosus (Freyn)Kuprian. [1058].
>Not climbing; Herb; Perennial.
>West Asia: Afghanistan(N), Iran(N); South Asia: Pakistan(N).
>Description: 1058; Illustration: 1058.

subsp. **fruticosus** Chrtkova-Zert. [1058]
: Not climbing; Shrub; Perennial.
West Asia: Afghanistan(N), Iran(N).
Description: 1058; Illustration: 1058.

L. garcinii DC. [1058]
L. pumilus Parsa [1058]; *L. sharifii* Rech.f. & Esfand. [1058].
Not climbing; Herb; Perennial.
West Asia: Iran(N), Oman(N), Qatar(N), Saudi Arabia(N), South Yemen(N); United Arab Emirates(N); Africa; South Asia: Pakistan(N).
Description: 1014, 1015, 1058.

L. gebelia Vent. [1009]
L. aleppicus Boiss. [1058].
Not climbing; Herb; Perennial.
West Asia: Iran(N), Iraq(N), Saudi Arabia(N); Africa; Middle East: Lebanon(N), Syria(N), Turkey(N); U.S.S.R.
Description: 1009, 1015, 1058; Illustration: 1058.
Forage 1009; Misc 1009.
Chrtkova-Zertova, in Rechinger 1985, recognised three varieties, *gebelia, villosus,* & *lanatus* [1058], but Townsend takes a broader view of the species, recognising two varieties, *gebelia* & *hirsutissimus* [1009]. The flowers are considered valuable for bees [1009].

L. glinoides Del. [1007]
L. arabicus var. *trigonelloides* (Webb)Webb [1007]; *L. trigonelloides* Webb & Berth. [1007].
Not climbing; Herb; Annual.
West Asia: Saudi Arabia(N).
Description: 1007.

L. halophilus Boiss. & Spruner [1009]
L. pusillus Viv. [1009]; *L. villosus* Forsskal [1009].
Not climbing; Herb; Annual.
West Asia: Bahrain(N), Iran(N), Iraq(N), Kuwait(N), Qatar(N), Saudi Arabia(N); United Arab Emirates(N); Africa; Europe; Middle East: Israel(N), Lebanon(N), Sinai(N), Syria(N), Turkey(N).
Description: 1001, 1009, 1058; Illustration: 1001, 1007, 1058.
Forage 1009.
L. halophilus Boiss. & Spruner is the correct name according to Flora of Turkey [1102].

L. krylovii Schischkin & Serg. [1058]
Not climbing; Herb; Perennial.
West Asia: Afghanistan(N), Iran(N); South Asia: Pakistan(N); U.S.S.R.
Description: 1058; Illustration: 1058.
Townsend treats this as a synonym of *L. corniculatus* var. *tenuifolius* [1009].

L. lalambensis Penzig [1007]
L. arabicus var. *glabrescens* Schweinf. [1007]; *L. brachycarpus* var. *lalambensis* (Schweinf.)Brand [1007].
Not climbing; Herb; Perennial.
West Asia: Saudi Arabia(N).
Description: 1007.

L. lanuginosus Vent. [1009]
Not climbing; Herb; Perennial.
West Asia: Iraq(N), Saudi Arabia(N); Middle East: Israel(N), Jordan(N), Syria(N).
Description: 1001, 1009; Illustration: 1001, 1009.

L. laricus Rech.f. [1058]
Not climbing; Herb; Perennial.
West Asia: Iran(N).
Description: 1058; Illustration: 1058.
L. schimperi Steudel forma *laricus* (Rech.f.)Parsa is a synonym of this.
Endemic to Iran [1058].

L. michauxianus Ser. [1058]
L. gebelia var. *tomentosus* Boiss.
Not climbing; Herb; Perennial.
West Asia: Iran(N).
Description: 1058; Illustration: 1058.
Townsend treats this taxon as a synonym of *L. gebelia* var. *hirsutissimus* [1009].

L. palustris Willd. [1081]
West Asia: South Yemen(U)
Identification uncertain.

L. quinatus (Forsskal)J.B.Gillett [1001]
Not climbing; Herb; Annual.
West Asia: North Yemen(N), Saudi Arabia(N); Africa.
Description: 1001; Illustration: 1001.

L. schimperi Boiss. [1001]
Not climbing; Herb; Perennial.
West Asia: Bahrain(N), Iran(N), Oman(N), Saudi Arabia(N); United Arab Emirates(N); Africa.
Description: 1001, 1058; Illustration: 1001, 1058.

L. tenuis Waldst. & Kit. [1058]
L. corniculatus subsp. *tenuis* (Kit.)Briq. [1058]; *L. tenuifolius* (L.)Reichb. [1058].
Not climbing; Herb; Perennial.
West Asia: Afghanistan(N), Iran(N); Africa; Europe.
Description: 1058; Illustration: 1058.
Townsend treats this taxon as a synonym of *L. corniculatus* var. *tenuifolius*; see p.206 of Flora of Iraq for a discussion of this [1009].

L. sp.aff.arabicus S.Collenette (provisional) [1001]
Not climbing; Herb
West Asia: Saudi Arabia(N)
Description: 1001; Illustration: 1001.
Collenette 2085 at Kew.

ORNITHOPUS L.

A genus of small herbs, with about six species. Some are cultivated as fodder and have been introduced well outside the Mediterranean centre of the genus.

O. compressus L. [1058]
Not climbing; Herb; Annual.
West Asia: Iran(N); Middle East: Turkey(N); U.S.S.R.
Description: 1015, 1058
Iranian record very old [1058].

PODOLOTUS Benth.

P. hosackiodes Benth. [1058]
Astragalus hosackioides (Benth.) Benth. [1058]; *Kerstania nuristanica* Rech.f. [1058].
Not climbing; Herb; Perennial.
West Asia: Iran(N), Afghanistan(N); South Asia: India; Pakistan.
Description: 1058; Illustration: 1058; Distribution Map: 1263.
The systematic position of this plant has been much discussed. Lassen (pers. comm.) considers that it is a good member of the Loteae.

PSEUDOLOTUS Rech.f.

P. makranicus (Rech.f. & Esfand.) Rech.f. [1058]
: *Lotus makranicus* Rech.f. & Esfand. [1058].
Not climbing; Herb; Perennial.
West Asia: Afghanistan(N), Iran(N), Oman(N); South Asia: Pakistan(N).
Description: 1052, 1058; Illustration: 1052, 1058.
Gunn (1983) regarded *Pseudolotus* as synonomous with *Lotus*; Lassen (pers. comm.) regards it as a very distinct genus within the Loteae.

SCORPIURUS L.

A herbaceous genus in which the definition of species is difficult because of great variation in fruit form. About four species, mainly Mediterranean.

S. muricatus L. [1009].
: *S. subvillosus* L. [1009]; *S. sulcatus* L. [1009].
Not climbing; Herb; Annual.
West Asia: Iran(N), Iraq(N), Kuwait(N), Qatar(N), Saudi Arabia(N), United Arab Emirates(N); Africa; Europe; Middle East: Israel(N), Jordan(N), Lebanon(N), Sinai(N), Syria(N), Turkey(N).
Description: 1009, 1015, 1058; Illustration: 1009, 1014, 1058.
Forage 1009.
Very variable; taxonomic treatment also varies.

SECURIGERA DC.

A Mediterranean genus of about four species of small herbs, formerly treated as part of *Coronilla*.

S. orientalis (Miller) Lassen
: *Coronilla cappadocica* Willd. [1058]; *C. orientalis* Miller [1262].
Not climbing; Herb; Perennial.
West Asia: Iran(N); Middle East: Turkey(N); U.S.S.R.
Description: 1058; Illustration: 1058.

S. parviflora (Desv.) Lassen
: *C. parviflora* Willd. [1058]
Not climbing; Herb; Annual.
West Asia: Iran(N); Europe; Middle East: Lebanon(N), Syria(N), Turkey(N).
Description: 1058; Illustration: 1058.

S. securidaca (L.)Degen & Doerfler [1005]
: *C. securidaca* L. [1005, 1009]; *S. coronilla* DC. [1009].
Not climbing; Herb; Annual.
West Asia: Iran(N), Iraq(N); Africa; Europe; Middle East: Lebanon(N), Turkey(N); U.S.S.R.
Description: 1009, 1015, 1058; Illustration: 1009, 1058.
Environmental 1009; Medicine 1009.

S. varia (L.) Lassen
: *C. varia* L. [1009]
Not climbing; Herb; Perennial.
West Asia: Iran(U), Iraq(U); Europe; Middle East; Lebanon(N), Syria(N), Turkey(N);U.S.S.R.
Description: 1009, 1058; Illustration: 1058.
Environmental 1009; Forage 1009; Medicine 1009.
Exact native range uncertain [1009].
Two subspecies, *varia* and *hirta* have been recognized [1058].

VERMIFRUX J.B.Gillett

A monospecific herbaceous genus from countries bordering the Red Sea. Sometimes considered not to be distinct from *Lotus*.

V. abyssinica (A.Rich.)J.B.Gillett [1001]
Helminthocarpum abyssinicum A.Rich. [1005]
Not climbing; Herb; Annual.
West Asia: North Yemen(N), Saudi Arabia(N); Africa.
Description: 1001; Illustration: 1001.

MILLETTIEAE

PTYCHOLOBIUM Harms

A small herbaceous genus of three species in the drier parts of Africa and Arabia. Close to *Tephrosia* but differing in its contorted fruits and palmately compound leaves. See Brummitt (1980 — ref. 1095).

P. plicatum (Oliver)Harms [1095]
Not climbing; Herb; Perennial.
West Asia: Oman(N), South Yemen(N); Africa.
Description: 1095; Illustration: 1095.

subsp. **arabicum** Brummitt [1095]
Not climbing; Herb; Perennial.
West Asia: Oman(N), South Yemen(N)
Description: 1095; Illustration: 1095.

TEPHROSIA Pers.

A genus of about 400 species in the drier tropics. Most are herbs or soft-wooded shrubs. Some yield fish-poisons; the range of a few species has therefore been extended by cultivation.

T. apollinea (Del.)Link [1001]
Not climbing; Herb/Shrub; Perennial.
West Asia: Iran(N), North Yemen(N), Oman(N), Saudi Arabia(N), South Yemen(N); United Arab Emirates(N); Africa; South Asia: Pakistan(N).
Description: 1001, 1015, 1058; Illustration: 1001, 1007, 1058.

T. desertorum Scheele [1001]
Not climbing; Herb.
West Asia: Iran(N), Saudi Arabia(N).
Description: 1001, 1015; Illustration: 1001.

T. dura Baker [1091]
Not climbing; Herb/Shrub; Perennial.
West Asia: South Yemen(N).
Description: 1091.

T. elata Defl. [1093]
Not climbing; Herb/Shrub; Annual.
West Asia: North Yemen(N); Africa.
Description: 1093; Distribution Map:1093.

subsp. **elata** [1093]
    Not climbing; Herb/Shrub; Annual.
    West Asia: North Yemen(N).
    Description: 1093.

T. geminiflora Baker [1092]
    Not climbing; Herb; Perennial.
    West Asia: South Yemen(N).
    Description: 1092.

T. haussknechtii Bornm. [1012]
    Not climbing; Shrub; Perennial.
    West Asia: Iran(N), Oman(N).
    Description: 1015.

T. heterophylla Vatke [1251]
    West Asia: North Yemen(N).

T. humilis Guillemin & Perrottet
    Not climbing; Herb; Annual.
    West Asia: Oman(N).

T. nubica (Boiss.)Baker [1001]
    Not climbing; Herb/Shrub; Perennial.
    West Asia: North Yemen(N), Oman(N), Saudi Arabia(N), South Yemen(N).
    Description: 1001, 1007; Illustration: 1001.

    subsp. **arabica** (Boiss.)J.B.Gillett [1094]
        Not climbing; Herb/Shrub; Perennial.
        West Asia: North Yemen(N), Oman(N), Saudi Arabia(N), South Yemen(N).
        Description: 1007.

    subsp. **nubica** [1094]
        Not climbing; Herb/Shrub; Perennial.
        West Asia: South Yemen(N).

T. pentaphylla (Roxb.)G.Don [1064]
    T. senticosa sensu auctt. [1064]
    Not climbing; Herb; Annual.
    West Asia: Iran(N), North Yemen(N), Oman(N).

T. persica Boiss. [1012]
    Tephrosia apollinea subsp. *persica* (Boiss.)Bornm. [1058].
    Not climbing; Herb/Shrub; Perennial.
    West Asia: Iran(N), Oman(N), United Arab Emirates(N).
    Description: 1015, 1058; Illustration: 1058.

T. pumila (Lam.)Pers. [1001]
    Not climbing; Herb; Annual.
    West Asia: North Yemen(N), Saudi Arabia(N); Africa.
    Description: 1001; Illustration: 1001.

T. purpurea (L.)Pers. [1001]
    Not climbing; Herb; Perennial.
    West Asia: Iran(N), North Yemen(N), Oman(N), Saudi Arabia(N), South Yemen(N); Africa;
        Australasia; South Asia: Pakistan(N).
    Description: 1001, 1015, 1058; Illustration: 1001, 1058.
    Rechinger does not mention Brummitt's subspecies [1058].

    subsp. **leptostachya** (DC.)Brummitt [1001]
        Not climbing; Herb/Shrub; Perennial.
        West Asia: North Yemen(N), Saudi Arabia(N).
        Description: 1001; Illustration: 1001.
        Includes vars. *leptostachya* & *pubescens* (Baker)Brummitt.

T. quartiniana Greuter & Burdet [1252]
> *T. vicioides* A. Rich. [1252]
> Not climbing; Herb; Perennial.
> West Asia: Oman(N), Saudi Arabia(N); Africa.
> Description: 1001, 1007; Illustration: 1001.

T. schweinfurthii Defl.
> Not climbing; Herb; Perennial.
> West Asia: North Yemen(N).

T. strigosa (Dalz.)Santapau & Maheshw. [1058]
> Not climbing; Herb; Annual.
> West Asia: Oman(N), United Arab Emirates(N); South Asia: India(N), Pakistan(N).
> Description: 1058.

T. subtriflora Baker [1001]
> Not climbing; Herb; Perennial.
> West Asia: North Yemen(N), Oman(N), Saudi Arabia(N); Africa; South Asia: India(N), Pakistan(N).
> Description: 1001, 1058; Illustration: 1001.

T. tomentosa (Forsskal) Pers. [1251] (Provisional)
> West Asia: North Yemen(N), South Yemen(N).
> Identity uncertain; no material at Kew.

T. uniflora Pers. [1001]
> Not climbing; Herb/Shrub; Perennial.
> West Asia: Iran(N), North Yemen(N), Saudi Arabia(N), United Arab Emirates(N); Africa; South Asia: India(N), Pakistan(N).
> Description: 1001, 1007, 1058; Illustration: 1001.

>subsp. **petrosa** J.B.Gillett & Ali [1001]
>> Not climbing; Herb; Perennial.
>> West Asia: Iran(N), North Yemen(N), Saudi Arabia(N), South Yemen(N); South Asia: India(N), Pakistan(N).
>> Description: 1001, 1058; Illustration: 1001, 1058.

T. villosa (L.)Pers.
> Not climbing; Herb; Annual.
> West Asia: North Yemen(N), Saudi Arabia(N); Africa.

PHASEOLEAE

CAJANUS DC.

A genus of about 30 species of herbs, soft-wooded shrubs or climbers, most of which were formerly placed in the genus *Atylosia*. One species is an important and widely planted pulse crop. See Van der Maesen (1985 — ref. 1135).

C. cajan (L.)Millsp. [1058]
> *C. indicus* Sprengel [1134].
> Not climbing; Herb/Shrub; Perennial.
> West Asia: Afghanistan(N), North Yemen(I), Oman(U).
> Description: 1015, 1058; Illustration: 1135.
> Food or drink 1134; Medicine 1134.
> Pigeon Pea 1134
> An important pulse crop in S.Asia and Africa [1246].

CANAVALIA DC.

C. africana Dunn [1252]
C. virosa sensu auctt. [1252]
Climbing; Herb; Perennial.
West Asia: North Yemen(N); Africa.
Description: 1064.
The nomenclature of this taxon is discussed and clarified by Verdcourt, Kew Bull. 42: 658.

CLITORIA L.

About 70 species, mainly in tropical America, mostly herbaceous climbers. The flowers are held inverted. Some are cultivated as ornamentals.

C. ternatea L. [1009]
Climbing; Herb; Perennial.
West Asia: Iran(I), Iraq(I), North Yemen(I), Saudi Arabia(I), South Yemen(I).
Description: 1001, 1007, 1009; Illustration: 1001.
Domestic 1009, 1107; Environmental 1009, 1107; Forage 1009; Medicine 1009, 1107; Toxins 1107.
Exact native range obscured by cultivation [1009].

DOLICHOS L.

About 60 species in the Old World tropics, mostly in more seasonal areas. Herbs, some climbing.

D. sericeus E. Meyer [1252]
subsp. **formosus** (A. Rich.) Verdc. [1252]
Climbing; Herb; Perennial.
West Asia: North Yemen(N); Africa.
Description: 1064.

D. trilobus L. [1001]
Climbing; Herb; Perennial.
West Asia: North Yemen(N);Saudi Arabia(N);South Yemen(N); Africa.
Description: 1001; Illustration: 1001.

ERIOSEMA Desv.

A tropical genus of about 130 species, mostly perennial herbs which may be woody at the base.

E. longipedunculatum (A.Rich.)Baker
Not climbing; Herb; Perennial.
West Asia: North Yemen(N); Africa.
Includes the variety *hirsutum* Verdc.

FLEMINGIA Aiton f.

F. grahamiana Wight & Arn. [1252]
Not climbing; Herb/Shrub; Perennial.
West Asia: North Yemen(N); Africa.
Description: 1064.

GLYCINE Willd.

A small Asian and Australian genus of herbs. One species is very widely cultivated for its seeds and the oil obtained from them. Another widespread taxon is now placed in *Neonotonia*.

G. max (L.)Merr. [1009]
Not climbing; Herb; Annual.
West Asia: Afghanistan(I), Iraq(I).
Description: 1009; Illustration: 1009.
Cover crop 1009; Domestic 1009; Food or drink 1009.
Soy Bean 1009; Soya Bean 1009

G. ussuriensis Regel & Maack [1106]
Glycine soja Siebold & Zucc. [1106].
Climbing; Herb; Perennial.
West Asia: Afghanistan(I); East Asia: China(N); U.S.S.R.
Description: 1106; Illustration: 1106.
The wild soya bean.

LABLAB Adans.

One species, a herbaceous twiner probably native only in Africa, where it is widespread. Varieties derived from this are cultivated throughout the Old World tropics, though usually on a small scale.

L. purpureus (L.)Sweet [1009]
Dolichos lablab L. [1009]; *L. niger* Medikus [1109].
Climbing; Herb; Perennial.
West Asia: Bahrain(I), Iran(I), Iraq(I), North Yemen(I), Oman(I), Saudi Arabia(I), South Yemen(I); Africa; South Asia.
Description: 1001, 1009; Illustration: 1001.
Cover crop 1009; Environmental 1009; Forage 1009; Medicine 1009.
Bonavist Bean 1244; Egyptian Kidney Bean 1009; Hyacinth Bean 1009.
Often grown as an annual [1009].

MACROTYLOMA (Wight & Arn.)Verdc.

About thirty species of small herbs in the Old World tropics. Formerly included in *Dolichos*. Some are cultivated for fodder or their edible seeds.

M. axillare (E.Meyer)Verdc. [1001]
Dolichos axillaris E.Meyer [1110]; *Dolichos biflorus* sensu auctt. [1110].
Climbing; Herb; Perennial.
West Asia: North Yemen(N), Saudi Arabia(N); Africa.
Description: 1001, 1110; Illustration: 1001, 1110.
Includes the varieties *axillare* & *glabrum* (E.Meyer)Verdc.

NEONOTONIA J.A.Lackey

A genus of two species in the Old World tropics, formerly included in *Glycine*. One species is a widespread and variable herbaceous climber, sometimes cultivated as fodder.

N. wightii (Wight & Arn.)J.A.Lackey [1112]
> *Glycine javanica* L. [1064]; *Glycine wightii* (Wight & Arn.)Verdc. [1064].
> Climbing; Herb/Shrub; Perennial.
> West Asia: Saudi Arabia(N); Africa.
> Description: 1001; Illustration: 1001.
> Food or drink 1112; Forage 1112

subsp. **wightii** [1113]
> *Glycine wightii* subsp. *longicauda* (Schweinf.)Verdc. [1001]; *Glycine wightii* subsp. *wightii* (Wight & Arn.)Verdc. [1113].
> Climbing; Herb/Shrub; Perennial.
> West Asia: North Yemen(N);Saudi Arabia(N); Africa.
> Description: 1001; Illustration: 1001.

PHASEOLUS L.

A tropical American genus of about 50 species, of which several species are very widely cultivated in the tropical and temperate zones as pulse and green vegetable crops.

P. lunatus L. [1009]
> Climbing; Herb; Annual.
> West Asia: Iran(I), Iraq(I); South America.
> Description: 1009.
> Cover crop 1009; Food or drink 1009.
> Butter Bean 1009; Civet Bean 1009; Duffin Bean 1009; Lima Bean 1009; Sieva Bean 1009.

P. vulgaris L. [1009]
> Climbing; Herb; Annual.
> West Asia: Afghanistan(I), Iran(I), Iraq(I); South America.
> Description: 1009.
> Food or drink 1009; Forage 1009.
> Common Bean 1009; French Bean 1009; Haricot 1009; Haricot Bean 1009; Kidney Bean 1009.

RHYNCHOSIA Lour.

A tropical genus containing about 200 species. Most are small herbs, but some are tall climbers.

R. elegans A.Rich. [1005]
> *R. buramensis* Hutch. & Bruce [1005].
> Climbing; Shrub; Perennial.
> West Asia: North Yemen(N), Saudi Arabia(N).
> Description: 1001; Illustration: 1001.

R. flava (Forsskal)Thulin [1114]
> Not climbing; Shrub; Perennial.
> West Asia: North Yemen(N).
> Description: 1114.

R. malacophylla (Sprengel)Bojer [1001]
> Climbing; Herb; Perennial.
> West Asia: North Yemen(N);Saudi Arabia(N); Africa.
> Description: 1001, 1115; Illustration: 1001.
> Some Saudi Arabian material at Kew (Collenette 1327,3307,3509) is intermediate between this species and *R.minima* [1001]

R. minima (L.)DC. [1001]
> *R. memnonia* (Del.)DC. [1007]
> Climbing; Herb; Perennial.
> West Asia: Afghanistan(N), Iran(N), North Yemen(N), Oman(N), Qatar(N); Saudi Arabia(N), South Yemen(N); Africa; Australasia; South Asia: India(N), Pakistan(N).
> Description: 1001, 1015, 1058; Illustration: 1001, 1007, 1058.
> Includes the varieties *memnonia* (Del.)Cooke, *prostrata* (Harv.)Meikle & *minima*.
> Collenette 3586 from Saudi Arabia may be a related species [1001].

R. pulverulenta Stocks [1001]
> Climbing; Herb; Perennial.
> West Asia: Iran(N), Oman(N), Saudi Arabia(N), South Yemen(N); Africa; South Asia: India(N), Pakistan(N).
> Description: 1001, 1058; Illustration: 1001, 1058.

R. schimperi Boiss. [1009]
> Not climbing; Herb; Perennial.
> West Asia: Iran(N), North Yemen(N), Oman(N), Saudi Arabia(N); Africa; South Asia: Pakistan(N).
> Description: 1001, 1058; Illustration: 1001, 1058.

R. sublobata (Schum.)Meikle [1115]
> Climbing; Herb; Perennial.
> West Asia: North Yemen(N); Africa.
> Description: 1115; Illustration: 1115.

R. totta (Thunb.)DC. [1001]
> Climbing; Herb; Perennial.
> West Asia: North Yemen(N), Saudi Arabia(N); Africa.
> Description: 1001; Illustration: 1001.

R. usambarensis Taubert [1001]
> Climbing; Herb; Perennial.
> West Asia: North Yemen(N), Saudi Arabia(N); Africa.
> Description: 1001; Illustration: 1001.
> The West Asian material (Collenette 3152A & 3585, & J.R.I.Wood 1920) is not identical to the type & may represent another taxon.

R. sp. of S.Collenette (provisional) [1058]
> Not climbing; Herb/Shrub; Perennial.
> West Asia: Saudi Arabia(N).
> Description: 1001; Illustration: 1001.
> The illustration is of Collenette 1384.

TERAMNUS P.Browne

A small genus of herbs, some climbing, in the seasonal tropics of the Old and New Worlds. Some of the species are variable.

T. labialis (L.f.)Sprengel [1001]
> Climbing; Herb; Perennial.
> West Asia: North Yemen(N), Saudi Arabia(N); Africa.
> Description: 1001; Illustration: 1001.
> Includes the variety *abyssinicus* (A.Rich.)Verdc.

subsp. **arabicus** Verdc. [1001]
> Climbing; Herb; Perennial.
> West Asia: North Yemen(N), Saudi Arabia(N); Africa.
> Description: 1001; Illustration: 1001.

T. repens (Taubert)Baker f. [1001]
> Climbing; Herb; Perennial.
> West Asia: Saudi Arabia(N); Africa.
> Description: 1001; Illustration: 1001.

VATOVAEA Chiov.

A monospecific genus found in the dry regions of north-eastern Africa and southern Arabia. A vigorous perennial climber.

V. pseudolablab (Harms)J.B.Gillett [1016].
Vatovaea biloba Chiov. [1111]; *Vigna pseudolablab* Harms [1064].
Climbing; Herb; Perennial.
West Asia: Oman(N); Africa.
Description: 1016; Illustration: 1016, 1111.
Food or drink 1016; Forage 1016.

VIGNA Savi

A complex and difficult genus of about 150 species in the tropics of the Old and New Worlds. Most are herbaceous climbers; some are cultivated as fodder and pulse crops.

V. aconitifolia (Jacq.)Marechal [1058]
Phaseolus aconitifolius Jacq. [1058].
Not climbing; Herb; Annual.
West Asia: Saudi Arabia(I).
Description: 1001, 1058; Illustration: 1001, 1244.
Food or drink 1244; Forage 1244
Mat Bean 1244; Moth Bean 1244.

V. ambacensis Baker [1001]
Climbing; Herb; Perennial.
West Asia: Saudi Arabia(N); Africa.
Description: 1001; Illustration: 1001.

V. heterophylla A.Rich.
Climbing; Herb; Annual.
West Asia: North Yemen(N).

V. luteola (Jacq.)Benth. [1116]
V. nilotica (Del.)Hook.f. [1116].
Climbing; Herb; Perennial.
West Asia: Iran(U); Africa.
Not mentioned by Rechinger or Townsend.

V. membranacea A.Rich. [1001]
Climbing; Herb; Annual.
West Asia: North Yemen(N), Saudi Arabia(N); Africa.
Description: 1001; Illustration: 1001.

V. mungo (L.)Hepper [1117]
Phaseolus mungo L. [1117]
Not climbing; Herb; Annual.
West Asia: Afghanistan(I).
Description: 1015; Illustration: 1244.
Food or drink 1244.
Black Gram 1244; Urd Bean 1244.

V. radiata (L.)Wilczek [1009]
Azukia radiata (L.)Ohwi [1009,1116]; *Phaseolus aureus* Roxb. [1009,1116]; *Phaseolus mungo* sensu auctt. [1009,1116]; *Phaseolus radiatus* L. [1009,1116]; *Rudua aurea* (Roxb.)F.Maek. [1116].
Climbing; Herb; Annual.
West Asia: Iran(I), Iraq(I), North Yemen(I), Oman(I).
Description: 1009, 1058; Illustration: 1009, 1244.
Cover crop 1009; Food or drink 1009; Forage 1009; Medicine 1009.
Green Gram 1009; Mung Bean 1009.
An important pulse and forage crop [1244].

V. unguiculata (L.)Walp. [1009]
V. cylindrica (L.)Skeels [1009]; *Vigna sinensis* (L.)Hassk. [1009].
Climbing; Herb; Annual.
West Asia: Afghanistan(I), Iran(I), Iraq(I), North Yemen(I); Africa.
Description: 1009, 1058; Illustration: 1244.
Food or drink 1009; Forage 1009; Medicine 1009.
Black-Eye Pea 1009; Catjang 1009; Catjang Cowpea 1009.

V. vexillata (L.)A.Rich.
Climbing; Herb; Perennial.
West Asia: North Yemen(N); Africa.
Illustration: 1244.

PSORALEAE

CULLEN Medikus

About 35 species in the Old World tropics and sub-tropics. Formerly included in *Psoralea*; for the distinctions, see Stirton (1981).

C. corylifolia (L.)Medikus [1082]
Psoralea corylifolia L. [1082].
Not climbing; Herb; Annual.
West Asia: Iraq(N), North Yemen(N), Oman(N), South Yemen(N); South Asia: India(N); South-East Asia.
Description: 1009, 1016, 1058; Illustration: 1016.
Food or drink 1016; Medicine 1009, 1016.

C. drupacea (Bunge)Stirton [1082]
Psoralea drupacea Bunge [1082].
Not climbing; Herb; Perennial.
West Asia: Afghanistan(N), Iran(N); South Asia: U.S.S.R.
Description: 1015, 1052, 1058; Illustration: 1058.

C. jaubertiana (Fenzl)Stirton [1082]
Psoralea jaubertiana Fenzl [1082].
Not climbing; Herb/Shrub; Perennial.
West Asia: Iran(N), Iraq(N); Middle East: Syria(N), Turkey(N).
Description: 1009; Illustration: 1009.
The records for Iran and Iraq are both doubtful [1009], and the taxon is not mentioned by Rechinger (1984) [1058].

C. plicata (Del.)Stirton [1082]
Psoralea plicata Del. [1082].
Not climbing; Herb/Shrub; Perennial.
West Asia: Oman(N), Qatar(N), Saudi Arabia(N); Africa; South Asia.
Description: 1001, 1014, 1058; Illustration: 1001, 1007, 1014.
Forage 1014.

ROBINIEAE

ROBINIA L.

North American trees or shrubs; one species of the four is planted in many warm temperate regions for timber and ornament, and often becomes naturalised.

R. pseudacacia L. [1009]
Not climbing; Tree; Perennial.
West Asia: Afghanistan(I), Iran(I), Iraq(I); North America.
Description: 1009; Illustration: 1009.
Domestic 1009; Environmental 1009; Food or drink 1009; Forage 1009; Medicine 1009; Wood 1009.
Black Locust 1009; False Acacia 1009.

SESBANIA Adans.

A genus of fast-growing soft-wooded herbs, shrubs or small trees, usually found in seasonally or permanently wet sites. See Gillett (1963 — ref. 1055).

S. bispinosa (Jacq.)W.Wight [1009]
Sesbania aculeata (Willd.)Pers. [1009].
Not climbing; Herb/Shrub; Perennial.
West Asia: Afghanistan(U), Iran(U), Iraq(U), Oman(U); Africa; East Asia: China(N); Indian Ocean; South America; South Asia: Pakistan(N); South-East Asia.
Description: 1009,1058.
Cover crop 1009; Fibre 1009; Weed 1009.
Hadac, Haines & Waleed W.1885 (K) is intermediate between this and S. sesban [1009].

S. cannabina (Retz.)Poiret [1009]
Not climbing; Herb; Annual.
West Asia: Iraq(I); Africa; Australasia; East Asia: China(N); South Asia: India(N); South-East Asia.
Description: 1009; Illustration: [1009].
Cover crop 1009; Environmental 1009; Fibre 1009, 1055.
Widely cultivated & exact native area obscure 1009.
The application of this name is complex; see Gillett (1963).

S. concolor J.B.Gillett [1055]
Not climbing; Herb; Annual.
West Asia: South Yemen(N).
Description: 1055.

S. grandiflora (L.)Poiret [1044]
Not climbing; Shrub/Tree; Perennial.
West Asia: North Yemen(I).

S. leptocarpa DC. [1055]
Not climbing; Herb/Shrub; Perennial.
West Asia: North Yemen(N).

S. pachycarpa DC. [1055]
S. sinuo-carinata Ali [1055].
Not climbing; Herb/Shrub; Annual.
West Asia: North Yemen(U); Africa.
Description: 1055.

subsp. **pachycarpa** [1055]
Not climbing; Herb/Shrub; Annual.
West Asia: North Yemen(U).

S. sericea (Willd.)Link [1055]
Not climbing; Herb; Annual.
West Asia: Oman(N).

S. sesban (L.)Merr. [1009]
Not climbing; Shrub/Tree; Perennial.
West Asia: Afghanistan(N); Bahrain(N); Iran(N); Iraq(N); North Yemen(N); Oman(N); Saudi Arabia(N); Africa; Australasia; South Asia: India(N), Pakistan(N); South-East Asia.
Description: 1001,1009; Illustration: 1001,1009.
Cover crop 1009; Domestic 1009; Environmental 1009; Fibre 1009; Forage 1009; Medicine 1009; Wood 1009.
Hadac, Haines & Waleed W.1885 (K) is intermediate between this & *S. bispinosa* [1009].

subsp. **sesban** [1055]
Not climbing; Shrub/Tree; Perennial.
West Asia: Iran(N);Iraq(N).

SOPHOREAE

AMMODENDRON Fischer

A small genus of about six species, all shrubs or small trees growing in the dry regions of West and Central Asia.

A. karelinii Fischer & Meyer [1240]
Ammodendron persicum Boiss. [1248].
Not climbing; Shrub/Tree; Perennial.
West Asia: Afghanistan(N), Iran(N); U.S.S.R.
Description: 1058; Illustration: 1058.
Two varieties, *karelinii* & *conollyi* [1248].

CADIA Forsskal

A small genus of shrubs, all from Madagascar except for one species in the dry highlands of north-east Africa and southern Arabia. See Van der Maesen (1970).

C. purpurea (Picciv.)Aiton [1001]
Not climbing; Shrub; Perennial.
West Asia: North Yemen(N),Oman(N),Saudi Arabia(N), South Yemen(N); Africa.
Description: 1001; Illustration: 1001.

SOPHORA L.

50 species of trees and shrubs, mainly in the north temperate zones. Some are cultivated as ornamentals; a few yield timber.

S. alopecuroides L. [1009]
Goebelia alopecuroides (L.)Boiss. [1009]; *Sophora jauberti* Spach [1011].
Not climbing; Herb; Perennial.
West Asia: Afghanistan(N), Iran(N), Iraq(N); Europe; Middle East: Turkey(N); South Asia: Pakistan(N); U.S.S.R.
Description: 1009, 1015, 1058; Illustration: 1009.
2 subspecies.
Hybrids with *S. pachycarpa* recorded [1058].

subsp. **alopecuroides** [1009,1058]
Goebelia alopecuroides (L.) Bunge var. *alopecuroides* [1052].
Not climbing; Herb; Perennial.
West Asia: Afghanistan(N), Iran(N), Iraq(N); Europe; South Asia: Pakistan(N); U.S.S.R.
Description: 1009, 1058; Illustration: 1058.

subsp. **tomentosa** (Boiss.)Bornm. [1009]
Goebelia alopecuroides (L.)Bunge var. *tomentosa* Boiss. [1011]; *Sophora alopecuroides* var. *tomentosa* (Benth.)Brenan [1009].
Not climbing; Herb; Perennial.
West Asia: Afghanistan(N), Iran(N), Iraq(N); Middle East: Turkey(N); South Asia: Pakistan(N).
Description: 1009, 1058.
Environmental 1009; Forage 1009.

S. gibbosa (DC.)Yakovlev
Ammothamnus gibbosus (DC.)Boiss. [1009].
Not climbing; Herb/Shrub; Perennial.
West Asia: Iraq(N); Middle East: Syria(N).
Description: 1009, 1058; Illustration: 1009, 1058.
Medicine 1009.

S. japonica L. [1009]
Not climbing; Tree; Perennial.
West Asia: Iran(I), Iraq(I).
Description: 1009.
Domestic 1009; Environmental 1009; Medicine 1009; Wood 1009.

S. lehmanni (Bunge)Yakovlev [1119]
Ammothamnus lehmanni Bunge [1058].
Not climbing; Shrub; Perennial.
West Asia: Afghanistan(N), Iran(N); U.S.S.R.
Description: 1015, 1058; Illustration: 1058.
Rechinger retains this taxon in *Ammothamnus* [1058].
Gunn regards *Ammothamnus* as a synonym of *Sophora* [1090].

S. mollis (Royle)Baker [1058]
Not climbing; Shrub; Perennial.
West Asia: Afghanistan(N), Iran(N); South Asia: India(N), Pakistan(N).
Description: 1015, 1058; Illustration: 1058.

subsp. griffithii (Stocks)Ali [1058]
Keyserlingia griffithii (Stocks)Boiss.; *Sophora griffithii* Stocks; *Sophora korolkowii* Koehne.
Not climbing; Shrub; Perennial.
West Asia: Afghanistan(N), Iran(N); South Asia: Pakistan(N).
Description: 1058; Illustration: 1058.

S. pachycarpa C.Meyer [1058]
Ammothamnus intermedius Kuntze [1058]; *Goebelia pachycarpa* (C.Meyer)Boiss.
Not climbing; Herb; Perennial.
West Asia: Afghanistan(N), Iran(N); U.S.S.R.
Description: 1015, 1058; Illustration: 1058.
Hybrids with *S. alopecuroides* have been recorded [1058].

THERMOPSIDEAE

ANAGYRIS L.

A Mediterranean genus with one or two species, both shrubby. All parts of the plant are toxic and have an unpleasant smell.

A.foetida L. [1009]
Not climbing; Shrub; Perennial.
West Asia: Iran(N), Iraq(N), North Yemen(N), Saudi Arabia(N); Africa; Europe; Middle East: Israel(N), Jordan(N), Lebanon(N), Sinai(N), Syria(N), Turkey(N).
Description: 1009, 1015, 1058; Illustration: 1001, 1009, 1058.
Environmental 1009; Medicine 1009, 1057; Toxins 1009.
Bean Clover 1009; Bean Trefoil 1009; Mediterranean Stinkbush 1009; Stinking Wood 1009.

TRIFOLIEAE

MEDICAGO L.

A genus of about 60 species, mainly in the Mediterranean area but very widely introduced elsewhere. Most are small herbs with coiled pods. Some are cultivated for fodder. Taxonomic treatments vary considerably in the number of species recognized.

M. arabica (L.)Hudson [1009]
M. maculata Willd. [1009].
Not climbing; Herb; Annual.
West Asia: Iran(U); Africa; Europe; Middle East: Israel(N), Lebanon(N), Syria(N), Turkey(N); U.S.S.R.
Description: 1009, 1068; Illustration: 1009.
Cover crop 1009; Forage 1009.
Spotted Medick 1009, 1067.

M. astroites (Fischer & C.Meyer)Trautv. [1069]
Trigonella astroites Fischer & C.Meyer [1069].
Not climbing; Herb; Annual.
West Asia: Iran(N), Iraq(N); Middle East: Israel(N), Jordan(N), Lebanon(N), Syria(N), Turkey(N); U.S.S.R.
Description: 1009, 1015, 1058; Illustration: 1009, 1058.

M. biflora (Griseb.) E. Small [1069]
Trigonella lunata Boiss. [1069].
Not climbing; Herb; Annual.
West Asia: Iran (N); Middle East: Syria (N); U.S.S.R.
Description: 1015, 1072.

M. brachycarpa (M.Bieb.) Moris [1069]
Trigonella brachycarpa M.Bieb. [1069].
Not climbing; Herb; Annual.
West Asia: Iran(N), Iraq(N); Middle East: Lebanon(N), Syria(N), Turkey(N); U.S.S.R.
Description: 1058; Illustration: 1058.

M. constricta Durieu [1009]
M. globosa sensu auctt. [1009].
Not climbing; Herb; Annual.
West Asia: Iran(N), Iraq(N); Europe; Middle East: Lebanon(N), Syria(N), Turkey(N); U.S.S.R.
Description: 1009, 1058, 1068; Illustration: 1009, 1058.
Globe Medick 1009.

M. coronata (L.)Bartal. [1009]
M. coronata var. *brevipedunculata* Eig [1067]; *M. coronata* var. *multiflora* Eig [1067]; *M. polymorpha* var. *coronata* L. [1067].
Not climbing; Herb; Annual.
West Asia: Iran(N), Iraq(N); Africa; Europe; Middle East: Israel(N), Lebanon(N), Sinai(N), Syria(N), Turkey(N).
Description: 1009, 1058, 1068; Illustration: 1009, 1058.
Crown Medick 1009.

M. crassipes (Boiss.)E.Small [1069]
Trigonella crassipes Boiss. [1069].
Not climbing; Herb; Annual.
West Asia: Iran(N); Middle East: Lebanon(N), Syria(N), Turkey(N); U.S.S.R.
Description: 1015, 1058; Illustration: 1058.

M. edgeworthii Sirj. [1069]
M. pubescens Aylmer [1069]; *Trigonella pubescens* Aylmer [1069].
Not climbing; Herb; Perennial.
West Asia: Afghanistan(U).
Kew material determined by E.Small, 1988.

M. falcata L. [1009]

M. borealis Grossh. [1070]; *M. difalcata* Sinsk. [1070]; *M. erecta* Kotov [1070]; *M. quasifalcata* Sinsk. [1070]; *M. tenderlensis* Opperman [1070].
Not climbing; Herb; Perennial.
West Asia: Afghanistan(N), Iran(N); East Asia: China(N); Europe; Middle East: Lebanon(N), Syria(N), Turkey(N); South Asia: India(N); U.S.S.R.
Description: 1009, 1015; Illustration: 1009.
Forage 1009.
Blue Alfalfa 1009; Sickle Alfalfa 1009; Sickle Medick 1009; Yellow Lucerne 1009.
Of economic importance as a parent of *M.* × *varia* [1009].
A component of meadow-pasture rotations in steppe and forest zones [1009].

M. fischeriana Ser. [1069]

Trigonella fischeriana (Ser.)Trautv. [1069].
Not climbing; Herb; Annual.
West Asia: Iran(N), Iraq(N); Middle East: Turkey(N); U.S.S.R.
Description: 1009, 1058; Illustration: 1009.

M. hypogaea E.Small [1071]

Factorovskya aschersoniana (Urban)Eig [1071]; *Trigonella aschersoniana* Urban [1071].
Not climbing; Herb; Annual.
West Asia: Iraq(U); Africa; Middle East: Israel(N), Lebanon(N), Sinai(N), Syria(N), Turkey(N).
Description: 1009, 1071; Illustration: 1009.
Geocarpic; the fruits develop underground following burial of the ovaries.

M. intertexta (L.)Miller [1009]

M. ciliaris (L.)All. [1009].
Not climbing; Herb; Annual.
West Asia: Iran(I), Iraq(I); Africa; Europe; Middle East: Israel(N), Lebanon(N), Syria(N), Turkey(N).
Description: 1009; Illustration: 1009.
Some authorities treat *M. ciliaris* as a good species.

M. laciniata (L.)Miller [1009]

M. aschersoniana Urban [1009].
Not climbing; Herb; Annual.
West Asia: Afghanistan(N), Bahrain(N), Iran(N), Iraq(N), Kuwait(N); Africa; Europe; Middle East: Israel(N), Jordan(N), Sinai(N), Syria(N).
Description: 1009, 1058; Illustration: 1001, 1009, 1058.
Forage 1009.
Cut-Leaved Medick 1009.
Includes two varieties, *laciniata* & *brachyantha* [1009].

M. lanigera Winkler & O.Fedtsch. [1058]

Not climbing; Herb; Annual.
West Asia: Afghanistan(N), U.S.S.R.
Description: 1058, 1068; Illustration: 1058.

M. laxispira Heyn [1058]

Not climbing; Herb; Annual.
West Asia: Iraq(N).
Description: 1058; Illustration: 1058.
Endemic to Iraq [1058].

M. littoralis Lois. [1067]

Not climbing; Herb; Annual.
West Asia: Iran(U); Africa; Europe; Middle East: Israel(N), Lebanon(N), Syria(N), Turkey(N).
Description: 1067, 1068, 1070; Illustration: 1067.
Includes the varieties *littoralis* & *inermis* Moris [1067].

M. lupulina L. [1009]
Not climbing; Herb; Perennial.
West Asia: Afghanistan(N), Iran(N), Iraq(N), North Yemen(N), Saudi Arabia(N); Africa; East Asia: China(N); Europe; Middle East: Israel(N), Lebanon(N), Syria(N), Turkey(N); South Asia: India(N), Pakistan(N), U.S.S.R.
Description: 1009, 1058; Illustration: 1001, 1009, 1058.
Cover crop 1009; Forage 1009.
Black Medick 1009, 1067.
Sometimes used to adulterate alfalfa [1009].

M. medicaginoides (Retz.)E.Small [1069]
Trigonella arcuata C.Meyer [1069]; *Trigonella cancellata* Pers. [1069]; *Trigonella striata* L.f. [1069]; *Trigonella tenuis* M.Bieb. [1069].
Not climbing; Herb; Annual.
West Asia: Iran(N); Europe; U.S.S.R.
Description: 1058; Illustration: 1058.

M. minima (L.)Bartal. [1009]
Not climbing; Herb; Annual.
West Asia: Afghanistan(N), Iran(N), Iraq(N), Saudi Arabia(N); Africa; Australasia; Europe; Middle East: Israel(N), Lebanon(N), Syria(N), Turkey(N); U.S.S.R.
Description: 1009, 1058; Illustration: 1001, 1009, 1058.
Forage 1009.
Bur-Medick 1009, 1067; Little Medick 1009.
Includes the varieties *minima* & *brevispina* Benth.

M. monantha (C.Meyer)Trautv. [1069]
Trigonella brahuica Boiss. [1011]; *Trigonella geminiflora* Bunge [1069]; *Trigonella incisa* Benth. [1069]; *Trigonella monantha* C.Meyer [1069]; *Trigonella monantha* subsp. *geminiflora* (Bunge)Rech.f.1[[1058]; *Trigonella monantha* subsp. *incisa* (Benth.)Ali [1058]; *Trigonella monantha* subsp. *noeana* (Boiss.)Huber-Mor. [1058]; *Trigonella noeana* Boiss. [1069].
Not climbing; Herb; Annual.
West Asia: Afghanistan(N), Iran(N), Iraq(N), Qatar(N), Saudi Arabia(N); Middle East; South Asia.
Description: 1009, 1058; Illustration: 1009, 1058.
Rechinger (1984) distinguishes four subspecies which are lumped together by Small, Lassen & Brookes (1987).

M. monspeliaca L. [1069]
Trigonella monspeliaca (L.)Trautv. [1069]
West Asia: Iran(N), Iraq(N); Africa; Europe; Middle East.
Description: 1009, 1015, 1058; Illustration: 1009, 1058.

M. noeana Boiss. [1009]
Not climbing; Herb; Annual.
West Asia: Iraq(N); Middle East: Turkey(N).
Description: 1009, 1068, 1070; Illustration: 1009.
Note that *Trigonella noeana* Boiss. is a synonym of *Medicago monantha* (q.v.).

M. orbicularis (L.)Bartal. [1009]
Not climbing; Herb; Annual.
West Asia: Afghanistan(N), Iran(N), Iraq(N), Oman(I);Saudi Arabia(I); Africa; Europe; Middle East: Israel(N), Jordan(N), Lebanon(N), Syria(N), Turkey(N); South Asia: India(N); U.S.S.R.
Description: 1009, 1058; Illustration: 1001, 1009, 1058.
Cover crop 1009, 1067; Forage 1009, 1067.
Button Clover 1009, 1067; Button Medick 1009.

M. orthoceras (Karelin & Kir.)Trautv. [1069]
Trigonella orthoceras Karelin & Kir. [1069]; *Trigonella orthoceras* var. *anatolica* (Boiss. & Bal.)Boiss. [1058]; *Trigonella orthoceras* var. *baylissii* Blakelock [1009].
Not climbing; Herb; Annual.
West Asia: Iran(N), Iraq(N); Middle East: Turkey(N); South Asia: Pakistan(N); U.S.S.R.
Description: 1009, 1058; Illustration: 1009, 1058.

M. persica Boiss. [1069]
Trigonella persica (Boiss.)E.Small [1069].
Not climbing; Herb; Annual.
West Asia: Iran(N).
Description: 1015, 1058; Illustration: 1058.
Endemic to Iran [1058].

M. phrygia (Boiss. & Bal.)E.Small [1069]
Trigonella aurantiaca Boiss. [1069]; *Trigonella mareschiana* Hand.-Mazz. [1058]; *Trigonella mareschina* Hand.-Mazz. [1015]; *Trigonella phrygia* Boiss. & Bal. [1069].
Not climbing; Herb; Annual.
West Asia: Iran(N), Iraq(N); Middle East: Syria(N), Turkey(N).
Description: 1009; Illustration: 1009.
Environmental 1009.

M. polyceratia L. [1069]
Trigonella polyceratia (L.)Trautv. [1069].
Not climbing; Herb; Annual.
West Asia: Iran(U), Saudi Arabia(N); Africa; Europe.
Description: 1015, 1072.

M. polychroa Grossh. [1058]
West Asia: Iran(U).
Description: 1058; Illustration: 1058.

M. polymorpha L. [1009]
M. denticulata var. *apiculata* (Willd.)Posp. [1009]; *M. hispida* Gaertner [1009]; *M. hispida* var. *apiculata* (Willd.)E.Burnat [1009]; *M. hispida* var. *denticulata* (Willd.)E.Burnat [1009]; *M. polymorpha* subsp. *denticulata* (Del.)Boiss. [1009]; *M. polymorpha* var. *apiculata* A.al-Rawi [1009]; *M. serratifolia* A.al-Rawi [1009].
Not climbing; Herb; Annual.
West Asia: Afghanistan(U), Iran(U), Iraq(U), Kuwait(U), North Yemen(U), Qatar(U), Saudi Arabia(U); Africa; East Asia: China(N); Europe; Middle East: Israel(N), Jordan(N), Lebanon(N), Syria(N), Turkey(N); South Asia: Pakistan(N): U.S.S.R.
Description: 1009, 1058; Illustration: 1009, 1058.
Cover crop 1009; Food or drink 1009; Forage 1009, 1014; Medicine 1009; Weed 1014.
A common weed in gardens, lawns & cultivated lands [1014]; it forms a light ground cover and helps to protect the soil from erosion [1009].
Includes three varieties, *polymorpha*, *brevispina* (Benth.emend.Heyn)Heyn, and *vulgaris* (Benth.) Shinn. The variety *polymorpha* is the rarest form in Iraq, and is close to var. *vulgaris* [1009].

M. radiata (L.) Boiss. [1009]
Trigonella radiata L. [1009].
Not climbing; Herb; Annual.
West Asia: Afghanistan(N), Iran(N), Iraq(N); Middle East: Israel(N), Lebanon(N), Syria(N), Turkey(N); U.S.S.R.
Description: 1009, 1058; Illustration: 1009, 1058.
Forage 1009.

M. retrorsa Boiss. [1069]
Trigonella retrorsa (Boiss.)E.Small [1069].
West Asia: Afghanistan(N).
Description: 1072.

M. rigidula (L.)All. [1009]
M. gerardi Willd. [1009].
Not climbing; Herb; Annual.
West Asia: Afghanistan(N), Iran(N), Iraq(N); Africa; Europe; Middle East: Israel(N), Jordan(N), Lebanon(N), Syria(N), Turkey(N); U.S.S.R.
Description: 1009, 1058; Illustration: 1009.
Forage 1009.
Includes the varieties *rigidula*, *submitis* (Boiss.)Heyn, *agrestis* (Ten.)Burn. and *cinerascens* (Jord.)Rouy.

M. rotata Boiss. [1009]
Not climbing; Herb; Annual.
West Asia: Iraq(N); Africa; Middle East: Israel(N), Jordan(N), Lebanon(N), Syria(N), Turkey(N).
Description: 1009; Illustration: 1009.
Weed 1009, 1070.

M. sativa L. [1009]
Not climbing; Herb; Perennial.
West Asia: Afghanistan(N), Iran(N), Iraq(N), North Yemen(N), Oman(N), Qatar(N), Saudi Arabia(N); Africa; Europe; Middle East: Israel(N), Jordan(N), Lebanon(N), Syria(N), Turkey(N); South Asia: India(N), Pakistan(N), U.S.S.R.
Description: 1009, 1058; Illustration: 1009.
Environmental 1009; Food or drink 1009; Forage 1009, 1014, 1044; Weed 1067.
Alfalfa [1009, 1067; Lucerne 1009, 1067.
Probably native in the Mediterranean region and W. Asia [1014].
A very widespread forage and seed crop; it tolerates drought & salinity [1014].

subsp. microcarpa Urban [1067]
M. coerulea Ledeb. [1067]; *M. sativa* subsp. *coerulea* (Ledeb.)Schmalh. [1005].
Not climbing; Herb; Perennial.
West Asia: Iran(N).
Description: 1067.

subsp. sativa [1067]
Not climbing; Herb; Perennial.
West Asia: Iran(N), Iraq(N).
Description: 1058; Illustration: 1058.

M. truncatula Gaertner [1009]
M. tribuloides Desr. [1009]
Not climbing; Herb; Annual.
West Asia: Iran(U), Iraq(U); Africa; Europe; Middle East: Israel(N), Jordan(N), Lebanon(N), Syria(N), Turkey(N); U.S.S.R.
Description: 1009; Illustration: 1009.
Forage 1009.
Includes the varieties *truncatula* & *longiaculeata* Urban.

M. turbinata (L.)All. [1009]
M. tuberculata Willd. [1009].
Not climbing; Herb; Annual.
West Asia: Iran(N), Iraq(N); Africa; Europe; Middle East: Israel(N), Jordan(N), Lebanon(N), Syria(N), Turkey(N).
Description: 1009; Illustration: 1009.
Includes the variety *chiotica* Urban.

M. × varia Martyn [1058]
M. ladak Vassilcz.
Not climbing; Herb; Perennial.
West Asia: Iran(N); Europe; Middle East: Syria(N), Turkey(N); U.S.S.R.
Description: 1058; Illustration: 1058.
Forage 1070.
This is the artificial hybrid, *M. sativa* × *M. falcata*.

MELILOTUS Miller

A genus of about 20 species of tall annual herbs, mainly from southern Europe and western Asia, but the natural range of some species is obscured by introductions. Some are cultivated for fodder.

M. alba Medikus [1009]
Not climbing; Herb; Perennial.
West Asia: Afghanistan(N), Iran(N), Iraq(N), North Yemen(N), Qatar(N), Saudi Arabia(N); Africa; East Asia: China(N); Europe; Middle East: Israel(N), Lebanon(N), Syria(N), Turkey(N); South Asia: India(N), Pakistan(N); U.S.S.R.
Description: 1009, 1058; Illustration: 1001, 1009, 1058.
Cover crop 1009; Forage 1009; Misc. 1009; Weed 1014.
White Melilot 1009.
An excellent bee plant; also salt-tolerant [1009].
The exact native range of this taxon is obscure, but appears to embrace most of Europe, and west and central Asia [1009].
Includes the variety *parviflora* Boiss.

M. altissima Thuill. [1011]
Not climbing; Herb; Annual.
West Asia: Afghanistan(U); Europe.
The species is not otherwise recorded from W. Asia; the records are based on specimens without fruits and must be doubtful.

M. elegans Ser. [1015]
Not climbing; Herb; Annual.
West Asia: Iran(U), North Yemen(U).
Description: 1015.

M. indica (L.)All. [1009]
Not climbing; Herb; Annual.
West Asia: Afghanistan(N), Bahrain(N), Iran(N), Iraq(N), Kuwait(N), North Yemen(N), Oman(N), Qatar(N), Saudi Arabia(N); Africa; Europe; Middle East: Israel(N), Jordan(N), Lebanon(N), Sinai(N), Syria(N), Turkey(N); South Asia: India(N), Pakistan(N); U.S.S.R.
Description: 1009, 1058; Illustration: 1001, 1009, 1058.
Cover crop 1009; Forage 1009; Medicine 1009; Weed 1009, 1014.
Mosquitos dislike the plant's odour and so *Melilotus*, especially *M. indica*, may control the spread of malaria [1009].
A winter weed of cultivated areas; also a useful cover crop in rotations as it occupies the ground for a shorter period than lucerne or berseem [1009].

M. messanensis (L.)All. [1009]
Melilotus sicula (Vitman)B.D.Jackson [1009]
Not climbing; Herb; Annual.
West Asia: Iran(N), Iraq(N); Africa; Europe; Middle East: Israel(N), Jordan(N), Lebanon(N), Sinai(N), Syria(N), Turkey(N)
Description: 1009, 1058; Illustration: 1009.
Sicilian Melilot 1009; Small Sweetclover 1009.
Only one dubious record of this species in Iraq [1009].

M. officinalis (L.)Pallas [1009]
Not climbing; Herb; Annual.
West Asia: Afghanistan(U), Iran(U), Iraq(U); Europe; Middle East: Lebanon(N), Syria(N), Turkey(N); U.S.S.R.
Description: 1009, 1058; Illustration: 1009, 1058.
Domestic 1009; Food or drink 1009; Medicine 1009; Misc.1009; Weed 1064.
Common Yellow Melilot 1009; Medicinal Sweetclover 1009; Yellow Sweetclover 1009.
Includes the two varieties, *officinalis* & *micranthus* O.Schulz.
The roots are a component of tobacco snuff & of moth repellent [1009].
The exact native area of this species is obscure [1009].

M. suaveolens Ledeb. [1011]
Not climbing; Herb; Annual.
West Asia: Afghanistan(N); East Asia; U.S.S.R.
Description: 1052.

ONONIS L.

A genus of about 70 species, mostly in southern Europe. Many are shrubs but others are small annual herbs. Species delimitation is controversial in some groups.

O. afghanica Sirj. & Rech.f.
Ononis spinosa subsp. *afghanica* (Sirj. & Rech.f.)Kitam.
Not climbing; Shrub; Perennial.
West Asia: Afghanistan(N); South Asia: Pakistan(N).
Description: 1052, 1058; Illustration: 1052, 1058.

O. biflora Desf. [1009]
Not climbing; Herb; Annual.
West Asia: Iraq(N); Africa; Europe; Middle East: Israel(N), Lebanon(N), Syria(N), Turkey(N).
Description: 1009.
Neither of the collections supposed to be from Iraq can be definitely localised there [1009].

O. chorassanica Bunge [1015]
Not climbing; Herb; Annual.
West Asia: Iran(N).
Description: 1015.

O. hirta Poiret [1005]
Not climbing; Herb; Annual.
West Asia: Iraq(N); Africa; Europe; Middle East: Israel(N), Lebanon(N), Syria(N), Turkey(N).
Description: 1009, 1058.

O. mitissima L. [1009]
Not climbing; Herb; Annual.
West Asia: Iraq(N); Africa; Europe; Middle East: Israel(N), Lebanon(N), Turkey(N).
Description: 1009, 1058; Illustration: 1009.

O. natrix L. [1009]
Not climbing; Herb/Shrub; Perennial.
West Asia: Iraq(N), Saudi Arabia(N).
Description: 1001, 1009; Illustration: 1001.
EnvironmentaL 1009.
Collenette's material (C. 447, 4370, 4528) may not be identical, and has been named as *O. natrix* sensu lato. [1001].

subsp. stenophylla (Boiss.)Sirj. [1009]
Not climbing; Herb; Perennial.
West Asia: Iraq(U); Middle East: Israel(N), Lebanon(N), Syria(N).
Description: 1009.
Environmental 1009.

O. nuristanica Podl. [1058]
Not climbing; Herb/Shrub; Perennial.
West Asia: Afghanistan(N); South Asia: Pakistan(N).
Description: 1058, 1073; Illustration: 1058.
Endemic to Afghanistan [1058].

O. pubescens L. [1009]
Not climbing; Herb; Annual.
West Asia: Iraq(N); Africa; Europe; Middle East: Israel(N), Jordan(N), Lebanon(N), Sinai(N), Syria(N), Turkey(N).
Description: 1009.
Dubiously found once in desert region of Iraq; the specimen appears to have been lost and there may have been a misidentification [1009].

O. pusilla L. [1009]
Ononis columnae All. [1009].
Not climbing; Herb; Perennial.
West Asia: Iran(N), Iraq(N); Africa; Europe; Middle East: Lebanon(N), Syria(N), Turkey(N); U.S.S.R.
Description: 1009, 1058; Illustration: 1058.

O. reclinata L. [1009]
Ononis reclinata var. *minor* Moris [1009]; *Ononis reclinata* var. *mollis* (Savi)Heldr. [1009].
Not climbing; Herb; Annual.
West Asia: Bahrain(N), Iran(N), Iraq(N), Kuwait(N), North Yemen(N), Qatar(N), Saudi Arabia(N); Africa; Europe; Middle East: Israel(N), Lebanon(N), Sinai(N), Syria(N), Turkey(N).
Description: 1009, 1058; Illustration: 1001, 1058.

O. repens L. [1074]
Not climbing; Herb/Shrub; Perennial.
West Asia: Afghanistan(U), Iran(U), Iraq(N).
Description: 1009, 1058; Illustration: 1009, 1058.

subsp. antiquorum (L.)Greuter [1074]
Ononis antiquorum L. [1074]; *Ononis spinosa* subsp. *antiquorum* Franchet [1074].
Not climbing; Herb/Shrub; Perennial.
West Asia: Afghanistan(U), Iran(U), Iraq(U).
Description: 1015.

subsp. arvensis (L.)Greuter [1074]
Ononis arvensis L. [1074]; *Ononis arvensis* subsp. *arvensis* [1074]; *Ononis hircina* Jacq. [1074].
Not climbing; Herb/Shrub; Perennial.
West Asia: Afghanistan(N), Iran(N).
Description: 1058.

subsp. leiosperma (Boiss.)Greuter [1074]
Ononis leiosperma Boiss. [1074]; *Ononis spinosa* subsp. *leiosperma* Willd. [1074].
Not climbing; Shrub; Perennial.
West Asia: Iran(N), Iraq(N); Europe; Middle East.
Description: 1009, 1058; Illustration: 1058.

subsp. spinosa Greuter [1074]
Ononis campestris Koch & Ziz [1074]; *Ononis spinosa* Benth. [1074].
Not climbing; Shrub; Perennial.
West Asia: Afghanistan(N), Iran(N), Iraq(N).
Description: 1009, 1015, 1058; Illustration: 1058.

O. serrata Forsskal [1009]
Not climbing; Herb; Annual.
West Asia: Bahrain(N), Iran(N), Iraq(N), Kuwait(N), Saudi Arabia(N), United Arab Emirates(N); Africa; Europe; Middle East: Israel(N), Lebanon(N), Sinai(N), Syria(N), Turkey(N)
Description: 1009, 1058; Illustration: 1009, 1058.
Forage 1009.
Yellow Ononis 1009.

O. sicula Guss. [1009]
Ononis sicula subsp. *microcarpa* Milne-Redh. [1009]; *Ononis viscosa* subsp. *sicula* (Guss.)Huber-Mor. [1009].
Not climbing; Herb; Annual.
West Asia: Bahrain(N), Iran(N), Iraq(N), Qatar(N), Saudi Arabia(N); Africa; Europe; Middle East: Israel(N), Jordan(N), Lebanon(N), Sinai(N), Syria(N), Turkey(N).
Description: 1009, 1058; Illustration: 1009, 1058.

O. viscosa L. [1009]
Not climbing; Herb; Annual.
West Asia: Iran(N), Iraq(N); Africa; Europe; Middle East: Israel(N), Jordan(N), Lebanon(N), Syria(N), Turkey(N).
Description: 1009.
Recorded from "Arabia" [1009].

subsp. **breviflora** (DC.)Nyman [1009]
: Not climbing; Herb; Annual.
West Asia: Iran(N), Iraq(N).
Description: 1009, 1058; Illustration: 1009, 1058.
Recorded from "Arabia" [1058].

TRIFOLIUM L.

A genus of about 240 species of annual and perennial herbs, mainly in the north temperate regions but also in tropical upland and montane regions. Very important as fodder and as bee plants; many species have been extensively planted outside their natural range. See Zohary & Heller (1984).

T. alexandrinum L. [1009]
: *Trifolium maritimum* sensu Rawi [1009].
Not climbing; Herb; Annual.
West Asia: Iran(I), Iraq(I); Africa; Europe; Middle East: Israel(N), Lebanon(N), Syria(N), Turkey(N).
Description: 1009, 1058, 1077; Illustration: 1058, 1077.
Cover crop 1009; Forage 1009, 1077.
Berseem Clover 1009; Egyptian Clover 1009.

T. alpestre L. [1058]
: Not climbing; Herb; Perennial.
West Asia: Iran(N); Europe; Middle East: Turkey(N); U.S.S.R.
Description: 1058, 1077; Illustration: 1077.
3 varieties of which only var. *alpestre* occurs in West Asia [1077].

T. ambiguum M.Bieb. [1009]
: *Trifolium ambiguum* var. *majus* Hossain [1009]; *Trifolium ruprechtii* Tomasch. & Fed. [1077].
Not climbing; Herb; Perennial.
West Asia: Iran(N), Iraq(N); Europe; Middle East: Turkey(N); U.S.S.R.
Description: 1009, 1058, 1077; Illustration: 1058, 1077.
Forage 1009, 1077.

T. angustifolium L. [1009]
: Not climbing; Herb; Annual.
West Asia: Iran(N), Iraq(N); Africa; Europe; Middle East: Israel(N), Lebanon(N), Syria(N), Turkey(N); U.S.S.R.
Description: 1009, 1058, 1077; Illustration: 1058, 1077.

T. arvense L. [1009]
: Not climbing; Herb; Annual.
West Asia: Afghanistan(N), Iran(N), Iraq(N), Saudi Arabia(N); Africa; Europe; Middle East: Israel(N), Lebanon(N), Syria(N), Turkey(N); U.S.S.R.
Description: 1009, 1058, 1077; Illustration: 1001, 1058, 1077.
Forage 1009; Medicine 1009.
Haresfoot Clover 1009; Haresfoot Trefoil 1009.

T. aureum Pollich [1058]
: *Trifolium agrarium* L.
Not climbing; Herb; Perennial.
West Asia: Iran(N); Europe; Middle East.
Description: 1058, 1077; Illustration: 1077.
T. agrarium L. is a *nomen ambiguum*; *T. aureum* & *T. campestre* are both involved.

T. badium Schreber [1009]
Trifolium badium subsp. *rivulare* A.Rich. [1077]; *Trifolium badium* subsp. *rytidosemium* Benth. [1058, 1077]; *Trifolium rivulare* Boiss. & Bal. [1077]; *Trifolium rytidosemium* Boiss. & Hohen. [1077]; *Trifolium rytidosemium* var. *rivulare* (Boiss. & Hohen.)Zoh. [1058, 1077]; *Trifolium rytidosemium* var. *rytidosemium* Boiss. & Hohen. [1058].
Not climbing; Herb; Perennial.
West Asia: Iran(N), Iraq(N); Europe; Middle East: Turkey(N); U.S.S.R.
Description: 1009, 1058, 1077; Illustration: 1058, 1077.
There is much debate over the status of *T. rivulare* & *T. rytidosemium*.

T. boissieri Soyer-Will. & Godron [1009]
Not climbing; Herb; Annual.
West Asia: Iraq(N); Europe; Middle East: Israel(N), Lebanon(N), Syria(N), Turkey(N).
Description: 1009, 1077; Illustration: 1009, 1077.
Very close to *T. grandiflorum* [1058].

T. bullatum Boiss. & Hausskn. [1058]
Trifolium tomentosum subsp. *bullatum* (Boiss. & Hausskn.)Oppenh. [1058, 1077].
Not climbing; Herb; Annual.
West Asia: Iran(N), Iraq(N); Middle East: Israel(N), Jordan(N), Syria(N), Turkey(N).
Description: 1058, 1077; Illustration: 1058, 1077.
Includes 2 vars.; *bullatum* & *macrosphaerium* Zohary, both recorded from W. Asia [1077].
Townsend (1974) [1009] reduces this taxon to a variety of *T. tomentosum*.

T. campestre Schreber [1009]
Trifolium glaucescens Hausskn. [1009]; *Trifolium procumbens* sensu auctt. [1009]; *Trifolium pumilum* Hossain [1077].
Not climbing; Herb; Annual.
West Asia: Afghanistan(N), Iran(N), Iraq(N), Saudi Arabia(N); Africa; Europe; Middle East: Israel(N), Jordan(N), Lebanon(N), Syria(N), U.S.S.R.
Description: 1009, 1058; Illustration: 1001, 1058.
Forage 1009.

T. canescens Willd. [1058]
Not climbing; Herb; Perennial.
West Asia: Iran(N); Middle East: Turkey(N); U.S.S.R.
Description: 1015, 1058, 1077; Illustration: 1058, 1077.

T. caucasicum Tausch [1058]
Trifolium pannonicum sensu auctt. [1058].
Not climbing; Herb; Perennial.
West Asia: Iran(N); Middle East: Syria(N), Turkey(N); U.S.S.R.
Description: 1058, 1077; Illustration: 1058, 1077.

T. cherleri L. [1009]
Not climbing; Herb; Annual.
West Asia: Iran(N), Iraq(N); Africa; Europe; Middle East: Israel(N), Jordan(N), Lebanon(N), Syria(N), Turkey(N).
Description: 1009, 1058, 1077; Illustration: 1009, 1058, 1077.

T. clusii Godron & Gren. [1058]
Not climbing; Herb; Annual.
West Asia: Iran(N), Iraq(N); Africa.
Description: 1058, 1077; Illustration: 1058, 1077.
Zohary & Heller (1984) recognise 3 varieties, *clusii*, *gossypinum*, & *kahiricum*, of which only var. *kahiricum* recorded from W. Asia [1058].
Townsend [1009] considers that *T. clusii* may be *T. resupinatum* × *T. tomentosum*.

T. dasyurum C.Presl [1077]
Trifolium formosum Urv. [1077]; *Trifolium pamphylicum* sensu Rawi [1009].
Not climbing; Herb; Annual.
West Asia: Iran(N), Iraq(N); Africa; Europe; Middle East: Israel(N), Jordan(N), Lebanon(N), Syria(N), Turkey(N).
Description: 1009, 1058; Illustration: 1009, 1058.

T. dubium Sibth. [1077]
Trifolium filiforme L. [1058].
Not climbing; Herb; Annual.
West Asia: Iraq(N); Europe; Middle East: Turkey(N); U.S.S.R.
Description: 1077; Illustration: 1077.

T. echinatum M.Bieb. [1058]
Trifolium supinum Savi [1058].
Not climbing; Herb; Annual.
West Asia: Iran(N), Iraq(N); Europe; Middle East: Israel(N), Lebanon(N), Syria(N), Turkey(N); U.S.S.R.
Description: 1009, 1058, 1077; Illustration: 1058, 1077.
Forage 1009.

T. fragiferum L. [1009]
Trifolium fragiferum var. *pulchellum* Lange [1009].
Not climbing; Herb; Perennial.
West Asia: Afghanistan(N), Iran(N), Iraq(N), North Yemen(N), Saudi Arabia(N), South Yemen(N); Africa; Europe; Middle East: Israel(N), Jordan(N), Lebanon(N), Syria(N), Turkey(N); U.S.S.R.
Description: 1009, 1058, 1077; Illustration: 1001, 1058, 1077.
Strawberry Clover 1009; Strawberry Trefoil 1009.

T. glomeratum L. [1058]
Not climbing; Herb; Annual.
West Asia: Iran(N); Africa; Europe; Middle East: Turkey(N).
Description: 1015, 1058, 1077; Illustration: 1058, 1077.

T. grandiflorum Schreber [1009]
Trifolium speciosum Willd. [1009].
Not climbing; Herb; Annual.
West Asia: Iran(N), Iraq(N).
Description: 1009, 1058, 1077; Illustration: 1009, 1058, 1077.

T. haussknechtii Boiss. [1009]
Not climbing; Herb; Annual.
West Asia: Iraq(N); Middle East: Syria(N), Turkey(N).
Description: 1009, 1058, 1077; Illustration: 1077.
Includes the varieties *candollei* (Post.)Hossain & *haussknechtii* [1077].

T. hirtum All. [1009]
Not climbing; Herb; Annual.
West Asia: Iran(N), Iraq(N); Africa; Europe; Middle East: Lebanon(N), Syria(N), Turkey(N); U.S.S.R.
Description: 1009, 1058, 1077; Illustration: 1009, 1058, 1077.
Forage 1009.

T. hybridum L. [1009]
Trifolium elegans Ser. [1009].
Not climbing; Herb; Perennial.
West Asia: Iran(N), Iraq(N); Africa; Europe; U.S.S.R.
Description: 1009, 1058, 1077; Illustration: 1058.
Forage 1009, 1077; Misc. 1009.
Alsike Clover 1009.
The exact native range obscured by frequent cultivation & subsequent naturalization in almost the whole temperate region in both hemispheres [1009].
Includes the varieties *hybridum*, *elegans* (Savi)Boiss., and *anatolicum* (Boiss.)Boiss [1058].
Prized as a honey plant [1009].

T. lappaceum L. [1009]
Trifolium lappaceum var. *rhodense* (Pampan.)Rech.f. [1058, 1077].
Not climbing; Herb; Annual.
West Asia: Iran(N), Iraq(N), Kuwait(N); Africa; Europe; Middle East: Israel(N), Lebanon(N), Sinai(N), Syria(N), Turkey(N); U.S.S.R.
Description: 1009, 1058, 1077; Illustration: 1077.
Includes the varieties *lappaceum* & *zoharyi* Eig [1077].

T. leucanthum M.Bieb. [1009]
Trifolium leucanthum var. *declinatum* Boiss. [1009].
Not climbing; Herb; Annual.
West Asia: Iran(N), Iraq(N); Africa; Europe; Middle East: Israel(N), Syria(N), Turkey(N); U.S.S.R.
Description: 1009, 1058, 1077; Illustration: 1058, 1077.

T. mazanderanicum Rech.f. [1058]
Not climbing; Herb; Perennial.
West Asia: Iran(N)
Description: 1015, 1058, 1077; Illustration: 1058, 1077.
Endemic to Iran [1058].

T. medium L. [1058].
Trifolium medium subsp. *banaticum* (Heuffel)Hendrych [1077]; *Trifolium medium* subsp. *majus* Baker [1077].
Not climbing; Herb; Perennial.
West Asia: Iran(N); Europe; Middle East: Turkey(N); U.S.S.R.
Description: 1015, 1058, 1077; Illustration: 1058, 1077.
Four varieties, *medium, pseudomedium, banaticum* & *sarosiense*; only var. *medium* in West Asia [1077].

T. micranthum Viv. [1058]
Not climbing; Herb; Annual.
West Asia: Iran(N); Africa; Europe; Middle East.
Description: 1058, 1077; Illustration: 1058, 1077.

T. montanum L. [1058]
Not climbing; Herb; Perennial.
West Asia: Iran(N); Europe; Middle East: Turkey(N); U.S.S.R.
Description: 1015, 1058, 1077.
Three subspecies, *montanum, humboldtianum,* & *rupestre*, of which only subsp. *humboldtianum* occurs in West Asia [1077].

subsp. humboldtianum (A.Brown & Asch.)Hossain
Trifolium humboldtianum A.Brown & Asch. [1058].
Not climbing; Herb; Perennial.
West Asia: Iran(N); Middle East: Turkey(N); U.S.S.R.
Description: 1058, 1077; Illustration: 1058, 1077.

T. nigrescens Viv. [1009]
Not climbing; Herb; Annual.
West Asia: Iran(N), Iraq(N); Africa; Europe; Middle East: Israel(N), Jordan(N), Lebanon(N), Syria(N), Turkey(N); U.S.S.R.
Description: 1009, 1058, 1077; Illustration: 1009, 1058.
Two subspecies, *nigrescens* & *petrisavii* [1077].

subsp. nigrescens [1009]
Not climbing; Herb; Annual.
West Asia: Iraq(N); Africa; Europe; Middle East: Turkey(N).
Description: 1009, 1077.

subsp. petrisavii (Clem.)Holmboe [1009]
Trifolium meneghinianum Clem. [1009].; *Trifolium petrisavii* Clem. [1077].
Not climbing; Herb; Annual.
West Asia: Iran(N), Iraq(N); Europe; Middle East: Israel(N), Lebanon(N), Syria(N), Turkey(N); U.S.S.R.
Description: 1009, 1058, 1077; Illustration: 1058, 1077.
Hossain, in Flora of Turkey, distinguishes a var. *meneghinianum*.
Not recognised by Townsend & Guest (1974) or Zohary & Heller (1984).

T. ochroleucum Hudson [1058]
Not climbing; Herb; Perennial.
West Asia: Iran(N); Africa; Europe; Middle East: Turkey(N); U.S.S.R.
Description: 1015, 1058, 1077; Illustration: 1077.

T. pauciflorum Urv.
    Not climbing; Herb; Annual.
    West Asia: Iraq(U).
    Description: 1077.

T. phleoides Willd. [1058]
    Not climbing; Herb; Annual.
    West Asia: Iran(N); Africa; Europe; Middle East: Turkey(N); U.S.S.R.
    Description: 1058, 1077; Illustration: 1077.

T. physodes M.Bieb. [1009]
    Trifolium amani Dinsm. [1058]; *Trifolium amani* var. *glabrescens* Thieb. [1058]; *Trifolium physodes* var. *psilocalyx* Boiss. [1009]; *Trifolium rechingeri* Vassilcz. [1077].
    Not climbing; Herb; Perennial.
    West Asia: Afghanistan(N), Iran(N), Iraq(N); Europe; Middle East: Lebanon(N), Syria(N), Turkey(N); U.S.S.R.
    Description: 1009, 1058, 1077; Illustration: 1009, 1058, 1077.
    Forage 1009.
    Two varieties, *physodes* & *psilocalyx* [1077].

T. pilulare Boiss. [1009]
    Trifolium pilulare var. *longipedunculatum* M.Evenari [1077].
    Not climbing; Herb; Annual.
    West Asia: Iran(N), Iraq(N); Europe; Middle East: Israel(N), Jordan(N), Lebanon(N), Syria(N), Turkey(N).
    Description: 1009, 1058, 1077; Illustration: 1009, 1058, 1077.

T. pratense L. [1009]
    Trifolium fontanum Bobrov [1058].
    Not climbing; Herb; Perennial.
    West Asia: Afghanistan(N), Iran(N), Iraq(N); Africa; Europe; Middle East: Israel(N), Lebanon(N), Turkey(N); U.S.S.R.
    Description: 1009, 1058, 1077; Illustration: 1058, 1077.
    Cover crop 1009; Domestic 1009; Forage 1009; Medicine 1009.
    Red Clover 1009.
    Zohary & Heller (1984) [1077] provisionally recognise 6 vars., of which the vars. *pratense*, *americanum* & *sativum* recorded from W. Asia [1077].

T. purpureum Lois. [1009]
    Not climbing; Herb; Annual.
    West Asia: Iran(N), Iraq(N); Africa; Europe; Middle East: Israel(N), Jordan(N), Lebanon(N), Sinai(N), Syria(N), Turkey(N); U.S.S.R.
    Description: 1009, 1058, 1077; Illustration: 1009, 1058, 1077.
    Forage 1009.
    Purlpe Clover 1009.
    Three varieties, *purpureum*, *desvauxii* & *pamphylicum*, of which only var. *purpureum* definitely occurs in W. Asia [1077].

T. radicosum Boiss. & Hohen. [1058]
    Trifolium guestii Blakelock [1077]; *Trifolium tumens* var. *rechingeri* Zoh. & Heller [1077].
    Not climbing; Herb; Perennial.
    West Asia: Iran(N), Iraq(N); Middle East: Turkey(N).
    Description: 1009, 1058, 1077; Illustration: 1058, 1077.
    2 varieties, *radicosum* & *guestii* [1077]. Townsend (1974) regards *T. guestii* as a good species.

T. repens L. [1009]
    Trifolium pachypodium O.Schwarz [1058].
    Not climbing; Herb; Perennial.
    West Asia: Afghanistan(N), Iran(N), Iraq(N); Africa; Europe; Middle East: Israel(N), Lebanon(N), Syria(N), Turkey(N); South Asia: Pakistan(N), U.S.S.R.
    Description: 1009, 1058, 1077; Illustration: 1058, 1077.
    Cover crop 1077; Forage 1009, 1077; Medicine 1009, 1077.
    Dutch Clover 1009, 1077; White Clover 1009.
    Zohary & Heller (1984) recognise eight varieties [1058].

T. resupinatum L. [1009]

Trifolium resupinatum var. *microcephalum* Zoh. [1009].
Not climbing; Herb; Annual.
West Asia: Afghanistan(U), Iran(U), Iraq(U), Kuwait(U), Qatar(U); Africa; Europe; Middle East: Israel(N), Jordan(N), Lebanon(N), Sinai(N), Syria(N); South Asia: India(N), Pakistan(N), U.S.S.R.
Description: 1009, 1058, 1077; Illustration: 1009, 1058, 1077.
Forage 1009.
Its exact native area has been obscured by cultivation, but may be in the southeast Mediterranean; tolerates some salinity [1009].
3 varieties, *resupinatum, majus, microcephalum*, all occurring in West Asia [1077].

T. retusum L. [1009]

Trifolium parviflorum Ehrh. [1009].
Not climbing; Herb; Annual.
West Asia: Iraq(N); Africa; Europe; Middle East: Turkey(N); U.S.S.R.
Description: 1009, 1058, 1077; Illustration: 1009, 1077.

T. scabrum L. [1009]

Trifolium compactum Post [1009].
Not climbing; Herb; Annual.
West Asia: Afghanistan(N), Iran(N), Iraq(N); Africa; Europe; Middle East: Israel(N), Lebanon(N), Syria(N), Turkey(N); U.S.S.R.
Description: 1009, 1058, 1077; Illustration: 1058, 1077.
Forage 1009.
Rough Clover 1009.
Zohary & Heller (1984) treat *T. compactum* as a synonym of *T. lucanicum*.
Townsend regarded *T. lucanicum* as closely related to *T. scabrum* and doubtfully distinct [1009].

T. scutatum Boiss. [1077]

Not climbing; Herb; Annual.
West Asia: Iraq(N); Africa; Middle East: Israel(N), Lebanon(N), Syria(N), Turkey(N).
Description: 1077; Illustration: 1077.
Not mentioned by Townsend (1974) for Iraq.
Very similar to *T. plebeium* [1077].

T. sebastianii Savi [1058]

Not climbing; Herb; Annual.
West Asia: Iran(N); Europe; Middle East: Turkey(N); U.S.S.R.
Description: 1058, 1077; Illustration: 1077.

T. semipilosum Fresen. [1077]

Not climbing; Herb; Perennial.
West Asia: North Yemen(N); Africa
Description: 1077, 1079; Illustration: 1077.
Two varieties, *semipilosum* & *glabrescens* Gillett [1077], both recorded from North Yemen [1077].

T. spadiceum L. [1058]

Not climbing; Herb; Perennial.
West Asia: Iran(N); Europe; Middle East: Turkey(N); U.S.S.R.
Description: 1058, 1077; Illustration: 1058, 1077.

T. spumosum L. [1009]

Not climbing; Herb; Annual.
West Asia: Iran(N), Iraq(N); Africa; Europe; Middle East: Israel(N), Lebanon(N), Syria(N), Turkey(N); U.S.S.R.
Description: 1009, 1058, 1077; Illustration: 1058, 1077.
Forage 1009.

T. squamosum L. [1058]

Not climbing; Herb; Annual.
West Asia: Iran(N); Africa; Europe; Middle East: Lebanon(N), Syria(N), Turkey(N); U.S.S.R.
Description: 1058, 1077; Illustration: 1077.

T. stellatum L. [1009]
Not climbing; Herb; Annual.
West Asia: Iran(N), Iraq(N); Africa; Europe; Middle East: Israel(N), Jordan(N), Lebanon(N), Sinai(N), Syria(N), Turkey(N); U.S.S.R.
Description: 1009, 1058, 1077; Illustration: 1009, 1077.
Forage 1009.
Star Clover 1009; Starry Clover 1009.
Four varieties, *stellatum, adpressum, longiflorum* & *xanthinum*; all but *longiflorum* occur in West Asia [1058].

T. striatum L. [1009]
Not climbing; Herb; Annual.
West Asia: Iran(N), Iraq(N); Africa; Europe; Middle East: Turkey(N); U.S.S.R.
Description: 1009, 1058, 1077; Illustration: 1009, 1058, 1077.

T. subterraneum L. [1009]
Not climbing; Herb; Annual.
West Asia: Iran(N), Iraq(N); Africa; Europe; Middle East: Israel(N), Lebanon(N), Syria(N), Turkey(N); U.S.S.R.
Description: 1009, 1058, 1077; Illustration: 1058, 1077.
Cover crop 1009.
Subterranean Clover 1009.
Infraspecific classification disputed; eight varieties in two subspecies (*subterraneum* and *brachycalycinum*). West Asian material is subssp. *subterraneum* [1077].

subsp. subterraneum [1058]
Not climbing; Herb; Annual.
West Asia: Iran(N); Africa; Europe; Middle East: Israel(N), Syria(N), Turkey(N); U.S.S.R.
Description: 1058, 1077; Illustration: 1077.
2 varieties, *subterraneum* & *brachycladum*, of which only var. *subterraneum* occurs in West Asia [1077].

T. suffocatum L. [1058]
Not climbing; Herb; Annual.
West Asia: Iran(N); Africa; Europe; Middle East: Israel(N), Lebanon(N), Syria(N), Turkey(N); U.S.S.R.
Description: 1058, 1077; Illustration: 1058, 1077.

T. sylvaticum Lois. [1009]
Trifolium bonnevillei Mout. [1058]; *Trifolium smyrnaeum* Boiss. [1009].
Not climbing; Herb; Annual.
West Asia: Iraq(N); Europe; Middle East: Syria(N), Turkey(N).
Description: 1009, 1058, 1077; Illustration: 1009, 1077.
Superficially very similar to *T. striatum* & *T. scabrum* [1009].
The illustrations in Townsend [1009] & Zohary & Heller [1077] are remarkably different.

T. tomentosum L. [1009]
Not climbing; Herb; Annual.
West Asia: Iran(N), Iraq(N), Saudi Arabia(N); Africa; Europe; Middle East: Israel(N), Jordan(N), Lebanon(N), Sinai(N), Syria(N), Turkey(N); U.S.S.R.
Description: 1009, 1058, 1077; Illustration: 1001, 1009, 1077.
Forage 1009.
Six varieties, of which two, *tomentosa* & *curvisepalum* are recorded for West Asia [1077].

T. trichocephalum M.Bieb. [1058]
Trifolium trichocephalum var. *lonchophyllum* Hossain [1058]; *Trifolium trichocephalum* var. *macrophyllum* Hossain [1058, 1077].
Not climbing; Herb; Perennial.
West Asia: Iran(N); Middle East: Turkey(N); U.S.S.R.
Description: 1015, 1058, 1077; Illustration: 1077.

T. tumens M.Bieb. [1011]
Trifolium curvisepalum Tackh. [1058]; *Trifolium talyschense* Chalilov [1058].
Not climbing; Herb; Perennial.
West Asia: Afghanistan(N), Iran(N); Middle East: Turkey(N); U.S.S.R.
Description: 1015, 1058, 1077; Illustration: 1058, 1077.
Two varieties, *tumens* & *talyschense* [1077]; Kitamura (1960) also recognises a var. *majus* from Afghanistan.

T. vavilovii Eig [1058]
Not climbing; Herb; Annual.
West Asia: Iran(N), Iraq(N); Middle East: Israel(N), Syria(N), Turkey(N).
Description: 1058, 1077; Illustration: 1058, 1077.

TRIGONELLA L.

About 80 species, mostly herbs, found mainly in southern Europe, north Africa and central Asia. Many species are desert annuals. One species (*T. foenum-graecum*) is widely cultivated for its spicey seeds.

T. adscendens (Nevski)Afan. & Gontch. [1058]
Melissitus adscendens (Nevski)Ikonn. [1058].
Not climbing; Perennial.
West Asia: Afghanistan(N).
Description: 1058; Illustration: 1058.

T. afghanica Vassilcz. [1011]
Not climbing; Shrub; Perennial.
West Asia: Afghanistan(N).
Description: 1058; Illustration: 1058.
Endemic to Afghanistan [1058].

T. anguina Del. [1009]
Not climbing; Herb; Annual.
West Asia: Bahrain(N), Iran(N), Iraq(N), Kuwait(N), Qatar(N), Saudi Arabia(N); Africa; Middle East: Jordan(N), Sinai(N); South Asia: Pakistan(N).
Description: 1009, 1058; Illustration: 1001, 1009, 1058.
Forage 1009, 1014.

T. aphanoneura Rech.f. [1058]
Not climbing; Shrub; Perennial.
West Asia: Iran(N).
Description: 1058; Illustration: 1058.
Endemic to Iran [1058].

T. bactriana Vassilcz. [1011].
T. emodi sensu Aitch. [1011].
Not climbing; Shrub; Perennial.
West Asia: Afghanistan(N).
Description: 1058; Illustration: 1058.
Endemic to Afghanistan [1058].

T. badachschanica Afan. [1058]
Not climbing; Perennial.
West Asia: Afghanistan(N).
Description: 1058.

T. calliceras Fischer [1058]
Not climbing; Herb; Annual.
West Asia: Iran(N); U.S.S.R.
Description: 1015, 1058, 1072; Illustration: 1058.

T. capitata Boiss. [1058]
   Not climbing; Herb; Annual.
   West Asia: Iran(N); Middle East: Turkey(N); U.S.S.R.
   Description: 1015, 1058, 1072; Illustration: 1058.

T. cedretorum Vassilcz. [1075]
   Not climbing; Herb; Perennial.
   West Asia: Afghanistan(N).
   Description: 1075.

T. coelesyriaca Boiss. [1009]
   Not climbing; Herb; Annual.
   West Asia: Iran(N), Iraq(N); Middle East: Israel(N), Jordan(N), Lebanon(N), Syria(N), Turkey(N).
   Description: 1009, 1058, 1072; Illustration: 1009.
   Environmental 1009.

T. coerulescens (M.Bieb.)Hal. [1009]
   T. azurea C.Meyer [1009].
   Not climbing; Herb; Annual.
   West Asia: Iran(N), Iraq(N); Africa; Europe; Middle East: Syria(N), Turkey(N); U.S.S.R.
   Description: 1009, 1058, 1072; Illustration: 1009, 1058.

T. corniculata L. [1011]
   Not climbing; Herb; Annual.
   West Asia: Afghanistan(N); Europe; South Asia: India(N), Pakistan(N).
   Description: 1015, 1072.

T. cylindracea Desv. [1072]
   T. cylindrica Desv. [1007].
   Not climbing; Herb; Annual.
   West Asia: Iran(N), Saudi Arabia(N); Middle East: Israel(N), Syria(N).
   Description: 1007, 1072.
   Migahid's [1007] name is a spelling mistake.

T. edelbergii (Sirj. & Rech.f.)Rech.f. [1058]
   T. lipskyi var. *edelbergii* Sirj. & Rech.f. [1058].
   West Asia: Afghanistan(N).
   Description: 1058; Illustration: 1058.
   Endemic to Afghanistan [1058].

T. elliptica Boiss. [1009]
   T. disperma Vassilcz. [1072].
   Not climbing; Herb/Shrub; Perennial.
   West Asia: Afghanistan(N), Iran(N), Iraq(N).
   Description: 1009, 1015, 1058; Illustration: 1009, 1058.
   Includes the varieties *brachycarpa* Bornm. & *latialata* Bornm.

T. emodi Benth. [1011]
   T. cachemiriana Cambess. [1011]; *T. griffithii* Boiss. [1072].
   Not climbing; Shrub; Perennial.
   West Asia: Afghanistan(N); South Asia: India(N), Pakistan(N).
   Description: 1052, 1058, 1072; Illustration: 1058.

T. filipes Boiss. [1015]
   Not climbing; Herb; Annual.
   West Asia: Iran(N), Iraq(N); Middle East: Israel(N), Jordan(N), Lebanon(N), Syria(N), Turkey(N).
   Description: 1009, 1058, 1072; Illustration: 1009, 1058.

T. foenum-graecum L. [1009]
T. foenum-graecum var. *haussknechtii* Sirj. [1009].
Not climbing; Herb; Annual.
West Asia: Afghanistan(N), Iran(N), Iraq(N), Kuwait(N), North Yemen(N), Saudi Arabia(N), South Yemen(N); Africa; Europe; Middle East: Israel(I), Jordan(I), Lebanon(I), Sinai(I), Syria(I); U.S.S.R.
Description: 1007, 1009, 1058; Illustration: 1009, 1058.
Domestic 1009; Food or drink 1009, 1011; Forage 1009; Medicine 1009, 1011.
Common Fenugreek 1009; Fenugreek 1009.

T. freitagii Vassilcz. [1058]
Perennial.
West Asia: Afghanistan(N).
Description: 1058; Illustration: 1058.
Endemic to Afghanistan [1058].

T. gharuensis Rech.f. [1011]
Not climbing; Shrub; Perennial.
West Asia: Afghanistan(N); South Asia: Pakistan(N).
Description: 1052, 1058; Illustration: 1052, 1058.

T. gracilis Benth. [1011]
West Asia: Afghanistan(N); South Asia: India(N), Pakistan(N).
Description: 1072.

T. grandiflora Bunge [1011]
Not climbing; Herb; Annual.
West Asia: Afghanistan(N); Europe; U.S.S.R.
Description: 1058, 1072; Illustration: 1058.

T. hamosa L. [1009]
Not climbing; Herb; Annual.
West Asia: Afghanistan(N), Bahrain(N), Iran(N), Iraq(N), Kuwait(N), Oman(N), Qatar(N), Saudi Arabia(N); Middle East: Israel(N), Jordan(N), Lebanon(N), Sinai(N), Syria(N); South Asia: India(N).
Description: 1009, 1072; Illustration: 1001, 1007.
Medicine 1009.

subsp. **hamosa** [1009]
Not climbing; Herb; Annual.
West Asia: Iraq(N).
Description: 1009.
Egyptian Fenugreek 1009.
May not in fact occur in Iraq; occurs in "Arabia" [1009].

T. heratensis Rech.f. [1058]
Not climbing; Herb/Shrub; Perennial.
West Asia: Afghanistan(N).
Description: 1058; Illustration: 1058.
Endemic to Afghanistan [1058].

T. ionantha Rech.f. [1058]
Not climbing; Herb; Perennial.
West Asia: Afghanistan(N).
Description: 1058; Illustration: 1058.

T. kafirniganica Vassilcz. [1058]
Not climbing; Herb; Perennial.
West Asia: Afghanistan(N).
Description: 1058; Illustration: 1058.

T. koeiei Sirj. & Rech.f. [1058]
Not climbing; Herb/Shrub; Perennial.
West Asia: Afghanistan(N).
Description: 1052, 1058; Illustration: 1058.
Endemic to Afghanistan [1058].

T. kotschyi Boiss. [1072]
Not climbing; Herb; Annual.
West Asia: Iran(N); Middle East: Israel(N), Syria(N), Turkey(N).
Description: 1015, 1072.

T. laciniata L. [1007]
Not climbing; Herb; Annual.
West Asia: Saudi Arabia(N); Africa; Europe.
Description: 1007, 1072.

T. latialata (Bornm.)Vassilcz. [1058]
T. latealata (Bornm.)Vassilcz.
Not climbing; Shrub; Perennial.
West Asia: Iran(N).
Description: 1058; Illustration: 1058.
Endemic to Iran [1058].

T. laxiflora Aitch. & Baker [1058]
Perennial.
West Asia: Afghanistan(N).
Description: 1058, 1072.
Endemic to Afghanistan [1058].

T. laxissima Vassilcz. [1076]
Not climbing; Herb; Perennial.
West Asia: Afghanistan(N).
Description: 1076.

T. lipskyi Sirj. [1011]
Perennial.
West Asia: Afghanistan(N).
Description: 1052, 1072.
Includes the variety *edelbergii* Sirj. & Rech.f [1011].

T. macroglochin Durieu [1058]
Not climbing; Herb; Annual.
West Asia: Iran(N), Iraq(N); Middle East: Syria(N), Turkey(N).
Description: 1058; Illustration: 1058.

T. marco-poloi Vassilcz. [1075]
Not climbing; Herb; Perennial.
West Asia: Afghanistan(N).
Description: 1075.

T. pamirica Boriss [1058]
Not climbing; Shrub; Perennial.
West Asia: Afghanistan(N); U.S.S.R.
Description: 1058.

T. podlechii Vassilcz. [1075]
Not climbing; Herb; Perennial.
West Asia: Afghanistan(N).
Description: 1075.

T. pycnotricha Rech.f. [1058]
Not climbing; Herb/Shrub; Perennial.
West Asia: Afghanistan(N).
Description: 1058; Illustration: 1058.
Endemic to Afghanistan [1058].

T. rechingeri Vassilcz. [1075]
Not climbing; Herb; Perennial.
West Asia: Afghanistan(N).
Description: 1075.

T. salangensis Vassilcz. [1058]
Not climbing; Herb; Perennial.
West Asia: Afghanistan(N).
Description: 1058; Illustration: 1058.
Endemic to Afghanistan [1058].

T. spicata Sibth. & J.Smith [1009]
Not climbing; Herb; Annual.
West Asia: Iran(N), Iraq(N); Europe; Middle East: Israel(N), Lebanon(N), Syria(N), Turkey(N); U.S.S.R.
Description: 1009, 1058, 1072; Illustration: 1009, 1058.

T. spruneriana Boiss. [1009]
T. torulosa Griseb. [1009].
Not climbing; Herb; Annual.
West Asia: Iran(N), Iraq(N); Europe; Middle East: Israel(N), Lebanon(N), Syria(N), Turkey(N); U.S.S.R.
Description: 1009, 1058, 1072; Illustration: 1009, 1058.

T. stellata Forsskal [1009]
Not climbing; Herb; Annual.
West Asia: Bahrain(N), Iran(N), Iraq(N), Kuwait(N), Oman(N), Qatar(N), Saudi Arabia(N); Africa; Middle East: Israel(N), Jordan(N), Sinai(N), Syria(N).
Description: 1009, 1058, 1072; Illustration: 1001, 1009, 1058.
Forage 1014.

T. stenocarpa Rech.f. [1058]
Not climbing; Shrub; Perennial.
West Asia: Iran(N).
Description: 1058; Illustration: 1058.
Endemic to Iran [1058].

T. strangulata Boiss. [1009]
Not climbing; Herb; Annual.
West Asia: Iran(N), Iraq(N); Europe; Middle East: Lebanon(N), Syria(N), Turkey(N); U.S.S.R.
Description: 1009, 1058, 1072; Illustration: 1009, 1058.

T. subenervis Rech.f. [1058]
Not climbing; Shrub; Perennial.
West Asia: Iran(N).
Description: 1058; Illustration: 1058.
Endemic to Iran [1058].

T. teheranica Bornm. [1058]
T. theranica sensu Parsa
Not climbing; Herb; Annual.
West Asia: Iran(N).
Description: 1058, 1072; Illustration: 1058.
Endemic to Iran [1058].

T. turkmena Popov [1058]
Not climbing; Herb; Perennial.
West Asia: Iran(N).
Description: 1058; Illustration: 1058.
Endemic to Iran [1058].

T. uncata Boiss. & Noe [1058]
T. hamosa subsp. *uncata* (Boiss. & Noe)C.Towns. [1058].
Not climbing; Herb; Annual.
West Asia: Afghanistan(N), Iran(N), Iraq(N), Saudi Arabia(N), United Arab Emirates(N); Middle East: Israel(N), Jordan(N), Lebanon(N), Syria(N); South Asia: India(N), Pakistan(N).
Description: 1009, 1058, 1072; Illustration: 1001, 1009, 1058.
Forage 1009.

T. uncinata Banks & Sol. [1009]
T. brevidentata Blakelock [1009].
Not climbing; Herb; Annual.
West Asia: Iran(N), Iraq(N); Middle East: Syria(N), Turkey(N).
Description: 1009, 1058; Illustration: 1009, 1058.

T. verae Sirj. [1058]
Not climbing; Herb; Annual.
West Asia: Afghanistan(N).
Description: 1058, 1072; Illustration: 1058.

T. xeromorpha Rech.f. [1058]
T. griffithii var. *micrantha* Gilli [1058].
Not climbing; Herb; Perennial.
West Asia: Afghanistan(N).
Description: 1058; Illustration: 1058.
Endemic to Afghanistan [1058].

VICIEAE

LATHYRUS L.

About 150 species, mainly in the north temperate regions, but a few also in upland Africa and temperate South America. Some are cultivated as fodder and human food, but there can be toxicity problems.

L. annuus L. [1009]
Climbing; Herb; Annual.
West Asia: Afghanistan(N), Iran(N), Iraq(N); Africa; Europe; Middle East: Israel(N), Jordan(N), Lebanon(N), Syria(N), Turkey(N); U.S.S.R.
Description: 1009, 1015, 1129; Illustration: 1009, 1129.
Food or drink 1009.

L. aphaca L. [1009]
L. apaca L. [1011].
Climbing; Herb; Annual.
West Asia: Afghanistan(N), Iran(N), Iraq(N), Saudi Arabia(N); Africa; Europe; Middle East: Israel(N), Jordan(N), Lebanon(N), Sinai(N), Syria(N), Turkey(N); South Asia: India(N); U.S.S.R.
Description: 1009, 1015, 1129; Illustration: 1001.
Forage 1009; Medicine 1009.
Yellow Vetch 1009; Yellow Vetchling 1009.

L. bijugus Boiss. & Noe (provisional) [1009]
Not climbing; Herb; Annual.
West Asia: Iraq(N).
Description: 1009.
Endemic to Iraq; known only from the type, collected in 1850 [1009].
Recorded in error from Iran in Index Kewensis.

L. boissieri Sirj. [1009]
  L. nervosus (Boiss.)Boiss. [1009].
  Not climbing; Herb; Perennial.
  West Asia: Iran(N), Iraq(N); Middle East: Turkey(N).
  Description: 1009, 1015, 1129; Illustration: 1129.

L. cassius Boiss. [1009]
  Climbing; Herb; Annual.
  West Asia: Iran(N), Iraq(N); Middle East: Israel(N), Lebanon(N), Syria(N), Turkey(N).
  Description: 1009, 1129; Illustration: 1009, 1129.

L. chloranthus Boiss. [1009]
  Climbing or not; Herb; Annual.
  West Asia: Iran(N), Iraq(N); Middle East: Turkey(N); U.S.S.R.
  Description: 1009, 1015, 1129; Illustration: 1009, 1129.
  Townsend cites the authority as Boiss. & Bal.[1009].

L. cicera L. [1009]
  Climbing; Herb; Annual.
  West Asia: Iran(N), Iraq(N), Saudi Arabia(N); Africa; Europe; Middle East: Israel(N), Lebanon(N), Syria(N), Turkey(N); South Asia: Pakistan(N); U.S.S.R.
  Description: 1009, 1015, 1129; Illustration: 1001, 1129.
  Food or drink 1009; Forage 1009.
  Parsa mentions a var. *ciliatus* Lipsky [1015].
  Rechinger (1979) believes that *L. szowitsii* may be synonymous.[1129].

L. cyaneus (Steven)K.Koch [1129]
  Orobus cyaneus Steven [1015].
  Not climbing; Herb; Perennial.
  West Asia: Iran(N); Middle East: Turkey(N); U.S.S.R.
  Description: 1015, 1129; Illustration: 1129.

L. gorgoni Parl. [1009]
  L. amoenus Fenzl [1009]; *L. amoenus* var. *lineatus* Dinsm. [1132]; *L. cicera* var. *lineatus* Post [1009].
  Climbing; Herb; Annual.
  West Asia: Iran(N), Iraq(N); Africa; Europe; Middle East: Israel(N), Jordan(N), Lebanon(N), Sinai(N), Syria(N), Turkey(N)
  Description: 1009, 1015, 1129; Illustration: 1009, 1129.
  Townsend includes three varieties, *gorgoni*, *lineatus* & *pilosus* [1009].

L. hirsutus L. [1009]
  Climbing; Herb; Annual.
  West Asia: Afghanistan(N), Iran(N), Iraq(N); Africa; Europe; Middle East: Israel(N), Lebanon(N), Turkey(N); U.S.S.R.
  Description: 1009, 1015, 1129; Illustration: 1129.

L. humilis (Ser.)Sprengel [1131]
  L. altaicus var. *humilis* Ledeb. [1129]; *L. humilis* DC. [1131].
  Not climbing; Herb; Perennial.
  West Asia: Afghanistan(N); East Asia: China(N); Europe; South Asia: Pakistan(N); U.S.S.R.
  Description: 1129.

L. inconspicuus L. [1009]
  L. erectus Lagasca [1009]; *L. erectus* var. *stenophyllus* Boiss. [1009]; *L. hispidulus* Boiss. [1009].
  Climbing; Herb; Annual.
  West Asia: Afghanistan(N), Iran(N), Iraq(N); Europe; Middle East: Lebanon(N), Syria(N), Turkey(N); South Asia: Pakistan(N); U.S.S.R.
  Description: 1009, 1015, 1129; Illustration: 1009, 1129.
  Includes the varieties *inconspicuus* & *stenophyllus* (Boiss.)Rech.f.; intermediates between these occur in Iraq, Iran & Afghanistan.[1129].

L. incurvus (Roth)Willd. [1129]
  Climbing; Herb; Perennial.
  West Asia: Iran(N); U.S.S.R.
  Description: 1015, 1129; Illustration: 1129.

L. latifolius L. [1129]
Climbing; Herb; Perennial.
West Asia: Afghanistan(I), Iran(I); Africa; Europe; U.S.S.R.

L. laxiflorus (Desf.)Kuntze [1129]
L. laxiflora sensu Rech.f. [1129].
Not climbing; Herb; Perennial.
West Asia: Iran(N); Middle East: Turkey(N); U.S.S.R.
Description: 1129; Illustration: 1129.
2 subspecies, *laxiflorus* & *angustifolius* of which only *laxiflorus* in West Asia.

subsp. **laxiflorus** [1129]
Not climbing; Herb; Perennial.
West Asia: Iran(N); U.S.S.R.
Description: 1129.

L. marmoratus Boiss. & Blanche [1130]
Climbing; Herb; Annual.
West Asia: Iraq(N).
Not mentioned by Townsend (1974) or Rechinger (1979).

L. mulkak Lipsky [1129]
Climbing; Herb; Perennial.
West Asia: Afghanistan(N); U.S.S.R.
Description: 1129; Illustration: 1129.

L. neurolobus Boiss. & Heldr. [1015]
Not climbing; Herb; Perennial.
West Asia: Iran(I).
Description: 1015.
Iranian material sterile; the determination must be doubtful [1015].

L. nissolia L. [1009]
Not climbing; Herb; Annual.
West Asia: Iran(N), Iraq(N); Africa; Europe; Middle East: Israel(N), Lebanon(N), Syria(N), Turkey(N); U.S.S.R.
Description: 1009, 1129; Illustration: 1129.

L. nivalis Hand.-Mazz. [1131]
Not climbing; Herb; Perennial.
West Asia: Iran(N), Iraq(N).

L. ochrus (L.)DC. [1009]
Climbing; Herb; Annual.
West Asia: Iraq(N); Africa; Europe; Middle East: Israel(N), Lebanon(N), Syria(N), Turkey(N).
Description: 1009.

L. odoratus L. [1009]
Climbing; Herb; Annual.
West Asia: Iraq(I).
Description: 1009.
Domestic 1009; Environmental 1009.
Sweet Pea 1009.

L. pannonicus (Jacq.)Garcke [1131]
Not climbing; Herb; Perennial.
West Asia: Iran(N).
Determination of Kew material uncertain; locality also not definite [1102].

subsp. **multijugus** (Ledeb.)Bassler [1130]
L. multijugus (Ledeb.)Czefr. [1130]; *L. pannonicus* var. *paucijugus* (Ledeb.)Schischkin [1130]; *Orobus albus* var. *multijugus* (Ledeb.)Ledeb. [1130]; *Orobus lacteus* var. *multijugus* Ledeb. [1130].
Not climbing; Herb; Perennial.
West Asia: Iran(N).
Determination of Kew material uncertain; locality also not definite [1102].

L. pratensis L. [1009]
Climbing; Herb; Perennial.
West Asia: Afghanistan(N), Iran(N), Iraq(N), Saudi Arabia(N); Africa; East Asia: China(N); Europe; Middle East: Lebanon(N), Turkey(N); South Asia: India(N); U.S.S.R.
Description: 1009, 1124, 1129; Illustration: 1001, 1129.
Environmental 1009; Forage 1009.
Meadow Pea 1009.

L. pseudocicera Pampan. [1129]
L. pseudo-cicera Pampan. [1129].
Climbing; Herb; Annual.
West Asia: Iran(N), Iraq(N).
Description: 1129; Illustration: 1129.
Townsend (1974) regards this as a synonym of *L. gorgoni* var. *lineatus* [1009].

L. roseus Steven [1129]
Not climbing; Herb; Perennial.
West Asia: Iran(N); Middle East: Turkey(N); U.S.S.R.
Description: 1015, 1129; Illustration: 1129.

L. rotundifolius Willd. [1009]
Climbing; Herb; Perennial.
West Asia: Iran(N), Iraq(N); Middle East: Turkey(N); U.S.S.R.
Description: 1009, 1015, 1129; Illustration: 1129.
Environmental 1009.
Rechinger (1979) recognises 2 subspp., rotundifolius & miniatus [1129], which are not recognised by Townsend (1974).
Only subsp. *miniatus* in W. Asia; *rotundifolius* occurs in the Crimea.

subsp. **miniatus** (Steven)P.Davis [1129]
L. miniatus Steven [1009].
Climbing; Herb; Perennial.
West Asia: Iran(N), Iraq(N); U.S.S.R.
Description: 1129; Illustration: 1129.
Environmental 1009.

L. sativus L. [1009]
L. sativus var. *stenophyllus* Boiss. [1009].
Climbing; Herb; Annual.
West Asia: Afghanistan(U), Iran(U), Iraq(U); Africa; Europe; Middle East: South Asia: U.S.S.R.
Description: 1009, 1015, 1129; Illustration: 1009, 1129.
Cover crop 1009; Food or drink 1009; Forage 1009.
Chickling Pea 1009.

L. saxatilis (Vent.)Vis. [1001]
Not climbing; Herb; Annual.
West Asia: Saudi Arabia(N).
Description: 1001; Illustration: 1001.

L. sphaericus Retz. [1009]
Climbing; Herb; Annual.
West Asia: Afghanistan(N), Iran(N), Iraq(N); Africa; Europe; Middle East: U.S.S.R.
Description: 1009, 1124, 1129; Illustration: 1009, 1129.

L. szowitsii Boiss. (provisional) [1129]
Climbing; Herb; Annual.
West Asia: Iran(N).
Description: 1015.
Mentioned by Rechinger (1979) as doubtfully distinct from *L. cicera* [1129].

L. tuberosus L. [1129]
Climbing; Herb; Perennial.
West Asia: Iran(N), Iraq(N); Europe; U.S.S.R.
Description: 1015, 1129; Illustration: 1129.

L. vernus (L.)Bernh. [1129]
>Not climbing; Herb; Perennial.
>West Asia: Iran(N); Europe; U.S.S.R.
>Description: 1015, 1129; Illustration: 1129.

>subsp. **vernus** [1130]
>>Not climbing; Herb; Perennial.
>>West Asia: Iran(N)

L. vinealis Boiss. & Noe [1009]
>Climbing; Herb; Annual.
>West Asia: Iran(N), Iraq(N); Middle East: Lebanon(N), Syria(N), Turkey(N)
>Description: 1009, 1129; Illustration: 1129.

LENS Miller

L. culinaris Medikus [1130]
>*Ervum lens* L. [1009]; *L. culinare* Medikus [1102]; *L. esculenta* Moench [1009].
>Climbing; Herb; Annual.
>West Asia: Afghanistan(N), Iran(N), Iraq(N), North Yemen(N), Saudi Arabia(N); Africa; East Asia; Europe; Middle East: Israel(N), Jordan(N), Lebanon(N), Syria(N), Turkey(N); South Asia: India(N); U.S.S.R.
>Description: 1009, 1015, 1129; Illustration: 1009, 1129.
>Food or drink 1009; Forage 1009; Medicine 1009.
>Common Lentil 1009; Lentil 1009.
>Very similar to *L. orientalis*.

L. orientalis (Boiss.)Hand.-Mazz. [1129]
>*Ervum orientale* Boiss. [1009]; *L. cyanea* (Boiss. & Hohen.)Alef. [1130]; *L. cyaneum* (Boiss. & Hohen.)Alef. [1015]; *L. pygmaea* Grossh. [1129].
>Climbing; Herb; Annual.
>West Asia: Afghanistan(N), Iran(N), Iraq(N), Saudi Arabia(N); Europe; Middle East: Lebanon(N), Syria(N), Turkey(N); U.S.S.R.
>Description: 1009, 1015, 1129; Illustration: 1001, 1009, 1129.
>Forage 1009.
>Rechinger (1979) accepts *L. cyanea* as a good species [1129].

PISUM L.

Annual herbs; probably only two species but there is some disagreement about the rank of the various taxa. Extensively cultivated as a food and fodder crop.

P. sativum L. [1009]
>Climbing; Herb; Annual.
>West Asia: Afghanistan(N), Iran(N), Iraq(N), Saudi Arabia(U); Africa; East Asia: China(I); Middle East: Israel(N), Lebanon(N), Syria(N), Turkey(N); South Asia: India(I), Pakistan(I); U.S.S.R.
>Description: 1009, 1129.
>Common Pea 1009.
>Originated in W. or C. Asia, but exact native area obscured by cultivation [1009].

>subsp. **elatius** (M.Bieb.)Asch. & Graebner [1009]
>>*P. arvense* sensu Guest [1009]; *P. elatius* M.Bieb. [1009]; *P. humile* Boiss. & Noe [1009]; *P. sativum* subsp. *humile* (Holmboe)Greuter [1009].
>>Climbing; Herb; Annual.
>>West Asia: Afghanistan(N), Iran(N), Iraq(N); Europe; Middle East: U.S.S.R.
>>Description: 1009, 1015, 1129; Illustration: 1009.
>>Food or drink 1009; Forage 1009.
>>Some authorities recognise a third subsp., *P. sativum* subsp. *humile*.
>>Var. *pumilio* is used for food and fodder [1009]

subsp. **sativum** [1009]
Climbing; Herb; Annual.
West Asia: Afghanistan(N), Iran(N), Iraq(N), Saudi Arabia(U); Europe.
Description: 1001, 1009, 1129; Illustration: 1001.
Food or drink 1009; Forage 1009.
Field Pea 1009; Garden Pea 1009; Green Pea 1009.

VAVILOVIA Al.Fed.

A monospecific genus, sometimes included in *Pisum*. A herb, occurring in western Asia only.

V. formosa (Steven)Al.Fed. [1130]
Alophotropis formosa (Steven)Grossh. [1129]; *Pisum formosum* (Steven)Alef. [1130]; *Vicia aucheri* Boiss. [1129].
Not climbing; Herb; Perennial.
West Asia: Iran(N), Iraq(N); Middle East: Lebanon(N), Turkey(N); U.S.S.R.
Description: 1009, 1015, 1129.
Townsend distinguishes a var. *pubescens* of *Pisum formosum* [1009].

VICIA L.

140 species, many climbing herbs, mainly in the north temperate regions but also in upland tropical areas. Some species (particularly *V. faba*) are very important food crops; others provide fodder.

V. afghanica Chrtkova-Zert. [1129]
Climbing; Herb; Perennial.
West Asia: Afghanistan(N).
Description: 1129; Illustration: 1129.
Endemic to Afghanistan [1009].

V. aintabensis Boiss. [1129]
Climbing; Herb; Annual.
West Asia: Iran(N), Iraq(N); Middle East: Syria(N).
Description: 1129; Illustration: 1129.

V. anatolica Turrill [1129]
V. hajastana Grossh. [1129].
Climbing; Herb; Annual.
West Asia: Iran(N); Middle East: Turkey(N); U.S.S.R.
Description: 1015, 1129; Illustration: 1129.

V. articulata Hornem. [1009]
V. monantha (L.)Desf. [1102].
Climbing; Herb; Annual.
West Asia: Iraq(U); Africa; Europe.
Description: 1009.

V. assyriaca Boiss. [1129]
Climbing; Herb; Annual.
West Asia: Iran(N), Iraq(N); Middle East.
Description: 1009, 1129; Illustration: 1009, 1129.

V. bithynica (L.)L. [1130]
Lathyrus bithynicus L. [1130].
Climbing; Herb; Annual.
West Asia: Afghanistan(U).
Not mentioned by Rechinger (1979) [1129].

V. canescens Labill. [1009]
V. rechingeri Chrtkova-Zert. [1130].
Climbing; Herb; Perennial.
West Asia: Iran(N), Iraq(N); Europe; Middle East: Lebanon(N), Turkey(N); U.S.S.R.
Description: 1009, 1129; Illustration: 1129.

subsp. gregaria (Boiss. & Heldr.)P.Davis [1009]
V. armena Boiss.; *V. canescens* subsp. *latistipulata* P.Davis [1009]; *V. gregaria* Boiss. & Heldr. [1009]; *V. persica* var. *stenophylla* Boiss. [1009]; *V. variegata* var. *stenophylla* (Boiss.)C.Towns. [1009].
Climbing; Herb; Perennial.
West Asia: Iran(N), Iraq(N); Middle East: Turkey(N); U.S.S.R.
Description: 1009, 1129; Illustration: 1129.
Forage 1009; Toxins 1009.
Subsp. *latistipulata* accepted by Vicieae Database Project (1986) [1130].
V. armena treated as a good species in Rechinger (1979) [1129].

subsp. variegata (Willd.)P.Davis [1130]
V. akhmaganica Kazarjan [1130]; *V. aucheri* Jaub. & Spach [1130]; *V. persica* Boiss. [1130]; *V. variegata* Willd. [1130]; *V. variegata* var. *aucheri* (Jaub. & Spach)Bornm. [1130]; *V. variegata* subsp.*bornmulleri* Radzhi [1130].
Climbing; Herb; Perennial.
West Asia: Afghanistan(N), Iran(N), Iraq(N); U.S.S.R.
Description: 1015, 1129; Illustration: 1129.
V. variegata ssp. *bornmulleri* placed as synonym of *V. persica* by Rechinger 1979 [1129].
V. aucheri, V. persica & *V. akhmaganica* treated as good species by Rechinger 1979 [1129].

V. cappadocica Boiss. & Bal. [1129]
V. paucijuga (Trautv.)B.Fedtsch. [1130].
Climbing; Herb; Perennial.
West Asia: Iran(N); U.S.S.R.
Description: 1015, 1129; Illustration: 1129.

V. ciceroidea Boiss. [1009]
Climbing; Herb; Perennial.
West Asia: Iran(N), Iraq(N); U.S.S.R.
Description: 1009, 1015, 1129; Illustration: 1129.
Parsa mentions a var. *multijuga* Boiss. [1015].

V. cracca L. [1129]
Climbing; Herb; Perennial.
West Asia: Afghanistan(N), Iran(N); Europe; U.S.S.R.
Description: 1129; Illustration: 1129.

subsp. stenophylla Velen. [1130]
V. variabilis Freyn & Sint. [1130].
Climbing; Herb; Perennial.
West Asia: Afghanistan(N), Iran(N), Iraq(N); South Asia: Pakistan(N); U.S.S.R.
Description: 1129; Illustration: 1129.
V. variabilis treated as a good species by Rechinger (1979) [1129].

subsp. tenuifolia (Roth)Gaudin [1009]
V. boissieri Freyn [1130]; *V. tenuifolia* Roth [1130].
Not climbing; Herb; Perennial.
West Asia: Afghanistan(N), Iran(N), Iraq(N); Europe; U.S.S.R.
Description: 1009, 1015; Illustration: 1009.
Forage 1009.
Parsa lists three vars. of *V. tenuifolia* [1015].

V. crocea (Desf.)B.Fedtsch. [1130]
V. aurantia (M.Bieb.)Boiss. [1130].
Not climbing; Herb; Perennial.
West Asia: Iran(N); U.S.S.R.
Description: 1129; Illustration: 1129.

V. cuspidata Boiss. [1129]
Climbing; Herb; Annual.
West Asia: Iran(N); Europe; Middle East: Lebanon(N), Turkey(N).
Description: 1129; Illustration: 1129.

V. cypria Unger & Kotschy [1129]
Climbing; Herb; Perennial.
West Asia: Iraq(U); Middle East: Lebanon(N), Turkey(N).
Description: 1129; Illustration: 1129.
Not recorded from West Asia by Rechinger (1979) [1129].

V. ervilia (L.)Willd. [1009]
Not climbing; Herb; Annual.
West Asia: Afghanistan(N), Iran(N), Iraq(N); Africa; Europe; Middle East: Israel(N), Jordan(N), Lebanon(N), Syria(N), Turkey(N); U.S.S.R.
Description: 1009, 1015, 1129; Illustration: 1129.
Cover crop 1009; Forage 1009.
Bitter Vetch 1009.

V. faba L. [1009]
Faba vulgaris Moench [1009].
Not climbing; Herb; Annual.
West Asia: Afghanistan(I), Iran(I), Iraq(I), Kuwait(I), North Yemen(I); Africa; Europe; Middle East: South Asia: U.S.S.R.
Description: 1009, 1015, 1129; Illustration: 1129.
Cover crop 1009; Food or drink 1009; Forage 1009; Medicine 1009.
Broad Bean 1009; Field Bean 1009; Horse Bean 1009; Small Bean 1009; Tick Bean 1009.

V. galeata Boiss. [1009]
Climbing; Herb; Annual.
West Asia: Iran(N), Iraq(N); Middle East: Israel(N), Lebanon(N), Turkey(N).
Description: 1009; Illustration: 1009.

V. grandiflora Scop. [1129]
Climbing; Herb; Annual.
West Asia: Iran(N), Iraq(N); Europe; Middle East: Turkey(N); U.S.S.R.
Description: 1015, 1129; Illustration: 1129.

V. hirsuta (L.)Gray [1009]
Climbing; Herb; Annual.
West Asia: Afghanistan(N), Iran(N), Iraq(N); Africa; Australasia; East Asia; Europe; Middle East: Lebanon(N), Syria(N), Turkey(N); South Asia; U.S.S.R.
Description: 1009, 1015, 1129; Illustration: 1129.
Forage 1009.

V. hybrida L. [1009]
Climbing; Herb; Annual.
West Asia: Afghanistan(N), Iran(N), Iraq(N); Africa; Europe; Middle East: Israel(N), Jordan(N), Lebanon(N), Syria(N), Turkey(N); U.S.S.R.
Description: 1009, 1015, 1129; Illustration: 1129.

V. hyrcanica Fischer & C.Meyer [1009]
Climbing; Herb; Annual.
West Asia: Afghanistan(N), Iran(N), Iraq(N); U.S.S.R.
Description: 1009, 1015, 1129; Illustration: 1129.
No definite record from Iraq, according to Townsend (1974) [1009].

V. iranica Boiss. [1129]
Not climbing; Herb; Perennial.
West Asia: Afghanistan(N), Iran(N).
Description: 1015, 1129; Illustration: 1129.

V. koeieana Rech.f. [1009]
Anatropostylia koeieana (Rech.f.)Kupicha [1130]; *V. blakelockiana* C.Towns. [1009].
Climbing; Herb; Annual.
West Asia: Iran(N), Iraq(N); U.S.S.R.
Description: 1009, 1129; Illustration: 1009, 1129.

V. kotschyana Boiss. [1129]
Climbing; Herb; Perennial.
West Asia: Iran(N).
Description: 1015, 1129; Illustration: 1129.
Endemic to Iran [1129].

V. lathyroides L. [1009]
Not climbing; Herb; Annual.
West Asia: Iran(N), Iraq(N); Africa; Europe; Middle East: Israel(N), Jordan(N), Lebanon(N), Syria(N), Turkey(N); U.S.S.R.
Description: 1009, 1129; Illustration: 1129.
Spring Vetch 1009.

V. laxiflora Brot. [1130]
V. gracilis Lois. [1130]; *V. tenuissima* (M.Bieb.)Schinz & Thell. [1130].
Climbing; Herb; Annual.
West Asia: Iran(N).
Description: 1015.
Not mentioned by Rechinger (1979) [1129].

V. lutea L. [1129]
Climbing; Herb; Annual.
West Asia: Iran(N); Africa; Europe; Middle East: Turkey(N); U.S.S.R.
Description: 1015, 1129; Illustration: 1129.

subsp. vestita (Boiss.)Rouy [1130]
V. hirta DC. [1130].
Climbing; Herb; Annual.
West Asia: Iran(N); Europe; Middle East: Turkey(N).
Description: 1129; Illustration: 1129.
Rechinger (1979) treats *V. hirta* as a good species [1129].

V. michauxii Sprengel [1009]
V. aintabensis sensu auctt. [1009].
Climbing; Herb; Annual.
West Asia: Afghanistan(N), Iran(N), Iraq(N); Middle East: Israel(N), Lebanon(N), Syria(N), Turkey(N); U.S.S.R.
Description: 1009, 1015, 1129; Illustration: 1129.

V. mollis Boiss. [1009].
Climbing; Herb; Annual.
West Asia: Iran(N), Iraq(N); Middle East: Lebanon(N), Syria(N), Turkey(N).
Description: 1009, 1129; Illustration: 1009, 1129.

V. monantha Retz. [1130]
Climbing; Herb; Annual.
West Asia: Afghanistan(N), Bahrain(N), Iran(N), Iraq(N), Kuwait(N), Qatar(N), Saudi Arabia(N); Africa; Europe; Middle East: Israel(N), Jordan(N), Lebanon(N), Syria(N), Turkey(N); South Asia: Pakistan(N); U.S.S.R.
Description: 1009, 1015, 1129; Illustration: 1129.
Forage 1009, 1014.

subsp. monantha [1130]
V. calcarata sensu auctt. [1009]; *V. cinerea* M.Bieb. [1130]; *V. griffithii* Baker [1130]; *V. monantha* subsp. *cinerea* (M.Bieb.)Maire [1129]; *V. monantha* var. *cinerea* (M.Bieb.)Boiss. [1130].
Climbing; Herb; Annual.
West Asia: Afghanistan(N), Iran(N), Iraq(N); South Asia: Pakistan(N).
Description: 1015, 1129; Illustration: 1129.
Rechinger (1979) treats *V. cinerea* as a good species [1129].

V. montbretii Fischer & C.Meyer [1130]

Ervum kotschyanum Boiss. [1009, 1130]; *Lens kotschyanus* (Boiss.)Nab. [1009, 1130]; *Lens montbretii* (Fischer & C.Meyer)P.Davis & Plitm. [1130]; *V. bombycina* Post.
Not climbing; Herb; Annual.
West Asia: Iraq(N); Middle East: Turkey(N).
Description: 1009, 1129; Illustration: 1129.

V. multijuga (Boiss.)Rech.f. [1129]

V. ciceroidea var. *multijuga* Boiss. [1129].
Climbing; Herb; Perennial.
West Asia: Iran(N); U.S.S.R.
Description: 1129; Illustration: 1129.

V. narbonensis L. [1009]

Climbing; Herb; Annual.
West Asia: Afghanistan(N), Iran(N), Iraq(N); Africa; Europe; Middle East: Israel(N), Jordan(N), Lebanon(N), Sinai(N), Syria(N), Turkey(N); South Asia: India(N); U.S.S.R.
Description: 1009, 1015, 1129; Illustration: 1129.
Food or drink 1009; Forage 1009.

V. noeana Boiss. [1102]

Climbing; Herb; Annual.
West Asia: Iran(N), Iraq(N); Middle East: Syria(N), Turkey(N).
Description: 1102.
Fl.Turkey maintains this as a good species, and recognizes two varieties, *noeana* & *megalodonta* [1102].
Townsend placed this as a synonym of *V. assyriaca* (q.v.) [1009].

V. palaestina Boiss. [1009]

Climbing; Herb; Annual.
West Asia: Iraq(N); Middle East: Israel(N), Lebanon(N), Syria(N), Turkey(N).
Description: 1009, 1129; Illustration: 1129.

V. pannonica Crantz [1129]

Climbing; Herb; Annual.
West Asia: Iran(N); Africa; Europe.
Description: 1015, 1129; Illustration: 1129.
Forage 1129.

V. peregrina L. [1009]

Climbing; Herb; Annual.
West Asia: Afghanistan(N), Iran(N), Iraq(N), Saudi Arabia(N); Africa; Europe; Middle East: Israel(N), Jordan(N), Lebanon(N), Sinai(N), Syria(N); South Asia: Pakistan(N); U.S.S.R.
Description: 1009, 1015, 1129; Illustration: 1129.
Includes 2 varieties, *latifolia* & *peregrina*.

V. rigidula Royle [1130]

V. pseudocassubica Rech.f. [1130].
Climbing; Herb; Perennial.
West Asia: Afghanistan(N), Iran(N); South Asia: India(N), Pakistan(N).
Description: 1052, 1129; Illustration: 1052, 1129.

V. sativa L. [1009]

Climbing; Herb; Annual.
West Asia: Afghanistan(N), Iran(N), Iraq(N), North Yemen(U), Oman(U), Saudi Arabia(N); Africa; Europe; Middle East: Israel(N); South Asia: Pakistan(N); U.S.S.R.
Description: 1009, 1015, 1129; Illustration: 1001, 1129.
Cover crop 1009; Forage 1009; Weed 1009.
Common Vetch 1009.

subsp. amphicarpa (Boiss.)Asch. & Graebner

V. amphicarpa Lam. [1009]; *V. sativa* var. *amphicarpa* (Boiss.)Alef. [1009].
Climbing; Herb; Annual.
West Asia: Afghanistan(N), Iran(N), Iraq(N); Africa; Middle East: Israel(N); U.S.S.R.
Description: 1129; Illustration: 1129.
Rechinger treats this as a good species [1009, 1129].

subsp. **nigra** (L.)Ehrh. [1001]
V. angustifolia L. [1130].
Climbing; Herb; Annual.
West Asia: Afghanistan(N), Iran(N), Iraq(N), Saudi Arabia(N); Africa; Europe; Middle East: South Asia: U.S.S.R.
Description: 1001, 1015, 1129; Illustration: 1001, 1129.

subsp. **sativa** [1130]
Climbing; Herb; Annual.
West Asia: Afghanistan(N), Iran(N), Iraq(N).
Description: 1009; Illustration: 1129.

V. sericocarpa Fenzl [1009]
Climbing; Herb; Annual.
West Asia: Iran(N), Iraq(N); Middle East: Israel(N), Jordan(N), Lebanon(N), Syria(N), Turkey(N)
Description: 1009, 1015, 1129; Illustration: 1009, 1129.

V. serratifolia Jacq. [1130]
V. narbonensis var. *serratifolia* (Jacq.)Ser. [1130]; *V. narbonensis* subsp. *serratifolia* (Jacq.)Arcang. [1130].
Not climbing; Herb; Annual.
West Asia: Iran(N), Iraq(N).
Description: 1009, 1129.
Townsend (1974) & Rechinger (1979) both treat this as a variety of *V. narbonensis* [1009, 1129].

V. setidens Bornm. [1015]
West Asia: Iran(U).
Description: 1015.
Not mentioned by Rechinger (1979) [1129], or Vicieae Database Project (1986) [1130].

V. singarensis Boiss. & Hausskn. [1009]
Climbing; Herb; Annual.
West Asia: Iran(N), Iraq(N).
Description: 1009, 1015; Illustration: 1009.
Endemic to Iraq; known only from the type [1009].
Townsend ignores Parsa's Iran record (1974) [1009, 1015].

V. sojakii Chrtkova-Zert. [1129]
Not climbing; Herb; Perennial.
West Asia: Iran(N).
Description: 1129; Illustration: 1129.
Endemic to Iran; known only from the type.

V. subvillosa (Ledeb.)Boiss. [1130]
Lathyrus subvillosus (Ledeb.)Aitch. & Hemsley [1129]; *Orobus triflorus* Stapf.
Not climbing; Herb; Perennial.
West Asia: Afghanistan(N), Iran(N).
Description: 1015, 1129; Illustration: 1129.
Includes var. *stenophylla* Popov [1052].
The type of *Orobus triflorus* has been annotated as this by Chrtkova-Zertova.

V. tetrasperma (L.)Schreber [1015, 1130]
Climbing; Herb; Annual.
West Asia: Afghanistan(N), Iran(N); Africa; East Asia; Europe; U.S.S.R.
Description: 1015, 1129; Illustration: 1129.

V. truncatula M.Bieb. [1129]
Not climbing; Herb; Perennial.
West Asia: Iran(N); Europe; Middle East: Turkey(N); U.S.S.R.
Description: 1015, 1129; Illustration: 1129.

V. venulosa Boiss. & Hohen. [1129]
Not climbing; Herb; Perennial.
West Asia: Iran(N); U.S.S.R.
Description: 1015, 1129; Illustration: 1129.

V. villosa Roth [1009]
Climbing; Herb; Perennial.
West Asia: Afghanistan(N), Iran(N), Iraq(N); Africa; Europe; Middle East: Israel(N), Lebanon(N), Syria(N), Turkey(N); U.S.S.R.
Description: 1009, 1015, 1129; Illustration: 1009, 1129.
Cover crop 1009; Forage 1009.

subsp. eriocarpa (Hausskn.)P.Ball [1009]
V. dasycarpa sensu Blakelock [1009]; *V. varia* sensu Zoh. [1009]; *V. varia* var. *eriocarpa* Hausskn. [1130].
Climbing; Herb; Annual.
West Asia: Iran(N), Iraq(N); Europe; Middle East: Syria(N), Turkey(N).
Description: 1009.

subsp. villosa 1009
Climbing; Herb; Annual.
West Asia: Afghanistan(N), Iraq(N).
Description: 1009.

GEOGRAPHICAL BIBLIOGRAPHY

Afghanistan 1010; 1011; 1051; 1052; 1058; 1126; 1129

Bahrain 1251; 1254; 1255

Iran 1010; 1015; 1051; 1058; 1103; 1129; 1187; 1251

Iraq 1009; 1251

Kuwait 1251; 1253

North Yemen 1251

Oman 1012; 1016; 1251

Qatar 1014; 1251

Saudi Arabia 1001; 1007; 1251; 1252

South Yemen 1044; 1251

United Arab Emirates 1251

BIBLIOGRAPHY

1001. Collenette, S. (1985). An Illustrated Guide to the Flowers of Saudi Arabia. London: Scorpion Publishing Ltd.
1002. Anderson, D.M.W. (1978). Chemotaxonomic aspects of the chemistry of *Acacia* gum. Kew Bulletin 32: 529–536.
1003. Chaudhary, S.A. (1983). *Acacia* & other genera of Mimosoideae in Saudi Arabia. Riyadh: National Herbarium, Regional Agriculture and Water Research Center.
1004. Simpson, K.A. (1989). Specimen in Herb.Kew.
1005. Lock, J.M. (1989). Legumes of Africa: a check-list. Royal Botanic Gardens, Kew.
1006. Wickens, G.E. (1969). A study of *Acacia albida* Del. Kew Bulletin 23: 181–202.
1007. Migahid, A.M. (1978). Flora of Saudi Arabia. Ed.2,Vol.1. Riyadh: University Publication.
1008. Brenan, J.P.M. (1953). In: Tropical African Plants: XXIII. Kew Bulletin 8: 97–98.
1009. Townsend, C.C.& Guest, E. (1974). Flora of Iraq. Vol.3. Baghdad: Ministry of Agriculture & Agrarian Reform.
1010. Rechinger, K.H. (1986). Flora Iranica. 161: Mimosaceae. Graz,Austria: Akademische Druck- und Verlaganstalt.
1011. Kitamura, S. (1960). Flora of Afghanistan. Results of the Kyoto University Scientific Expedition to the Karakoram and Hindukush, 1955, Vol.II. Kyoto University.
1012. Mandaville, J.P. (1977). In: The Scientific Results of the Oman Flora & Fauna Survey 1975. Journal of Oman Studies, Special Report. Oman: Ministry of Information and Culture.
1013. Ross, J.H. (1979). A Conspectus of the African *Acacia* species. Mem.Bot.Surv. S.Africa, 44.
1014. Batanouny, K.H. (1981). Ecology & Flora of Qatar. Qatar: Alden Press, for University of Qatar.
1015. Parsa, A. (1948). Flore de l'Iran. Vol.2. Teheran: Imprimerie Mazaheri.
1016. Miller, A.G. & Morris, M. (1988). Plants of Dhofar. Oman: Office of the Adviser for Conservation of the Environment.
1017. Brenan, J.P.M. (1957). Notes on Mimisoideae III. Kew Bulletin 12: 75–96.
1018. Howes, F.N. (1946). Fence & Barrier plants in warm climates. Kew Bulletin 1: 51–87.
1019. Wood, J.R.I. (1983). The identity of *Acacia oerfota* (Forssk.) Schweinf. (Leguminosae — Mimosoideae). Kew Bulletin 37: 451–454.
1020. Parker, R.N. (1921). In: Decades Kewenses XXXVIII; Decas CIV. Bull. Misc. Inf., Kew 1921: 309.
1021. Burtt Davy, J. (1922). New or noteworthy South African plants — V. Bull. Misc. Inf., Kew 1922: 322–335.
1022. Willis, J.H. (1972). A Handbook to Plants in Victoria: Vol. 2. Melbourne University Press.
1023. Zohary, M. (1976). New analytical flora of Israel. Tel Aviv: Am Oved Publishers Ltd.
1024. Brenan, J.P.M. (1983). Manual on taxonomy of *Acacia* species. Rome: F.A.O.
1025. Townsend, C.C. (1968). Contributions to the flora of Iraq: V. Notes on the Leguminales. Kew Bulletin 21: 435–458.
1026. Brenan, J.P.M. (1967). Notes on Mimosoideae XI. Kew Bulletin 21: 477–483.
1027. Hunde, A. (1982). Two new species of *Acacia* (Leguminosae — Mimosoideae) from Ethiopia and Yemen. Nordic J. Bot. 2: 337–342.
1028. Southgate, B.J. (1978). Variation in the susceptibility of African *Acacia* (*Leguminosae*) to seed beetle attack. Kew Bulletin 32: 541–544.

1029. Burtt Davy, J. (1930). *Acacia tortilis* and *Acacia spirocarpa*. Bull. Misc. Inf., Kew 1930: 402–404.
1030. Brenan, J.P.M. (1955). Notes on Mimosoideae: I. Kew Bulletin 10: 161–192.
1031. Burkart, A. (1976). A monograph of the genus *Prosopis* (Leguminosae subfam. Mimosoideae). J. Arnold Arb. 57: 450–525.
1032. Leonard, J. (1986). Une variété nouvelle de *Prosopis koelziana* Burkart (Mimosaceé de la péninsule arabique). Bull. Jard. Bot. Nat. Belg. 56: 483–485.
1033. Blakelock, R.A. (1948). The Rustam Herbarium, Iraq: Part I. Kew Bulletin 3: 375–444.
1034. Ffolliott, P.F. & Thames, J.L. (1983). Handbook on taxonomy of *Prosopis* in Mexico, Peru and Chile. Rome: F.A.O.
1035. Isely, D. (1973). Leguminosae of the United States: I. Subfamily Mimosoideae. Mem. New York Bot. Gard. 25: 1–151.
1036. Sprague, T.A. (1929). The correct spelling of certain generic names: V. Bull. Misc. Inf., Kew 1929: 241–243.
1037. Kostermans, A.J.G.H. (1954). A monograph of the Asiatic, Malaysian, Australian and Pacific species of Mimosaceae, formerly included in *Pithecolobium* Mart. Bull. Organ. Natuurw. Onderz. Indonesi 20: 1–122.
1038. Brenan, J.P.M. (1959). Leguminosae subfamily Mimosoideae. In: Flora of Tropical East Africa. (C.E. Hubbard & E. Milne-Redhead, Eds.). London, Crown Agents.
1039. Ross, J.H. (1974). Notes on miscellaneous *Acacia* species from Tropical Africa. Bothalia, 11: 291–294.
1040. Keay, R.W.J. & Brenan, J.P.M. (1949). A note on *Acacia dudgeoni* Craib. Kew Bulletin 4: 129–131.
1041. Ross, J.H. (1968). *Acacia senegal* (L.) Willd. in Africa, with particular reference to Natal. Bol. Soc. Brot. 42: 207–246.
1042. Leonard, J. (1985). Note sur *Prosopis cineraria* (L.) Druce et *P. koelziana* Burkart (Mimosaceés asiatiques). Bull. Jard. Bot. Nat. Belg. 55: 491–92.
1043. Hutchinson, J. (1931). XXV — The Flora of the Libyan Desert. Bull. Misc. Inf., Kew, 1931: 163–166.
1044. Boulos, L. (1988). A contribution to the flora of South Yemen (PDRY). Candollea 43: 559–585.
1045. Ross, J.H. (1975). The naturalized and cultivated exotic *Acacia* species in South Africa. Bothalia, 11: 463–470.
1046. De Wit, H.C.D. (1956). A revision of Malaysian Bauhinieae. Reinwardtia 3: 381–541.
1047. Brenan, J.P.M. (1967). Leguminosae subfamily Caesalpiniodeae. In: Flora of Tropical East Africa (E. Milne-Redhead & R.M. Polhill, Eds.). London: Crown Agents.
1048. De Wit, H.C.D. (1955). Revision of the genus *Cassia* (Caesalp.) as occurring in Malaysia. Webbia 11: 197–292.
1049. Brenan, J.P.M. (1958). New and noteworthy Cassias from tropical Africa. Kew Bulletin 13: 231–252.
1050. Hillcoat, D., Lewis, G. & Verdcourt, B. (1980). A new species of *Ceratonia* (*Leguminosae – Caesalpinoideae*) from Arabia and the Somali Republic. Kew Bulletin 35: 261–271.
1051. Rechinger, K.H. (1986). Flora Iranica. 160: Caesalpineaceae. Graz, Austria: Akademische Druck- und Verlagsanstalt.
1052. Rechinger, K.H. (1957). Leguminosae. In: Symbolae Afghanicae 3. (M. Koie & K.H. Rechinger, Eds.). Biol. Skr. Dan. Vid. Selsk. 9,3.

1053. Hooper, D. (1931). XLV - Some Persian drugs. Bull.Misc.Inf., Kew 1931: 299-344.
1054. Lock, J.M. (1988). *Cassia* sens.lat. (*Leguminosae - Caesalpinoideae*) in Africa. Kew Bulletin 43: 333-342.
1055. Gillett, J.B. (1963). *Sesbania* in Africa (excluding Madagascar) and Southern Arabia. Kew Bulletin 17: 91-160.
1056. Shaw, W.B.K. (1934). XXXVI — The Flora of the Libyan Desert. Bull.Misc.Inf., Kew, 1934: 281-289.
1057. Browicz, K. (1978). Geographic distribution of some shrubs from the family Leguminosae in southwestern Asia. Arboretum Kornickie 23: 5-30.
1058. Rechinger, K.H. (1984). Flora Iranica. 157: Papilionaceae II. Graz, Austria: Akademische Druck- und Verlagsanstalt.
1059. Dummer, R.A. (1913). A synopsis of the species of *Lotononis* Eckl. & Zeyh., and *Pleiospora* Harvey. Trans.Royal Soc.S.Afr. 3: 275-335.
1060. Polhill, R.M. (1982). *Crotalaria* in Africa & Madagascar. Rotterdam: A.A.Balkema.
1061. Baker, E.G. (1914). The African species of *Crotalaria*. J.Linn.Soc.,Bot. 42: 241-425.
1062. Peltier, M.A.G. (1959). Notes sur les Légumineuses — Papilionoidées de Madagascar et des Comores. 1. — Le genre *Crotalaria*. J. Agric. Trop. Bot. Appl. 6.
1063. Senn, H.A. (1939). The North American species of *Crotalaria*. Rhodora 41: 317-367.
1064. Gillett, J.B., Polhill, R.M. & Verdcourt, B. (1971). Leguminosae Subfamily Papilionoideae, Parts 1 & 2. In: Flora of Tropical East Africa. (E.Milne-Redhead & R.M.Polhill, Eds.). London: Crown Agents.
1065. Thulin, M. (1989). New or noteworthy species of Leguminosae in NE tropical Africa. Nordic J.Bot. 8: 457-488.
1066. Polhill, R.M. (1968). Miscellaneous notes on African species of *Crotalaria* L.: II. Kew Bulletin 22: 169-348.
1067. Moussavi, M. (1977). Contribution to the knowledge of *Medicago* species in Iran. Tehran: Ministry of Agriculture and Rural Development, Department of Botany, Publ.No.12.
1068. Heyn, C.C. (1963). The annual species of Medicago. Scripta Hierosolymitana (Publ. Hebrew Univ., Jerusalem) 12.
1069. Small, E. Lassen, P. & Brookes, B. (1987). An expanded circumscription of *Medicago* (*Leguminosae, Trifolieae*) based on explosive flower tripping. Willdenowia 16: 415-437.
1070. Lesins, K.A. & Lesins, I. (1979). Genus Medicago (Leguminosae): A taxogenetic study. The Hague: Dr.W.Junk bv.
1071. Small, E. & Brookes, B.S. (1984). Reduction of the geocarpic *Factorovskya* to *Medicago*. Taxon 33: 622-635.
1072. Sirjaev, G. (1928-33). Generis *Trigonella* L. revisio critica, I-VI. Publ.Faculte Sci.Uni.Masaryk, 102, 110, 128, 136, 148, 170, 192.
1073. Podlech, D. (1981). Neue Arten aus Afghanistan (Beitrage zur flora von Afghanistan XIII). Mitt. Bot. Staats. München 17: 477-484.
1074. Greuter, W. & Raus, T. (ed.) (1986). Med-Checklist Notulae,13. Willdenowia 16: 103-116.
1075. Vassilczenko, I. (1985). Generis *Trigonella* L. (Fabaceae) species novae ex Afghanistania. Nov. Syst. Plant. Vasc. 22: 143-146.
1076. Vassilczenko, I. (1983). Generis *Trigonella* L. (Fabaceae) species nova ex Afghanistania. Nov. Syst. Plant Vasc. 20: 126-127.

1077. Zohary, M. & Heller, D. (1984). The Genus *Trifolium*. Jerusalem: Israel Academy of Sciences & Humanities.
1078. Moussavi, M. (1979). List of plants of Evin Herbarium. Family: Leguminosae (Genus: *Trifolium*, with plates). Tehran:.Ministry of Agric. & Rural Dev., Dept. of Botany Publication No.14.
1079. Thulin, M. (1982). New varieties of *Trifolium semipilosum* in Ethiopia. Nordic J. Bot. 2: 51-52.
1080. Moussavi, M. (1975). Study of different species of *Lotus* in Iran. Entom. & Phytopath. Appl. 39.
1081. Gillett, J.B. (1958). *Lotus* in Africa south of the Sahara (excluding the Cape Verde islands and Socotra) and its distinction from *Dorycnium*. Kew Bulletin 13: 361-381.
1082. Stirton, C.H. (1981). Studies in the Leguminosae — Papilionoideae of southern Africa. Bothalia 13: 317-325.
1083. Gillett, J.B. (1958). *Indigofera* (*Microcharis*) in Tropical Africa. Kew Bulletin Add.Ser. 1: 1-166.
1084. Browicz, K. (1963). The genus *Colutea* L. A monograph. Monographiae Botanicae XIV.
1085. Podlech, D. & Deml, I. (1970). Neue und Bemerkenswerte Fabaceae aus Nordost Afghanistan II. Mitt.Bot.München 7: 329-346.
1086. Meikle, R.D. (1977). Flora of Cyprus, Vol.1. Royal Botanic Gardens, Kew: The Bentham-Moxon Trust.
1087. Townsend, C.C. (1967). A remarkable new species of *Chesneya* Lindl. (*Leguminosae*) from Afghanistan. Kew Bulletin 21: 293-294.
1088. Royle, J.F. (1835). Illustrations of the Botany and other branches of the Natural History of the Himalayan Mountains. London: W.H.Allen & Co.
1089. Vassilczenko, I.T. (1978). Generis *Oxytropis* DC. species novae ex Afghanistania. Byul. Mos. Obs. Isp. Pri. 83(4): 149.
1090. Gunn, C.R. (1983). A Nomenclator of Legume Genera. U.S. Dept. Agric. Tech. Bull. 1680.
1091. Baker, J.G. (1894). Botany of the Hadramaut Expedition. Bull. Misc. Inf. Kew 1894: 328-343.
1092. Baker, J.G. (1895). Decades Kewenses: Decas XXII.
1093. Brummitt, R.K. (1968). New and little known species from the Flora Zambesiaca area. XX: *Tephrosia*. Bol. Soc. Brot. 41: 219-393.
1094. Gillett, J.B. (1958). Notes on *Tephrosia* in Tropical Africa. Kew Bulletin 13: 111-132.
1095. Brummitt, R.K. (1980). Reconsideration of the genera *Ptycholobium*, *Caulocarpus*, *Lupinophyllum* and *Requienia* in relation to *Tephrosia* (*Leguminosae - Papilionoideae*). Kew Bulletin 35: 459-473.
1096. Gillett, J.B. (1966). The species of *Ormocarpum* Beauv. and *Arthrocarpum* Balf.f. (*Leguminosae*) in south-western Asia and Africa (excluding Madagascar). Kew Bulletin 20: 323-356.
1097. Thulin, M. (1985). Revision of *Taverniera* (*Leguminosae - Papilionoideae*). Acta Univ. Upsal.; Symb. Bot. Upsal. 25: 44-95.
1098. Podlech, D. (1967). Neue und bemerkenswerte Fabaceae aus Nordost Afghanistan. Mitt.Bot. München 6: 547-591.
1099. Wendelbo, P. (1980). New species of *Dionysia*, *Kickxia* and *Onobrychis* from Iran. Notes Roy. Bot. Edin. 38: 105-110.
1100. Fedtschenko, B.A. (1908). Turkestanskie espartsety. (Sainfoins from Turkestan). J. Russe Bot.(Russk. Bot. Zurn.) 3: 55-59.
1101. Sirjaev, G. (1925). *Onobrychis* Generis Revisio Critica 1, 2 & 3. Publ. Fac. Sc. Univ. Masaryk 56, 76 & 242.

1102. Davis, P.H. (1970). Flora of Turkey and the East Aegean Islands; Vol. 3. Edinburgh: University Press.
1103. Parsa, A. (1966). Flore de l'Iran: Vol.9. (2nd Supplement to Vol.2). Teheran.
1104. Bornmuller, J. (1906). Plantae Straussianae: sive enumeratio plantarum a Th. Strauss annis 1889-1899 in Persia occidentali collectarum. Bot. Centralbl., Beih. 19: 195-270.
1105. Verdcourt, B. (1970). Studies in the *Leguminosae - Papilionoideae* for the Flora of Tropical East Africa: II. Kew Bulletin 24: 235-307.
1106. Hermann, F.J. (1962). A revision of the genus *Glycine* and its immediate allies. U.S. Dept. Agric. Tech. Bull. 1268.
1107. Fantz, P.R. (1979). The Butterfly Pea of Ternate. Fairchild Trop. Gard. Bull. 34: 13-16.
1108. Verdcourt, B. (1981). *Dolichos trilobus* in Arabia. Kew Bulletin 36: 84.
1109. Verdcourt, B. (1970). Studies in the *Leguminosae - Papilionoideae* for the Flora of Tropical East Africa: III. Kew Bulletin 24: 379-447.
1110. Verdcourt, B. (1982). A Revision of *Macrotyloma* (Leguminosae). Hook. Ic. Pl. 38: 1-138.
1111. Gillett, J.B. (1966). Notes on *Leguminosae* (*Phaseoleae*). Kew Bulletin 20: 103-111.
1112. Lackey, J.A. (1977). *Neonotonia*, a new generic name to include *Glycine wightii* (Arnott) Verdcourt (Leguminosae, Papilionoideae). Phytologia 37: 209-12.
1113. Lackey, J.A. (1981). Infraspecific combinations in *Neonotonia wightii* (Arnott) Lackey (Leguminosae, Papilionoideae). Iselya 2: 11-12.
1114. Thulin, M. (1981). Notes on some species of *Rhynchosia* from NE Africa and Yemen. Nordic J. Bot.1: 37-42.
1115. Meikle, R.D. (1951). The identification of *Rhynchosia caribaea* (Jacq.) DC. and allied species. Kew Bulletin 6: 171-180.
1116. Verdcourt, B. (1970). Studies in the *Leguminosae - Papilionoideae* for the Flora of Tropical East Africa: IV. Kew Bulletin 24: 507-569.
1117. Marechal, R., Mascherpa, J-M. & Stanier, F. (1978). Étude taxonomique d'un groupe complexe d'espèces des genres *Phaseolus* et *Vigna* (Papilionaceae) sur la base des données morphologiques et polliniques, traitees par l'analyse informatique. Boissiera 28: 1-273.
1118. Yakovlev, G.P. (1967). Genus *Sophora* L. sect. *Pseudosophora* and *Keyserlingia* (Bunge) Yakovlev in the Near and Middle East. Vop. Farm. Leningrad 4: 63-67.
1119. Yakovlev, G.P. (1967). Systematical and geographical studies of genus *Sophora*. Vop. Farm. Leningrad 4: 41-62.
1120. Baker, E.G. In: Schinz, H. (1904). Beitrage zür Kenntnis der afrikanischen Flora (XVII). Viert. Nat. Ges. Zurich 49: 171-242.
1121. Boissier, E. (1872). Flora Orientalis II. Geneva: H.Georg.
1122. Baker, J.G. (1879). In: Hooker,J.D., Flora of British India 2: 56-306.
1123. Rechinger, K.H. (1951). Papilionaceae novae Iranicae. Engl., Bot. Jahrb. 75: 331-341.
1124. Parsa, A. (1947). New species and varieties of the Persian Flora: 1. Kew Bulletin 2: 17-22.
1125. Rechinger, K.H. (1979). Three new species of Papilionaceae from Iran. Candollea 34: 235-239.
1126. Podlech, D. & Anders, O. (1977). Florula des Wakhan (Nordost-Afghanistan). Mitt. Bot. München 13: 361-502.
1127. Hutchinson, J. & Bruce, E.A. (1941). Enumeration of the plants collected

by Mr. J.B.Gillett in Somaliland and Eastern Abyssinia. Bull. Misc. Inf., Kew 1941: 76-122.
1128. Van der Maesen, L.J.G. (1972). *Cicer* L., a monograph of the genus, with special reference to the chickpea (*Cicer arietinum* L.), its ecology and cultivation. Med. Land. Wageningen 72(10): 1-342.
1129. Rechinger, K.H. (1979). Flora Iranica: Papilionaceae I — Vicieae. Graz: Akademische Druck- und Verlagsanstalt.
1130. Allkin, R., Goyder, D.J., Bisby, F.A. & White,R.J. (1986). Names and synonyms of species and subspecies in the Vicieae: Issue 3. Vicieae Database Project Publication No.7.
1131. Allkin, R., MacFarlane, T.D., White, R.J., Bisby, F.A. & Adey, M.E. (1985). The Geographical distribution of *Lathyrus*. Vicieae Database Project Publication No.6.
1132. Townsend, C.C. (1971). Contributions to the Flora of Iraq: X. Notes on the Leguminales: 2. Kew Bulletin 25: 447-471.
1133. Bornmuller, J. & Gauba, E. (1940). Florae Keredjensis fundamenta. (Plantae Gaubaeanae Iranicae). Supplementum. I. Species novae. Feddes Repert. 49: 253-272.
1134. Oza, G.M. (1972). What is the native home of the Pigeon Pea ? The Indian Forester 98: 477-8.
1135. Van der Maesen, L.J.G. (1985). *Cajanus* DC. and *Atylosia* Wight & Arn. (Leguminosae). Agric. Univ. Wageningen Papers 85-4: 1-225.
1136. Oliver, D. (1881). *Acacia hunteri*. Hook. Ic. Pl. 16: t.1350.
1137. Rechinger, K.H. (1964). Notizen zür Orient-Flora 36. *Acacia iraqensis* n. sp. Anz. Osterr. Akad. Wiss., Math.- Nat. Kl. 101: 16.
1138. Brenan, J.P.M. (1963). Notes on African *Caesalpinioideae*. Kew Bulletin 17: 197-214.
1139. Franchet, M.A. (1896). Note sur une collection de plantes rapportées du Pamir en 1894 par M. E. de Poncins. Bull. Mus. Hist. Nat. Paris 2: 342-347.
1140. Bentham, G. (1871). Revision of the genus *Cassia*. Trans. Linn. Soc., Lond. 27: 503-591.
1141. Kaur, H. (1964). Illustrations of Indian plants. *Crotalaria* — 1. Bull. Nat. Bot. Gardens, Lucknow, 87.
1142. Maassoumi, A.A.R. (1989). Revision of *Astragalus* L. section *Hemiphaca* Bunge (Leguminosae) in Iran. Mitt. Bot. Munchen 28: 501-512.
1143. Podlech, D. (1989). Personal communication.
1144. Komarov, V.L. (1937). Flora Tadzhikistanica. Vol.5. Leguminosae. Moscow & Leningrad: Academiae Scientarum URSS.
1145. Ovczinnikov, L.N. & Kinzikaeva, (1981). Flora Tadjikskoë SSR, Vol.6. Leningrad: Academy NAUK.
1146. Podlech, D. (1983). Zur Taxonomie und Nomenclatur der Tragacanthoiden Astragali. Mitt. Bot. Munchen 19: 1-23.
1147. Sirjaev, G. (1939). Conspectus Tragacantharum (*Astragalus* L. subgenus *Tragacantha* Bunge). Feddes Repert. 47: 194-208, 225-261.
1148. Maassoumi, A.A.R. (1987). Notes on *Astragalus* L. in Iran. II. New species from Iran. Iranian J. Bot.3: 189-195.
1149. Bunge, A. (1869). Generis *Astragali* species Gerontogeae. Mem. Acad. Imp. Sci. St. Petersburg, 7th Ser., 9,16 & 15,1.
1150. Goncharov, N.F. (1946). Flora of the U.S.S.R. Vol. 12. Leguminosae: Astragalus. Moscow & Leningrad: Academy Nauk USSR. (Also in translation by N.Landau (1965); Jerusalem: Israel Program for Scientific Translation Ltd.).
1151. Takhtajan,A.L.(Ed.) (1962). Flora Armenii. Acad.NAUK Armenskoi SSR.

1152. Lock, J.M. (1990). Pers.obs.
1153. Lock, J.M. (1990). Specimen in Herb. Kew.
1154. Bornmuller, J. (1914). Reliquiae Straussianae. Weitere Beitrage zür Kenntnis der Flora des westlichen Persiens. Bot.Centralbl. Beihefte 32: 349-419.
1155. Hedge, I.C. & Podlech, D. (1987). A first survey of *Astracantha* and *Astragalus* in the Arabian Peninsula. Bot. Jahrb. Syst. 108: 259-270.
1156. Sirjaev, G. & Rechinger, K.H. (1951). Astragali et Onobrychidei novi orientales. Ann. Naturhist. Mus. Wien 58: 2-16.
1157. Rechinger, K.H. (1955). Astragali novi iranici. VII. Additis synonymis novis. Anz. Ost. Akad. Wiss.: Math.- nat. Kl. 92: 109-115.
1158. Bornmuller, J. (1908). Novitiae Florae Orientalis. Mitt. Thur. Bot. Ver. NS 23: 1-27.
1159. Aitchison, J.E.T. (1888). The Botany of the Afghan Delimitation Commission. Trans. Linn. Soc.,ser.2,3: 1-139.
1160. Rawi,A.al- (1964). Wild Plants of Iraq with their distribution. Dep.Agr.Iraq Tech.Bull. 14, pp.248.
1161. [Not used].
1162. Aitchison, J.E.T. (1880). On the Flora of the Kuram Valley, Afghanistan. Part 2: List of the plants collected, with notes and descriptions of new species. J. Linn. Soc., Bot. 18: 29-113.
1163. Ali,S.I. (1961). Revision of the genus *Astragalus* L. from W Pakistan and NW Himalayas. Biologia 7: 7-92.
1164. Turrill, W.B. (1929). On the Flora of the Nearer East:IV. Bull. Misc. Inf., Kew 1929: 223-235.
1165. Gilliat-Smith, B. & Turrill, W.B (1930). On the Flora of the Nearer East.VII. A contribution to our knowledge of the flora of Azerbaijan, N. Persia. Bull. Misc. Inf., Kew 1930: 375-398.
1166. Gilli, A. (1941). Ein Beitrag zür Flora des Elburs-Gebirges in Nord-Iran. Feddes Repert. 50: 263-283.
1167. Sirjaev, G. (1943). Generis Astragali L. species et varietates novae. Feddes Repert. 52: 1-15.
1168. Rechinger, K.H. (1940). Plantae novae Iranicae: I. Feddes Repert. 50: 33-48.
1169. Pau, C. & Vicioso, C. (1918). Plantas de Persia y de Mesopotamia recogidas por D. Fernando Martinez de la Escalera. Trab. Mus. Nac. Cienc. Nat., Ser. Bot. 14: 1-48.
1170. Sirjaev, G. & Rechinger, K.H. (1953). Astragali novi Iranici: I. Anz. Ost. Akad. Wiss., Math. Nat. Kl. 90: 114-125.
1171. Deml, I. (1972). Revision der Sektionen *Acanthophace* Bunge und *Aegacantha* Bunge der Gattung *Astragalus* L. Boissiera 21: 1-235.
1172. Maassoumi, A.A.R. (1989). The genus *Astragalus* in Iran. Vol.2 — Perennials. Isl. Rep. Iran, Res. Inst. For. Rnglds Tech. Publ. 44-1989.
1173. Podlech, D. (1990). Pers.comm.
1174. Becht, R. (1978). Revision der Sektion *Alopecuroidei* der Gattung *Astragalus* L. Phanerog. Monog. 10: 1-227.
1175. Podlech, D. (1988). Revision von *Astragalus* L. sect *Caprini* DC. (Leguminosae). Mitt. Bot. Munchen 25: 1-924.
1176. Assadi, M. (1988). Plants of Arasbaran Protected Area, NW Iran, Part 2). Iran J. Bot. 4: 127-142.
1177. Bornmuller, J. (1938). Zür Flora Persiens. Notizbl. Bot. Gard. Berlin 14: 270-287.
1178. Podlech, D. (1988). Beitrage zür Kenntnis der Gattung *Astragalus* L.

(Leguminosae). III. Einige neue Arten aus dem Iran, aus Afghanistan und Turkestan. Mitt.Bot.München 27: 51-64.
1179. Rechinger, K.H., Dulfer, H. & Patzak, A. (1961). Sirjaevii fragmenta Astragalogica, XIII – XVII. Sitz. Ost. Akad. Wiss. math. nat. Kl. 170: 9-68.
1180. Bornmuller, J. & Gauba, E. (1935). Florulae Keredjensis fundamenta. (Plantae Gaubaeanae iranicae). Feddes Repert. 39: 73-124.
1181. Podlech, D. (1985). Beitrage zur Kenntnis der Gattung *Astragalus* L. 1. Neue und bemerkenswerte Arten aus Afghanistan. Bot. Jahrb. Syst. 107: 55-73.
1182. Rechinger, K.H. (1962). Revision einiger Typen von Velenovsky's Plantae Arabicae Musilianae. Bot. Notiser 115: 35-48.
1183. Danin, A. (1976). *Astragalus camelorum* Barbey, a rediscovered species from the Isthmic Desert (N Sinai). Israel J. Bot. 25: 214-215.
1184. Sirjaev, G. & Rechinger, K.H. (1954). Astragali novi Iranici IV. Anz. Ost. Akad. Wiss. math.- nat. Kl. 91: 159-165.
1185. Podlech, D. (1973). Neue und bemerkenswerte *Astragalus*-arten aus Afghanistan. Mitt. Bot. München 11: 259-321.
1186. Freitag, H. (1983). *Astragalus kavirensis*, eine neue Art von *Astragalus* section *Ammodendron* aus dem Iran. Willdenowia 13: 133-136.
1187. Leonard, J. (1986). Contribution à l'étude de la flore et de la vegetation des deserts d'Iran: 6. Bull. Jard. Bot. natn. Belg., Suppl.
1188. Bornmuller, J. (1941). Beitrage zür Kenntnis der *Astragalus*-Arten Persiens, einschliesslich einiger Arten der Flora Turkestans. Feddes Repert. 50. 151-177.
1189. Sirjaev, G. (1944). Conspectus praecursorius specierum subgeneris *Trimeniaeus* Bunge generis *Astragali* L. Feddes Repert. 53: 220-253.
1190. Kamelin, R.V., Kovalevskaya, S.S. & Nabijev, M.M. (Eds.) (1981). Conspectus Florae Asiae Mediae, Vol. 6. Taschkent: Edit. Acad. Sci. Uz.S.S.R.
1191. Agerer-Kirchhoff, C. (1976). Revision von *Astragalus* L. sect. *Astragalus*. Boissiera 25: 1-198.
1192. Eig, A. (1955). Systematic Studies on *Astragali* of the Near East. Jerusalem: Israel Scientific Press.
1193. Mouterde, P. (1966). Nouvelle Flore du Liban et de la Syrie. Beirut: Dar el-Machreq Editeurs.
1194. Sirjaev, G. (1944). *Astragalus stalinskyi* species nova. Feddes Repert. 53: 75-76.
1195. Tietz, S. (1988). Revision von *Astragalus* sect. *Campylanthus* Bunge, sect. *Microphysa* Bunge und sect. *Poterion* Bunge. Mitt. Bot. München 27: 135-380.
1196. Maassoumi, A.A.R. (1988). Notes on the genus *Astragalus* L. (Leguminosae) in Iran. III. New Records and new species. Iran J. Bot. 4: 127-142.
1197. Podlech, D. & Maassoumi, A.A.R. (1987). Nine new taxa of the genus *Astragalus*, sect. *Caprini* (Leguminosae) from Iran. Iran J. Bot. 3: 95-110.
1198. Maassoumi, A.A.R. & Podlech, D. (1987). Some new species and subspecies of *Astragalus* L. section *Caprini* DC. from Iran. Bot. Jahrb. Sys. 109: 261-278.
1199. Freitag, H. & Podlech, D. (1980). Zwei neue *Astragalus*-arten aus dem Touran - Schutzgebiet im Iran. Mitt. Bot. München 16 (Beihefte): 7-10.
1200. Maassoumi, A.A.R. & Podlech, D. (1988). Eleven new species and a new section from the Genus *Astragalus* (Leguminosae) in Iran. Iran J. Bot. 4: 71-90.

1201. Bornmuller, J. (1905). Beitrage zur Flora der Elbursgebirge Nord-Persiens. Bull. Herb. Boiss. Ser.II, 5: 752–767.
1202. Rechinger, K.H. (1940). Plantae novae Iranicae. II. Feddes Repert. 48: sgebir121–167.
1203. Podlech, D. (1986). Beitrage zur kenntnis der Gattung *Astragalus* L. (Leguminosae). II. *Astragalus renzianus* species nova aus dem Iran. Mitt. Bot. München 22: 1–3.
1204. Wendelbo, P. (1953). Plants from Tirich Mir. A contribution to the Flora of Hindukush. Nytt. Mag. Bot. 1: 1–70.
1205. Ali, S.I. (1958). Notes on the genus *Astragalus* L. from W Pakistan and the NW Himalayas. Kew Bulletin 13: 303–318.
1206. Podlech, D. (1975). Revision der Sektion *Caraganella* Bunge der Gattung *Astragalus* L. Mitt. Bot. München 12: 153–166.
1207. Grey-Wilson, C. (1974). Some notes on the flora of Iran and Afghanistan. Kew Bulletin 29: 19–81.
1208. Podlech, D. & Maassoumi, A.A.R. (1989). Two new species of *Astragalus* from Iran and their affinity to *A. weirianus* from Afghanistan. In: Kit Tan(Ed.) The Davis and Hedge Festschrift. Edinburgh: University Press.
1209. Wenniger, J. (1990). Mitt.Bot.Munchen, in press
1210. Ott, E. (1978). Revision der Sektion *Chronopus* der Gattung *Astragalus* L. Phan. Monog. 9: 1–142.
1211. Rechinger, K.H. (1941). Plantae novae iranicae et regionum adjacentium. III. Feddes Repert. 50: 255–262.
1212. Rechinger, K.H. & Dulfer, H. (1969). Sirjaevi fragmentae Astragalogiae 18. Sitz. Osterr. Akad. Wiss. Math. Nat Kl. 177: 96–125.
1213. Sirjaev, G. & Rechinger, K.H. (1953). Astragali novi Iranici. II. Anz. Osterr. Akad. Wiss. Math. nat. Kl. 90: 154–163.
1214. Ali, S.I. (1973). Contributions to the genus *Astragalus* (Leguminosae) from Pakistan. VI. Further additions and corrections. Kew Bulletin 28: 297–304.
1215. Bornmuller, J. & Gauba, E. (1942). Florae Keredjensis fundamenta (Plantae Gaubaeanae Iranicae) Supplementum. 2.Enumeratio specierum. Feddes Repert. 51: 33–48.
1216. Bornmuller, J. (1915). Plantae Brunsianae. Aufzahlung der von F.Bruns in nordlichen Persien gesammelten Pflanzen. Bot. Centralbl. (Beih). 33: 270–324.
1217. Podlech, D. (1984). Revision von *Astragalus* sect. *Herpocaulos* Bunge. Mitt. Bot. München 20: 441–449.
1218. Bornmuller, J. (1934). Aus der Flora Afghanistans. Bot. Jahrb. 66: 216–240.
1219. Sytin, A. (1986). Revisio generis *Astragalus* L. (Fabaceae) sectionis *Hololeuce* specierum Caucasicis. Nov. Syst. Vyss. Rast. 23: 79–86.
1220. Rechinger, K.H., Dulfer, H. & Patzak, A. (1959). Sirjaevii fragmenta Astragalogica 5–8. Sitz. Osterr. Akad. Wiss. math. nat. Kl. 168: 95–182.
1221. Greuter, W., Burdet, H.M. & Long, G. (Eds.) (1989). Med-Checklist Vol.4 Dicotyledones (Lauraceae — Rhamnaceae). Geneva: Conservatoire et Jardin botaniques de la Ville de Geneve.
1222. Maassoumi, A.A.R. (1988). Revision of *Astragalus* L. section *Hypoglottoidei* DC. in Iran. Mitt. Bot. München 27: 125–133.
1223. Agerer-Kirchhoff, C. & Agerer, R. (1977). Eine neue Sektion der Gattung *Astragalus* L.: *Laxiflori* Agerer-Kirchhoff. Mitt. Bot. München 13: 203–234.
1224. Rechinger, K.H., Dulfer, H. & Patzak, A. (1959). Sirjaevii fragmenta Astragalogia 9–11. Sitz. Osterr. Akad. Wiss. Math-nat. Kl. 168: 693–718.

1225. Sirjaev, G. & Rechinger, K.H. (1954). Astragali novi Iranici. V. Anz. Osterr. Akad. Wiss. Math-nat. Kl.91: 166–173.
1226. Sirjaev, G. & Rechinger, K.H. (1953). Astragali novi Iranici. III. Anz. Osterr. Akad. Wiss. Math-nat. Kl. 90: 180–184.
1227. Gilli, A. (1939). Neue Arten und Varietaten aus dem Elbursgebirge in Nord-Iran. Feddes Repert. 46: 43–48.
1228. Nikitin, V.V. (Ed.) (1949). Flora Turkmenii. Vol.4. Academy Nauk SSSR.
1229. Rechinger, K.H., Dulfer, H. & Patzak, A. (1958). Sirjaevii fragmenta Astragalogia 1–3. Sitz. Ost. Akad. Wiss. Math-nat. Kl. 167: 51–93.
1230. Kamelin, R.V. (1977). Notae de genero *Astragalus* L. Nov. Syst. Vyss. Rast. 14: 161–168.
1231. Rechinger, K.H., Dulfer, H. & Patzak, A. (1958). Sirjaevii fragmenta Astragalogia 4. Sitz. Ost. Akad. Wiss. math-nat. Kl. 167: 321–361.
1232. Podlech, D. & Deml, I. (1967). Eine interessante neue *Astragalus* -art aus Afghanistan. Mitt. Bot. München 6: 541–546.
1233. Rechinger, K.H. & Dulfer, H. (1969). Sirjaevii fragmenta Astragalogia. 18. Sitz. Ost. Akad. Wiss. Math-nat. Kl. 177: 89–132.
1234. Borissova, A. (1947). Leguminosae novae e flora URSS. Not. Syst. Inst. Bot. Ac. Sc. URSS 10: 43–84.
1235. Rechinger, K.H. (1949). Plantarum novarum orientalium descriptiones. Osterr. Bot. Zeitschr. 95: 422–427.
1236. Barneby, R.C. (1974). Dragma hippomanicum. 1. Brittonia 26: 109–114.
1237. Podlech, D. (1975). Revision der Sektion *Stipitella* G.Grig. ex Podlech der Gattung *Astragalus*. Mitt. Bot. München 12: 33–50.
1238. Podlech, D. & Kozik, E. (1983). Revision von *Astragalus* L. Sektion *Theiochrus* Bunge. Mitt. Bot. München 19: 351–362.
1239. Maassoumi, A.A.R. & Ghahreman, M. (1988). A new species of *Astragalus* section *Tricholobus* from Iran. Iran J. Bot. 4: 109–110.
1240. Aitchison, J.E.T. (1882). On the Flora of the Kuram Valley, &c., Afghanistan. Part II. J. Linn. Soc. Bot. 19: 139–200.
1241. Dymock, W. (1890). Pharmacographia Indica. Bombay.
1242. Post, G.E. (1932). Flora of Syria, Palestine and Sinai. Ed. 2. Beirut: American Press.
1243. Gargary, H. (1982). The National Programme of Iran. Plant Genet. Res. Newsl. 50: 51–52.
1244. Duke, J.A. (1981). Handbook of Legumes of World Economic Importance. New York: Plenum Press.
1245. Larsen, K., Larsen, S.S. & Vidal, J.G. (1980). 18. Leguminosae – Caesalpinioideae. In Flore du Cambodge, Laos et Viet-nam. Paris: Museum National d'Histoire Naturelle.
1246. Van der Maesen, L.J.G. (1987). In: The Chickpea. (Saxena, M.C. & Singh, K.B., eds.).
1247. Allkin, R., Macfarlane, T.D., White, R.J. & Bisby, F.A. (1984). The Geographical Distribution of *Vicia*. Vicieae Database Project Publ. No. 5.
1248. Yakovlev, G.P., Svijazeva, O.A. & Batchili, F. (1982). On the genus *Ammodendron* (Fabaceae). Bot. Zhurn. 67: 60–61.
1249. Vassilchenko, I.T. (1988). New Asiatic species of *Oxytropis*. Bjull. Mosk. Obsc. Isp. Prir., Otd. Biol. 93 (3): 97–102.
1250. Mozaffarian, V. (1988). New species & new plant records from Iran. Iran J. Bot. 4: 61–70.
1251. Heller, D. & Heyn, C.C. (1990). Conspectus Florae Orientalis. Fascicle 5 —Mimosaceae-Papilionaceae, Podostemaceae. Jerusalem: Israel Academy of Science & Humanities.

1252. Mandaville, J.P. (1990). Flora of Eastern Saudi Arabia. London & New York: Kegan Paul International.
1253. Daoud, H.S. (rev. A.Al-Rawi) (1985). Flora of Kuwait. Vol.1: Dicotyledons. London, Boston & Melbourne: KPI Ltd. & Kuwait University.
1254. Phillips, D.C. (1988). Wild Flowers of Bahrain. Bahrain: Manama.
1255. Cornes, M.D. & Cornes, C.D. (1989). The Wild Flowering Plants of Bahrain. London: IMMEL Publications.
1256. Teitz, S. & Zarre, M. (1990). Erganzungen zur Revision von *Astragalus* L. sect. *Microphysa* Bunge (Leguminosae). Mitt. Bot. München 29: 539–540.
1257. Maassoumi, A.A.R. & Podlech, D. (1990). New species of *Astragalus* L. (Leguminosae) from Iran. Mitt. Bot. München 29: 509–514.
1258. Maassoumi, A.A.R. (1990). Some new species of *Astragalus* L. section *Malacothrix* in Iran. Mitt. Bot. München 29: 503–508.
1259. Kress-Deml, I. (1990). Funf neue dornige *Astragalus*-Arten aus Afghanistan. Mitt. Bot. München 29: 573–588.
1260. Podlech, D. (1990). Revision von *Astragalus* L. sect. *Platyglottis* Bunge (Leguminosae). Mitt. Bot. München 29: 541–572.
1261. Podlech, D. (1990). Die Typifizierung der altweltlichen Sektionen der Gattung *Astragalus* L. (Leguminosae). Mitt. Bot. München 29: 461–494.
1262. Schwrtz, O. (1939). Flora des tropischen Arabia. Mitt. Inst. allg. Bot., Hamburg 10: 1–393.
1263. Ali, S.I. (1958). On the identity of *Kerstania* Rech.f. J. Bombay Nat. Hist. Soc. 55: 378–380.
1264 Gabali, S.A. & Al-Gifri, A.-N. (1990). Flora of South Yemen — Angiospermae. A provisional check-list. Feddes Repert. 101: 373–383.

INDEX

Abrus bottae Defl. 17
  precatorius L. 17
Acacia abyssinica Benth. 9
  abyssinica subsp. abyssinica 9
  adansonii Guillemin & Perrottet 12
  albida Del. 14
  arabica sensu Brenan 12
  arabica (L.)Willd.) 12
  arabica var. kraussiana Benth. 12
  asak (Forsskal)Willd. 9,13
  campoptila Schweinf. 9
  cornigera (L.)Willd. 9
  cyanophylla Lindley 13
  dekindtiana A.Chev. 11
  eburnea (L.f.)Willd. 9
  edgeworthii T.Anders. 9
  ehrenbergiana Hayne 10
  elatior Brenan 10
  elatior subsp.elatior 10
  erythraea Chiov. 9
  etbaica Schweinf. 9
  etbaica subsp. etbaica 10
  etbaica subsp. uncinata Brenan 10
  farnesiana (L.)Willd. 10
  flava (Forsskal)Schweinf. 10
  flava var. ehrenbergiana (Hayne) Roberty 10
  gerrardii sensu auctt. 11
  gerrardii subsp. negevensis Zoh. 11
  glaucophylla A.Rich. 9
  greggii A.Gray 10
  hamulosa Benth. 10,13
  hockii De Wild. 10
  horrida (L.)Willd.,p.p. 11
  humifusa Chiov. 9
  hunteri Oliver 11
  hydaspica R.Parker 11
  iraqensis Rech.f. 11,12
  iraquensis Rech.f. 11
  jacquemontii Benth. 11
  karroo Hayne 11
  laeta Benth. 11
  leprosa DC. 11
  macracantha Willd. 11
  mellifera (Vahl)Benth. 11
  mellifera subsp. mellifera 11
  menachensis Schweinf. 12
  modesta Wall. 11
  natalitia E.Meyer 11
  negevensis (Zoh.)Zoh. 11
  negrii sensu Hepper 12
  nilotica (L.)Del. 12
  nilotica var. adansonii (Guillemin & Perrottet)A.Chev. 12
  nilotica subsp. adstringens (Schum.& Thonn.)Roberty 12
  nilotica subsp. indica (Benth.)Brenan 12
  nilotica var. indica (Benth.)A.F.Hill 12
  oerfota sensu Brenan 10
  oerfota (Forsskal)Schweinf. 10
  orfota sensu auctt. 12
  origena A.Hunde 12
  pachyceras O.Schwartz 12
  paolii Chiov. 13
  pseudosocotrana Chiov. 9
  pterygocarpa Benth. 12
  raddiana Savi 13
  saligna (Labill.)Wendl. 13
  scorpioides (L.)W.Wight 12
  senegal (L.)Willd. 13
  seyal Del. 13
  socotrana Balf.f. 9
  spirocarpa A.Rich. 13
  spirocarpa var. major Schweinf. 13
  sultani Chiov. 9
  tortilis (Forsskal)Hayne 13
  tortilis var. pubescens Aylmer 13
  tortilis subsp. raddiana (Savi) Brenan 13
  tortilis subsp. spirocarpa (A.Rich.) Brenan 13
  tortilis subsp. tortilis 13
  triacantha A.Rich. 9
Aeschynomene indica L. 17
Albizia julibrissin Durazz. 14
  julibrissin sensu Blakelock 15
  lebbec sensu Migahid 14
  lebbeck (L.)Benth. 14
  lebbek sensu auctt. 14
Alhagi camelorum Fischer 27
  camelorum subsp. turcorum Sinsk. 27
  canescens (Regel)Keller & Shap. 26
  graecorum Boiss. 26
  kirghisorum Schrenk 27
  mannifera Desv. 26
  mannifera var. karduchorum Boiss. 26
  maurorum sensu auctt. 26
  maurorum Medikus 27
  maurorum var. assyriacum Nab. 26
  maurorum var. karduchorum Boiss. 26
  persarum Boiss. & Buhse 27
  pseudalhagi Desv. 27
  sparsifolium Shap. 27
  turcorum Boiss. 27
Alophotropis formosa (Steven)Grossh. 213
Alysicarpus glumaceus (Vahl)DC. 25
  longifolius Wight & Arn. 25

Alysicarpus (*cont.*)
  ovalifolius (Schum.)Leonard 25
  rugosus (Willd.)DC. 25
  vaginalis (L.)DC. 25
Ammodendron karelinii Fischer & Meyer 186
  persicum Boiss. 186
Ammothamnus gibbosus (DC.)Boiss. 187
  intermedius Kuntze 187
  lehmanni Bunge 187
Amorpha fruticosa L. 18
Amphinomia platycarpa (Viv.)Cuf. 24
Anagyris foetida L. 187
Anatropostylia koeieana (Rech.f.)Kupicha 216
Anthyllis boissieri Sagorski 169
  lachnophora Juz. 169
  vulneraria L. 170
  vulneraria subsp. boissieri (Sagorski) Bornm. 169
Arachis hypogaea L. 17
Argyrolobium abyssinicum Jaub. & Spach 148
  arabicum (Decne.)Jaub. & Spach 148
  biebersteinii P.Ball 149
  calycinum (M.Bieb.)Boiss. 149
  confertum Polh. 149
  crotalarioides Jaub. & Spach 149
  kotschyi Boiss. 149
  lotoides Trautv. 149
  prostratum Boiss. 149
  roseum (Cambess.)Jaub. & Spach 149
  roseum subsp. ornithopodioides (Jaub. & Spach)S.M.H.Jafri & Ali 149
  roseum subsp. roseum 149
  rupestre (E.Meyer)Walp. 149
  rupestre subsp. remotum (A.Rich.) Polhill 149
  stenophyllum Boiss. 149
  trigonelloides Jaub. & Spach 149
  uniflorum (Decne.)Jaub. & Spach 150
  virgatum Baker 149
Astracantha acetabulosa (C.Towns.) Podl. 27
  acmophylla (Bunge)Podl. 27
  adscendens (Boiss. & Hausskn.) Podl. 27
  albispina (Sirj. & Bornm.)Podl. 27
  alexandri (Sirj.)Podl. 27
  amblolepis (Fischer)Podl. 28
  atropatana (Bunge)Podl. 28
  aurea (Willd.)Podl. 28
  baghensis (Bunge)Podl. 28
  basianica (Boiss. & Hausskn.) Podl. 28
  bienerti (Bunge)Podl. 28
  brachycalyx (Fischer)Podl. 28

  caspica (M.Bieb.)Podl. 28
  cephalotes (Banks & Sol.)Podl. 29
  cerasocrena (Bunge)Podl. 29
  cesarensis (Sirj. & Bornm.)Podl. 29
  chorassanica (Bunge)Podl. 29
  crassinervia (Boiss.)Podl. 29
  crenophila (Boiss.)Podl. 29
  deinacantha (Boiss.)Podl. 29
  diphtherites (Fenzl)Podl. 29
  divaricata (Bunge)Podl. 30
  drymophila (Bornm.)Podl. 30
  dschuparensis (Freyn & Bornm.) Podl. 30
  echidna (Bunge)Podl. 30
  echidnaeformis (Sirj.)Podl. 30
  echinus (DC.)Podl. 30
  echinus subsp. arabica Hedge & Podl. 30
  elisabethae (Sirj. & Rech.f.)Podl. 30
  erinacea (Fischer)Podl. 30
  eriocephala (Willd.)Podl. 31
  eriosphaera (Boiss. & Hausskn.) Podl. 31
  eriostyla (Boiss. & Hausskn.) Podl. 31
  eschkerensis (Boiss. & Hausskn.) Podl. 31
  esfandiarii (Sirj. & Rech.f.)Podl. 31
  farakulumensis (Sirj. & Rech.f.) Podl. 31
  floccosa (Boiss.)Podl. 31
  florulenta (Boiss. & Hausskn.)Podl. 31
  garaensis (Sirj.)Podl. 31
  gemiana (Boiss. & Hausskn.) Podl. 31
  ghilanica (Fischer)Podl. 32
  gigantostrobus (Rech.f. & Aellen) Podl. 32
  glabrifolia (Bunge)Podl. 32
  glaucopsoides (Bornm.)Podl. 32
  globiflora (Boiss.)Podl. 32
  gossypina (Fischer)Podl. 32
  gummifera (Labill.)Podl. 32
  heratensis (Bunge)Podl. 32
  hypsogeton (Bunge)Podl. 33
  karabaghensis (Bunge)Podl. 33
  kuhistana (Bunge)Podl. 33
  kurdica (Boiss.)Podl. 33
  lagonyx (Fischer)Podl. 33
  lateritia (Boiss. & Hausskn.)Podl. 33
  lateritians (Freyn & Bornm.)Podl. 33
  leioclados (Boiss.)Podl. 33
  leiophylla (Freyn & Bornm.)Greuter 44
  lepidantha (Boiss.)Podl. 34
  leucoptila (Boiss. & Hausskn.) Podl. 34
  longifolia (Lam.)Podl. 34
  meschhedensis (Bunge)Podl. 34
  mesoleia (Boiss. & Hohen.)Podl. 34

Astracantha (*cont.*)
  meyeri (Boiss.)Podl. 34
  michauxiana (Boiss.)Podl. 34
  meyeri (Boiss.)Podl. 34
  michauxiana (Boiss.)Podl. 34
  microcephala (Willd.)Podl. 34
  morgani (Freyn)Podl. 35
  multispina (Freyn & Bornm.) Podl. 35
  muschiana (Kotschy & Boiss.)Podl. 33,35
  myriacantha (Boiss.)Podl. 35
  ochrobia (Bunge)Podl. 35
  ochrochlora (Boiss. & Hohen.)Greuter 44
  ochtoranensis (Freyn)Podl. 35
  octopus (C.Towns.)Podl. 35
  pachyacantha (Bunge)Podl. 35
  pachystachys (Bunge)Podl. 35
  parrowiana (Boiss. & Hausskn.)Podl. 36
  peristerea (Boiss. & Hausskn.)Podl. 36
  polyantha (Bunge)Podl. 36
  prusiana (Boiss.)Podl. 36
  pseudaurea (Sirj. & Rech.f.)Podl. 36
  psilacmos (Bunge)Podl. 28
  psilodontia (Boiss.)Podl. 36
  psilostyla (Bunge)Podl. 36
  ptilocephala (Baker)Podl. 36
  pulvinata (Bunge)Podl. 36
  pycnocephala (Fischer)Podl. 37
  radschirdensis (Sirj. & Bornm.)Podl. 37
  rayatensis (Eig)Podl. 32
  rhodochroa (Boiss. & Hausskn.)Podl. 37
  semipellita (Bunge)Podl. 37
  serpentinica (Sirj. & Rech.f.)Podl. 37
  sofica (Bunge)Podl. 37
  splendidissima (Sirj. & Rech.f.)Podl. 37
  stenonychioides (Freyn & Bornm.) Podl. 37
  strobilifera (Benth.)Podl. 37
  talarensis (Sirj. & Rech.f.)Podl. 38
  totschalensis (Bornm.)Podl. 38
  wartoensis (Boiss. & Kotschy)Podl. 28
  xanthogossypina (Hand.-Mazz.)Podl. 32
  zoharyi (Eig)Podl. 38
Astragalus aaronsohnianus *Eig* 63
  abbas-riazi Parsa 117
  abnormalis Rech.f. 96
  absentivus A.A.R.Maassoumi 62
  acantherioceras Rech.f. & Koie 43
  acanthochristianopsis Rech.f. & Koie 57
  acetabulosus *C.Towns.* 27
  achtalensis *Conrath & Freyn* 117
  acinaciferus *Boiss.* 81

aciphyllus Freyn 127
ackerbergensis Freyn & Sint. 96
acmophyllus Bunge 27
acutifolius Bunge 53
acutus Bunge 61
adpresse-pilosus Gontsch. 62
adpressipilosus Gontsch. 62
adraskanensis Podl. 68
adscendens Boiss. & Hausskn. 27
adulterinus Podl. 62
aduncus Willd. 114
aduncus M.Bieb. 114
aegobromus Boiss. & Hohen. 62,78
aegobromus var. hirsutus Boiss. 62
aegobromus var. longiscapus Bornm. 62
aeluropus Bunge 38
aff.chionobius sensu Rawi 43
affganus Boiss. 56
affghanus Boiss. 56
afganicus Bornm. 56
afghanicus Bornm. 56
afghano-persicus Kitam. 122
afghanomontanus Sirj. & Rech.f. 128
agrestis Freyn 118
aharicus A.A.R.Maassoumi & Podl. 62
ahmad-adlii Bornm. & Gauba 54
ahmed-adlii Bornm. & Gauba 54
ahouicus Parsa 105
aitchisonii Sirj. & Rech.f. 71
aitchisonii Baker 116
aiwadzhii B.Fedtsch. 54
ajfreidii Aitch. & Baker 46
ajfreidii subsp. ajfreidii 46
ajfreidii subsp. brevivexillatus Deml 46
ajubensis Bunge 50
aksuensis Bunge 90
aktauensis Gontsch. 62
al-hamedensis Rech.f. 128
alamliensis Rech.f. 128
albispinus Sirj. & Bornm. 27
albosetulosus Sirj. & Rech.f. 57
albovillosus Kitam. 57
aleppicus Boiss. 57
aleppicus sensu Rawi 58
aleppicus subsp. megaloceras (Eig) C.Agerer-Kirchhoff 57
alexandri Sirj. 27
alexandrinus Boiss. 64
alienus Podl. 62
aligudarzicus Parsa 62
aloisii Deml 46
alopecias Pallas 50
altimurensis Deml 46
altissimus Rech.f. 80
alyssoides Lam. 91
amadiensis Eig 28

235

Astragalus (*cont.*)
  amanus Boiss. 73
  amblolepis Fischer 28
  amblolepis var. *rayatensis* Eig 28
  amherstianus Benth. 86
  ammodendroides Bornm. 54
  ammophilus Karelin & Kir. 118
  ammophilus var. *obtusatus* Popov 118
  ammophilus var. *persepolitanus* (Boiss.)Ali 118
  anacamptoides Sirj. & Rech.f. 85
  anacamptus Bunge 85
  anacardius Bunge 50
  anachoreticus Podl. 54
  ancistrocarpus Boiss. & Hausskn. 96
  ancyleus Boiss. 120
  andalanicus Boiss. & Hausskn. 32
  andalanicus var. *elymaiticus* (Boiss. & Hausskn.)Sirj. 32
  andalanicus var. *validior* (Bornm.)Sirj. 32
  andarabicus Podl. 66
  andersianus Podl. 119
  andrachnaefolius sensu auctt. 29
  andrachne Bunge 29
  andrachne var. *sefinensis* (Bornm.)Sirj. 29
  andrachne var. *sintenisianus* Sirj. 29
  anfractuosus Bunge 131
  angustatus Boiss. 128
  angustatus Bunge 128
  angustidens Freyn & Sint. 65
  angustidens var. *recterostratus* Sirj. & Rech.f. 65
  angustidens var. *retusefoliolatus* Sirj. & Rech.f. 65
  angustidens var. *sessilis* Sirj. & Rech.f. 65
  angustidens var. *strictifolius* Sirj. & Rech.f. 65
  angustiflorus C.Koch 62
  angustiflorus subsp. *amanus* (Boiss.) Chamberlain 73
  angustiflorus subsp. angustiflorus 62
  anisacanthus Boiss. 121
  anisacanthus subsp. anisacanthus 121
  anisacanthus subsp. schurabicus (Bunge)S.Tietz 121
  ankylotus Fischer & Meyer 56
  anomalus Bunge 76
  anserinaefolius Boiss. 105
  anserinifolius Boiss. 105
  antheliophorus Deml 46
  anthosphaerus Rech.f. & Gilli 57
  aphananthos Kress-Deml 46
  aphthonus Rech.f. & Aellen 38
  applicatus Boriss. 117
  apricus Bunge 62
  apricus var. *rarus* (Sirj. & Rech.f.)Parsa 62

  aqrabatensis Podl. 63
  aquae-rubrae B.Fedtsch. 67
  arabicus Bunge 91
  arakensis Parsa 82
  arbelicus Bornm. 51
  arbusculinus Bornm. & Gauba 122
  archibaldii Podl. 63
  ardahelicus Parsa 112
  ardakensis Sirj. & Rech.f. 110
  arenarius Pallas 94
  argillosus Manden. 62
  argyroides G.Beck 128
  argyrophyllus Boiss. & Gaill. 29
  argyrostachys Boiss. 60
  armatus var. *libycus* Pampan. 122
  arnoceras Bunge 60
  ascophorus Fischer 63
  ashuricus Parsa 52
  askabadensis Kuntze 58
  askius Bunge 96
  askius var. *buhseanus* Boiss. 97
  aspadanus Bunge 87
  aspreticola Podl. 63
  assadii A.A.R.Maassoumi & Podl. 63
  assyriacus Freyn & Bornm. 58
  astarabadensis Sirj. & Rech.f. 95
  asterabadensis Sirj. & Rech.f. 95
  asterias Ledeb. 117
  atricapillus Bornm. 95
  atropatanus Bunge 28
  atropilosulus (Hochst.)Bunge 80
  atropilosulus subsp. abyssinicus (Hochst.)J.B.Gillett 80
  atropurpureus Boiss. 97
  aucheri Boiss. 128
  auganus Bunge 63
  aulacolobus Boiss. 59
  aureus Willd. 28
  avicennicus Parsa 63
  avromanicus Rech.f. 99
  azizii A.A.R.Maassoumi 89
  baba-alliar Parsa 121
  baba-alliar subsp. baba-alliar 121
  baba-alliar subsp. nudicarpus (Sirj. & Rech.f.)S.Tietz 121
  babakhanloui A.A.R.Maassoumi & Podl. 105
  babatagi Popov 63
  babatagii Popov 63
  babensis Sirj. & Rech.f. 125
  bachtiaricus Bunge 73
  bachtiaricus var. *leiocarpus* Bornm. 73
  badachschanicus Boriss. 80
  baeticus L. 83
  baghdadensis Rech.f. 82
  baghensis Bunge 28
  baghlanensis Deml 47
  bagramiensis Sirj. & Rech.f. 102
  bahrakianus Grey-Wilson 79
  baibakht Eig 63

Astragalus (cont.)
　baijiensis C.Towns. 103
　baissunensis Lipsky 57
　bakaliensis Bunge 86
　bakuensis sensu Baker 67
　bamianicus Podl. 63
　barbatus Lam. 124
　barbicalyx Ali 72
　bardsiricus Parsa 58
　barrowianus Aitch. & Baker 65
　bashgalensis Podl. 59
　bashmensis A.A.R.Maassoumi 89
　basianicus Boiss. & Hausskn. 28
　basilicus Podl. & A.A.R.Maassoumi 63
　basineri Trautv. 57
　bayattii Bornm. & Gauba 97
　bazmanicus Podl. 81
　becki Bornm. 105
　beckii Bornm. 105
　beersabeensis Eig 64
　beershabensis Rech.f. 64
　behboudii Sirj & Rech.f. 38
　behen Bertol. 122
　belangerianus Fischer 33
　belcheraghensis Podl. 63
　bethlehemiticus subsp. lepidanthus (Boiss.)Eig 34
　bezudensis Sirj. & Rech.f. 57
　biabanensis Sirj. & Rech.f. 55
　bibracteolatus Sirj. & Rech.f. 38
　bienertii Bunge 28
　bifoliolatus Sirj. & Rech.f. 129
　bijugus Sirj. & Rech.f. 129
　biovulatus Bunge 118
　birjandicus Parsa 105
　birjandieus Parsa 105
　biserrula Bunge 123
　bizgensis Rech.f. 68
　bludhistanicus Bunge 121
　bludhistanus Bunge 121
　bludshistanus Bunge 121
　bodeanus Fischer 110
　bodinieri Leul. 84
　boeticus L. 83
　bombycinus Boiss. 120
　bombycinus subsp. sultanensis (Bunge)Sirj. 120
　bordschensis Bornm. 97
　borraginaceus Rech.f. 57
　borujerdicus Parsa
　botryophorus A.A.R.Maassoumi & Podl. 87
　botschantzevii Kamelin & Rassul. 78
　bounophilus Boiss. 91
　brachyanthus Freyn & Sint. 95
　brachycalyx Fischer 28
　brachycentrus Fischer 38
　brachyceras Ledeb. 60
　brachycladus Boiss. 122
　brachylobus Fischer 129

brachyodontus Boiss. 116
brachypetalus Trautv. 95
brachypodus Boiss. 92
brachystachys DC. 63
brachystachys var. argentatus Eig 63
bracteosus Boiss. & Noe 103
brahuicus Boiss. 56
brahuicus Boiss. 97
brecklei Deml 50
brevidens Freyn & Sint. 114
brevidentatiformis Sirj. & Rech.f. 114
breviflorus DC. 31
brevipes Bunge 116
brevipetiolatus Sirj. & Rech.f. 54
brotherusii Podl. 64
brugieri Boiss. 121
bruguieri Boiss. 121
bruguieri var. leiocladus Bornm. 121
bruguieri var. nudicarpus Sirj. & Rech.f. 121
brunsianus Bornm. 91
buchtormensis var. pseudobuchtormensis (Sirj. & Rech.f.)Parsa 71
buhseanus Bunge 97
bulla Fischer 112
bungeanus Boiss. 114
bungei Winkler & B.Fedtsch. 120
bunophilus Boiss. 91
bunophilus Boiss. 91
buschirensis Sirj. & Rech.f. 50
butleri Dinsm. 103
cabulicus Boiss. 139
calamistratus Podl. 64
calcareus Sirj. & Rech.f. 45
calcicolus Podl. 80
callainus Podl. 64
calliphysa Bunge 121
calliphysa subsp. angustifolius S.Tietz 121
calliphysa subsp. calliphysa 121
callistachys Buhse 112
callistachys subsp. callistachys 112
callistachys subsp. porphyrobaphis (Fischer)S.Tietz 112
callystachys Buhse 112
calvescens Bunge 81
camelorum Barbey 54
camporum Boiss. 87
camptoceras Bunge 120
campylanthoides Bornm. 82
campylanthus Boiss. 60
campylanthus var. ebenidioides Bornm. 60
campylanthus var. subglobosus Bornm. 60
campylorrhynchus Fischer & Meyer 88
campylosema Boiss. 87
campylosema subsp. atropurpureus (Boiss.)Chamberlain 97

Astragalus (cont.)
  campylosema subsp. campylosema 97
  campylotrichus Bunge 61
  cancellatus Bunge 114
  candolleanus Boiss. 101
  candolleanus var. pindreensis Baker 76
  canispinus Boiss. 44
  canus Bunge 114
  capito Boiss. 124
  cappadocicus Boiss. 87
  caprinus L. 64
  caprinus subsp. caprinus 64
  caprinus subsp. lanigerus (Desf.)Maire 64
  caraganae Hohen. 58
  caraganae var. brevicalyx Eig 58
  caramanicus Bornm. 112
  carduchorum Boiss. 45
  carduchorum subsp. mandaliensis Eig 45
  carmanicus Bornm. 112
  caroli-henrici Deml 47
  cartilagineus Gontsch. 64
  cartilagineus subsp. cartilagineus 64
  cartilagineus subsp. honigbergeri (Sirj. & Rech.f.) Podl. 64
  caryolobus Bunge 58
  caspicus M.Bieb. 28
  caspius M.Bieb. 28
  catabostrychos Kress-Deml & Podl. 64
  catacamptus Bunge 85
  cbraehycalyx Fischer 28
  cemerinus G.Beck 112
  cephalanthus DC. 112
  cephalanthus var. schirasicus (Fischer)Bornm. 112
  cephalotes Banks & Sol. 29
  cephalotes sensu auctt. 114
  cephalotes var. brevicalyx Eig 29
  cerasinus Baker 97
  cerasocrenus Bunge 29
  cercidophacos Podl. & A.A.R.Maassoumi 79
  cesarensis Sirj. & Bornm. 29
  chaborasicus Boiss. & Hausskn. 114
  chadjanensis Franchet 95
  chaetopodus Bunge 97
  chahartaghensis A.A.R.Maassoumi & Podl. 105
  chalaranthus Boiss. & Hausskn. 61
  chamaerionotus Rech.f. 69
  chamaesarathron Rech.f. 82
  chardinii Boiss. 88
  charguschanus Freyn 64
  chartaceus Ledeb. 53
  chartostegius Boiss. & Hausskn. 45
  chaworth-musteri Sirj. & Rech.f. 49
  chionobiiformis C.Towns. 43
  chionobius Bunge 44

  chionobius var. hirtus Boiss. 44
  chionophilus Boiss. 87
  chitralensis Ali 66
  chlamydophorus Bornm. 121
  chlorosphaerus Boiss. & Noe 91
  chlorostachys Lindley 80
  chlorostegius Boiss. & Hausskn. 103
  chlorotaenius Freyn & Bornm. 53
  choicus Bunge 99
  chorassanicus Bunge 29
  chorizanthus Rech.f. & Gilli 82
  christophi Trautv. 53
  chromolepis Boiss. & Hohen. 28
  chrysanthus Boiss. & Hohen. 65
  chrysanthus var. elburzensis Parsa 65
  chrysostachys Boiss. 92
  chrysostachys var. villosus Bornm. 92
  chrysotrichus Boiss. 108
  cicerifolius Bunge 50
  cinereus Willd. 97
  cisdarwasicus Gontsch. 129
  cisoxanus Podl. 65
  citrinus Bunge 65
  citrinus subsp. barrowianus (Aitch. & Baker) Podl. 65
  citrinus subsp. citrinus 65
  citrinus subsp. khorasanicus Podl. 65
  coelicolor Sirj. & Rech.f. 87
  coeruleus Parsa 97
  collenettiae Hedge & Podl. 54
  coluteocarpus Boiss. 80
  coluteopsis Parsa 110
  commixtus Bunge 56
  comosus Bunge 105
  concinnus Bunge 65
  concinnus Boiss. 65
  concinus Bunge 65
  conduplicatus Bertol. 59
  confertiformis Sirj. & Rech.f. 90
  confertissimus Kitam. 47
  confiniorum Boriss. 98
  confusus Bunge 98
  congestus Baker 132
  conjecturalis Schischkin 126
  connectens Podl. 65
  conrathii Freyn 117
  constrictus Turrill 38
  contortuplicatus L. 83
  controversus A.A.R.Maassoumi & Podl. 65
  cordatus Bunge 92
  cornu-caprae Sirj. & Rech.f. 103
  cornutus Bunge 122
  cornutus Pallas 129
  cornutus var. glaber Parsa 121
  coronilla Bunge 118
  corrugatus Bertol. 88
  corrugatus subsp. tenuirugis (Boiss.)Eig 88

Astragalus (*cont.*)
  cottonianus Aitch. & Baker 98
  courvidens Freyn & Bornm. 132
  crassinervius Boiss. 29
  crassispinus Bunge 104
  crassus Bunge 81
  crenophilus Boiss. 29
  cretaceus Boiss. & Kotschy 83
  cretensis Pallas 28
  crispocarpus Nab. 124
  cruciatus Link 88
  cruciatus sensu auctt. 117
  cruentus Balbis 88
  cucullaris Boiss. 52
  cuneifolius Bunge 125
  curvidens Freyn & Bornm. 132
  curviflorus Boiss. 104
  curvipes Trautv. 65
  curvipes var. transitorius Sirj. & Rech.f. 65
  curvirostris Boiss. 98
  curvirostris subsp. curvirostris Boiss. 98
  curvirostris subsp. psilocarpus C.Towns. 98
  cuscutae Bunge 98
  cyclophyllon G.Beck 98
  cylindraceus C.Meyer 107
  cyrusianus Parsa 86
  czuiliensis Golosk. 51
  dactylocarpus Boiss. 81
  dactylocarpus subsp. acinaciferus (Boiss.)E.Ott 81
  dactylocarpus subsp. dactylocarpus Boiss. 81
  daenaensis Boiss. 89
  daenensis Boiss. 89
  dahuricus Koch 117
  damascenus Boiss. & Gaill. 88
  danicus Retz 94
  darii Sirj. & Rech.f. 115
  darlingtonii Podl. 66
  darmadanicus Podl. 66
  decemjugus Bunge 47
  declinatus Willd. 74
  declinatus subsp. pinetorum (Boiss.)Ponert 74
  declinatus var. superglaber Freyn & Bornm. 74
  declinatus var. suprahirsutus Freyn 74
  decurrens Boiss. 51
  deickianus Bornm. 83
  deinacanthus Boiss. 29
  dekazygus Sirj. & Rech.f. 98
  delicatulus Podl. 66
  deluensis Bunge 166
  demavendicolus Bornm. & Gauba 113
  demavendicolus subsp. demavendicolus 113
  demavendicolus subsp. microphysopsis S.Tietz 113
  demavendicus Boiss. & Buhse 98
  demawendicus Boiss. & Buhse 98
  dendridium Fischer 44
  dendroproselius Rech.f. 129
  denticulatus Podl. 66
  derbendicus Bunge 62
  deserti-syriaci Eig 64
  dichroanthus Freyn & Sint. 51
  dictamnoides sensu Podl. 79
  dictyolobus Bunge 103
  dictyolobus sensu auctt. 103
  didymophysus Bunge 83
  dieterlei Podl. 119
  dietrichii C.Agerer-Kirchhoff 58
  dignus Boriss. 66
  dilankuri Lipsky 98
  dinsmorei Mout. 63
  diopogon Bunge 132
  dipelta Bunge 84
  diphtherites Fenzl 29
  diphtherites subsp. brachyanthus Eig 29
  discernendus Sirj. & Rech.f. 47
  distans Fischer 110
  distantior Turrill 39
  distinctissimus Rech.f. & Edelb. 119
  divaricatus Bunge 30
  diversifolius Trautv. 72
  djadjerudensis Sirj. & Rech.f. 52
  djashmensis Sirj., Rech.f. & Aellen 74
  djenarensis Sirj. & Rech.f. 85
  djiroftensis Sirj. & Rech.f. 112
  djuparensis Freyn & Bornm. 30
  dolius Boiss. & Hausskn. 39
  dorcoceras Bunge 60
  doshman-ziariensis A.A.R.Maassoumi & Podl. 124
  drepanophora Bornm. 98
  drepanophorus Bornm. 98
  drymophilus Bornm. 30
  dscheratuensis Sirj. & Rech.f. 76
  dscheratuensis var. viridis Sirj & Rech.f. 76
  dschuparensis Freyn & Bornm. 30
  duplostrigosus Dinsm. 56
  durandianus Aitch. & Baker 85
  durhamii Turrill 50
  durudensis Sirj. & Rech.f. 113
  ebenoides Boiss. 88
  eburneus Bornm. & Gauba 129
  ecbatanus Bunge 61
  echanensis Podl. 66
  echidna Bunge 30
  echidnaeformis Sirj. 30
  echinops Boiss. 51
  echinus DC. 30
  edelbergianus Sirj. & Rech.f. 66
  edmondsonii Podl. 66

Astragalus (cont.)
  effusus Bunge 115
  eigii C.Agerer-Kirchhoff 58
  ekatherinae B.Fedtsch. 57
  ekbergii Podl. 66
  elatior Kitam. 58
  elbrusensis Boiss. 91
  elbursensis Boiss. 91
  elegans Bunge 105
  elegantua Bornm. & Gauba 86
  elegantulus Bornm. & Gauba 86
  elisabethae Sirj. & Rech.f. 30
  elwendicus Bornm. 66
  elymaiticus Boiss. & Hausskn. 39
  elymiaticus Boiss. & Hausskn. 39
  emarginatus Labill. 83
  endytanthus Podl. & Deml 47
  ensifer Nab. 81
  entomophyllus Boiss. & Hausskn. 108
  ephemeretorum Gontsch. 67
  ephemeretorum subsp. bilobulatus Rassul. 67
  eremophilus Boiss. 86
  erianthus Willd. 32
  erinaceus Fischer 30
  erinifolius Pau 61
  eriocarpus M.Bieb. 108
  eriocephalus Willd. 31
  erionotus Bunge 66
  eriopodus Boiss. 106
  eriosphaerus Boiss. & Hausskn. 31
  eriostomus Bornm. 110
  eriostylus Boiss. & Hausskn. 31
  erpocaulos Boiss. 119
  erubescens Podl. 66
  erwinii-gaubae Sirj. & Rech.f. 54
  erythrolepis Boiss. 132
  erythroseminus Boiss. 67
  erythrosemius Boiss. 67
  eschkerensis Boiss. & Hausskn. 31
  esfandiarii Sirj. & Rech.f. 31
  esferayanicus Podl. & A.A.R.Maassoumi 67
  etezadianus Parsa 39
  eugenii Grossh. 106
  eusarathron Kress-Deml & Podl. 67
  eustrophacanthus Rech.f. & Edelb. 125
  evanensis A.A.R.Maassoumi & Podl. 67
  eximius Bunge 51
  expansus Boiss. 114
  expectatus A.A.R.Maassoumi 106
  fabaceus M.Bieb. 67
  facetus A.A.R.Maassoumi & Podl. 89
  faghirensis Sirj. & Rech.f. 73
  faizabadensis Sirj. & Rech.f. 52
  falcinellus Boiss. 86
  falconeri Bunge 80
  farakulumensis Sirj. & Rech.f. 31
  faridanicus Parsa 99
  farkharensis Podl. 67
  farmanfarmajani Sirj. & Rech.f. 43
  farsicus Sirj. & Rech.f. 55
  faryabensis Podl. 72
  fasciculifolius sensu C.Towns. 121
  fasciculifolius Boiss. 122
  fasciculifolius subsp. arbusculinus (Bornm. & Gauba)S.Tietz 122
  fasciculifolius subsp. fasciculifolius 122
  fatmensis Chiov. 91
  feddei Sirj. 97
  ferdovsicus Parsa 51
  ferociformis Sirj.,Rech.f. & Aellen 39
  fieldianus Huber-Mor. 62
  filagineus Boiss. 32
  filamentosus Bunge 67
  filicaulis Fischer & C.Meyer 118
  filicaulis subsp. rytilobus (Bunge)Popov 118
  finitimus Bunge 52
  firuzkuhensis Podl. 67
  fissilis Freyn & Sint. 34
  fissus Freyn & Sint. 34
  flabellatus Podl. 86
  flavicomus Bunge 51
  flemingii Ali 67
  flexilipes Bornm. 110
  flexirachis Rech.f. & Edelb. 48
  flexus Fischer 67
  floccosifolius Sumn. 67
  floccosus Boiss. 31
  florulentus Boiss. & Hausskn. 31
  forskahlei Boiss. 122
  fragiferus Bunge 113
  fragilidens Freyn & Sint. 29
  fragrans Willd. 46
  franziskae Deml 47
  fraternellus Bornm. 86
  fraxinella Bunge 62
  freitagii Deml 47
  fridae Rech.f. 99
  fruticosus Pallas 129
  fuliginosus G.Beck 99
  fursei Podl. 114
  gagnieui A.A.R.Maassoumi & Podl. 68
  galbineus A.A.R.Maassoumi 106
  galiifolius Podl. 68
  gandjehicus Parsa 113
  garaensis Sirj. 31
  gaubae Bornm. 68
  gaudanensis B.Fedtsch. 68
  gautieri Battand. & Trabut 91
  gelectus Freyn 34
  gemellus Podl. 54
  gemianus Boiss. & Hausskn. 31
  geminanus Boiss. & Hausskn. 31
  genistoides Boiss. 48
  gerensis Boiss. 81

Astragalus (*cont.*)
gerruensis Bornm. & Sirj. 39
getschesarensis Sirj. & Bornm. 39
gharemanii A.A.R.Maassoumi & Podl. 106
ghaznianus Sirj. & Rech.f. 39
ghilanicus Fischer 32
ghoortapacensis A.A.R.Maassoumi 106
ghorbandicus Podl. 68
gifanicus A.A.R.Maassoumi & Podl. 68
gigantostrobus Rech.f. & Aellen 32
gilgitensis Ali 68
gillettii C.Towns. 99
gillii Sirj. 39
gjunaicus Grossh. 91
glabellus Podl. 54
glaberrimus Sirj. & Rech.f. 116
glabrifolius Bunge 32
glabristylus Turrill 39
glabriusculus Gontsch. 70
glandulosus G.Beck 136
glaucacanthos Fischer 122
glaucops Bornm. 39
glaucopsoides Bornm. 32
globiceps Bunge 51
globiceps subsp. globiceps 51
globiflorus Boiss. 32
glochideus Boriss. 116
glumaceus Boiss. 92
glycophyllos L. 87
glycyphylloides DC. 87
glycyphyllos L. 87
glycyphyllos subsp. glycyphylloides (DC.)V.Matthews 87
glycyphyllos subsp. glycyphyllos 87
golestanicus A.A.R.Maassoumi & Podl. 68
gompholobium Bunge 68
goreanus Aitch. & Baker 55
gorganicus Sirj. & Rech.f. 87
gossypinoides Hand.-Mazz. 32
gossypinoides var. sindscharensis Sirj. 32
gossypinus Fischer 32
gracilipes Bunge 56
grahamianus sensu auctt. 49
grammocalyx Boiss. & Hohen. 87
grandiflorus Bunge 73
grantii Bunge 70
graveolens Benth. 84
gregarius Deml 47
grey-wilsonianus Podl. 68
griffithii Bunge 55
grisebachianus Aitch. & Baker 53
griseus Boiss. 106
groetzbachii Podl. 99
gudrunensis Boiss. 99
gueldenstaedtiae Bunge 99
guestii Eig 37
gummifer Labill. 32

guttatus Banks & Sol. 59
gymnopodus Boiss. 84
gypsaceus G.Beck 68
gysensis Del. 89
gyzensis Aitch. 89
gyzensis Del. 89
hadjarianus Parsa 65
hadroacanthus Rech.f. & Gilli 47
haematinus Sirj. & Rech.f. 95
haematodes Forsskal 57
haematoemius Sirj. & Bornm. 40
haematosemius Sirj. & Bornm. 40
hafez-shirazi Parsa 132
hamadanus Boiss. 51
hamosus L. 60
hamosus var. brachyceras (Ledeb.)Ledeb. 60
hamrinensis Bornm. 56
harpilobus Karelin & Kir. 89
harpocarpus Meff. 58
harsukhianus Rech.f. 86
hauarensis Boiss. 89
hedgeanus Podl. 68
hedgei C.Agerer-Kirchhoff 58
hedysaroides Willd. 107
helgurdensis C.Towns. 44
hemsleyi Aitch. & Baker 44
heratensis Bunge 32
herbertii A.A.R.Maassoumi 95
hermannii Freitag & Podl. 69
heteracanthus Bornm. 122
heterochrous Bornm. 72
heterochrus Bornm. 72
heterodoxus Bunge 106
heteromorphus Boriss. 99
heteroxus Bunge 106
hindukushensis Wendelbo 76
hippocrepidis Bunge 87
hirsutus Vahl 91
hirticalyx Boiss. 92
hirtus Bunge 124
hispidus Labill. 124
hoffmeisteri (Klotzsch)Ali 80
hohenackeri Boiss. 127
holdichianus Aitch. & Baker 100
hololeios Bornm. 47
hololeucus Boiss. & Buhse 91
holopsilus Bunge 106
holosemius Bunge 106
honigbergeri Sirj. & Rech.f. 64
hormozabadensis Sirj. & Rech.f. 64
horrictissimus Sirj. & Bornm. 40
horridissimus Sirj. & Bornm. 40
horridus Boiss. 44
horwoodii Eig 117
hosackioides (Benth.)Benth. 174
humifusa Baker 87
humilis M.Bieb. 127
huthianus Freyn & Bornm. 82
hymenocalyx Boiss. 51

241

Astragalus (*cont.*)
hymenochlaenus Bunge 62
hymenocystis Fischer & Meyer 92
hymenostegis Fischer 92
hypoglottis sensu Parsa 94
hypsogeton Bunge 33
hyrcanus Pallas 55
hystrix Fischer & Meyer 96
ibicinus Boiss. & Hausskn. 69
icmadophilus Hand.-Mazz. 45
idoneus A.A.R.Maassoumi 106
imbecillus A.A.R.Maassoumi & Podl. 69
imitans Podl. 80
immersus Baker 141
impexus Podl. 69
indistinctus Podl. & A.A.R.Maassoumi 69
inexpectatus A.A.R.Maassoumi & Podl. 83
infestus Boiss. 47
infestus var. glabrostipulatus Sirj. & Rech.f. 48
inquilinus A.A.R.Maassoumi 69
interiectus Kress-Deml 47
intermedius Boiss. 56
involucratus Lipsky 98
iodotropis Boiss. 45
iranicus Bunge 108
iranshahrii A.A.R.Maassoumi & Podl. 69
irmingardis Podl. 110
ischredensis Bunge 69
ishkamishensis Podl. 69
isopetalus Boiss. 85
ispahanicus Boiss. 126
ispirensis Boiss. 117
jabbor-khailii Kitam. 102
jagnobicus Lipsky 59
jakkabagi Lipsky 51
janthinus Boiss. & Hausskn. 40
jarmolenkoi Gontsch. 69
jesdianus Boiss. & Buhse 81
jessenii Bunge 52
jijaensis Sirj. & Rech.f. 61
jodostachys Boiss. & Buhse 117
jodotrophis Boiss. 45
jodotropis Boiss. 45
johannis Boiss. 69
jolderensis B.Fedtsch. 99
jubatus Bunge 121
jubatus var. semiglaber Bornm. 122
julii-gautieri Sirj. & Rech.f. 52
juratzkanus Freyn & Sint. 129
kabadianus Lipsky 129
kabristanicus Grossh. 106
kadschorensis Bunge 115
kahiricus DC. 85
kandaharensis Sirj. & Rech.f. 129
kapherrianus Fischer 93
karabaghensis Bunge 33

karabaghensis sensu Rawi 34
karadaghicus Eig 32
karakugensis Grossh. 55
karateginii Gontsch. 69
karelinii Bunge 56
karetensis Bunge 110
karputanus Boiss. & Noe 115
kashafensis Podl. 70
kashgakius Parsa 127
kashmarensis A.A.R.Maassoumi & Podl. 70
kaswinensis Bornm. 70
kavirensis Freitag 55
kellalensis Boiss. & Hausskn. 113
kendewanensis Gilli 107
kentrodes Fischer 81
keratensis Bunge 110
kerkuensis Bornm. 118
kerkukiensis Bornm. 118
kermanicus Parsa 107
kermanschahanenicus Sirj. & Rech.f. 81
kermanschahensis Bornm. 70
ketzkhovelianus Manden. 91
keyserlingii Bunge 40
khajiboulaghensis A.A.R.Maassoumi 107
khadjouicus Parsa 129
khaneradarensis Sirj. & Rech.f. 44
khayamicus Parsa 113
khorramabadensis Bornm. 115
khoshjailensis Sirj. & Rech.f. 110
khuzistanicus Sirj. & Rech.f. 64
khwaja-muhammadensis Podl. 70
kirghizicus Stschegl. 119
kirkukensis Eig 120
kirpicznikovii Grossh. 70
kirrindicus Boiss. 52
knappii Bornm. 40
kneuckeri Freyn 122
kochianus Sosn. 117
kodzhorensis Grossh. 115
koelzii Barneby 124
kohrudicus Bunge 92
kopetdaghi Boriss. 70
korovinianus Barneby 72
koschukensis Boiss. 79
koshukensis Boiss. 79
kotschyanus Boiss. 114
kotschyanus Fischer 120
kourosianus Parsa 107
kucanensis Rech.f. 73
kuhikakaschanicus Sirj. & Rech.f. 62
kuhistanus Bunge 33
kukkonenii Podl. 70
kulabensis Lipsky 52
kullmannii Podl. 59
kunarensis Podl. 70
kuntzei E.Sheldon 102
kuramensis Baker 81

Astragalus (cont.)
  kurdicus Boiss. 33
  kurdicus subsp. kurdorum Eig 33
  kurdicus var. muschianus (Kotschy & Boiss.)Chamberlain & V.Matthews 35
  kurrumensis Bunge 72
  kuschakevitschii O.Fedtsch. 102
  kuschkensis Boriss. 70
  laetus Bunge 70
  lagonyx Fischer 33
  lagopodioides Willd. 94
  lagopoides Lam. 92
  laguriformis Freyn 92
  laguroides Freyn 92
  lagurus Willd. 92
  lalandei Podl. 71
  lalesarensis Bornm. 110
  lambinonii Podl. 71
  lanceolatus Bunge 71
  lancifolius Gontsch. 130
  lanigerus Desf. 64
  laricus Boiss. 84
  laristanicus Bornm. & Gauba 107
  lasiosemius Boiss. 48
  lasiosemius var. leiosemius 48
  lasiosericeus Hook. et al. 48
  lasius Blatter 87
  lateritians Freyn & Bornm. 33
  lateritius Boiss. & Hausskn. 33
  latifolius Lam. 99
  latifolius var. choicus Boiss. 99
  latistylus Freyn 48
  laxus Boiss. & Hausskn. 126
  ledinghamii Barneby 125
  leiocalyx Bunge 70
  leioclados Boiss. 33
  leiocladus Boiss. 33
  leiolobus Bunge 88
  leiophyllus Freyn & Bornm. 44
  leiosemius (Lipsky)Popov 48
  leiosemius sensu Rech.f. 49
  leonardii A.A.R.Maassoumi 71
  lepidanthus Boiss. 34
  lepidotrichus Bornm. & Gauba 81
  lepidus Podl. 71
  leptacanthus Boiss. & Buhse 44
  leptodendron Fischer 28
  leptodermus Bunge 118
  leptorhaphis Bornm. & Gauba 127
  leptus Boiss. 48
  leptus var. bamianicus Sirj. & Rech.f. 48
  leptus subsp. ghazniensis Deml 48
  leptus subsp. leptus 48
  leptus subsp. longipedunculatus Deml 48
  leucargyreus Bornm. 92
  leucocephalus Bunge 120
  leucomelas Bunge 58
  leucophanus Bornm. 99
  leucoptilus Boiss. & Hausskn. 34
  leucospermus Bunge 50
  levieri Freyn 46
  lilacinus Boiss. 115
  lineatus Lam. 87
  lippertii A.A.R.Maassoumi 107
  litostachys Boiss. & Hausskn. 40
  litostachyus Boiss. & Hausskn. 40
  lobophorus Boiss. 73
  lobophorus var. pilosus (Bornm.) C.Towns. 73
  longemucronulatus Sirj. & Rech.f. 40
  longepedunculatus Sirj., Rech.f. & Aellen 78
  longibracteatus Sommier & Levier 108
  longicuspidatus St.Lager 99
  longicuspis Bunge 99
  longidens Freyn 87
  longiflorus Del. 85
  longifolius Lam. 34
  longimucronulatus Sirj. & Rech.f. 40
  longirostratus Pau 107
  longistipitatus Boriss. 123
  longistylus Bunge 40
  lovensis Rech.f. 100
  lugubris Rech.f. & Koie 43
  lumsdenianus Aitch. & Baker 110
  lunatus Pallas 115
  luristanicus Freyn 40
  luristanicus Bornm. 113
  lurorum Bornm. 113
  lurorum var. chamchidensis (Sirj. & Rech.f.)Parsa 113
  luteo-coerulea Baker 143
  luteocarpus Baker 90
  lycioides Boiss. 44
  maaratensis Eig 63
  maassoumii Podl. 71
  macrobotrys Bunge 55
  macrobotrys var. camelorum Velen. 54
  macrocephalus Willd. 52
  macrocephalus subsp. cucullaris (Boiss.)Chamberlain 52
  macrocephalus subsp. finitimus (Bunge)Chamberlain 51,52
  macrocephalus subsp. macrocephalus 52
  macronyx Bunge 71
  macropelmatus Bunge 71
  macropelmatus subsp. macropelmatus 71
  macropelmatus subsp. pseudobuchtormensis (Sirj. & Rech.f.)Podl. 71
  macropterus DC. 90
  macrorus Fischer & Meyer 107
  macrosemius Boiss. 45
  macrostachys DC. 107
  macrostegius Rech.f. 60
  macrosyrinx Rech.f. 131
  macrotropis sensu auctt. 129

Astragalus (cont.)
  macrourus Fischer & Meyer 107
  maculatus Lam. 57
  maculatus Bunge 100
  magistratus A.A.R.Maassoumi et al. 127
  magnifolius Parsa 40
  magnus Sirj. & Rech.f. 68
  maharluensis Bornm. & Gauba 56
  malacophyllus Bunge 76
  malanogramma Bornm. 111
  managettae Sirj. & Rech.f. 71
  mangeri Bornm. 100
  manucehrii Sirj. & Rech.f. 92
  mardabadensis Bornm. & Gauba 124
  mareoticus Del. 89
  masanderanus Bunge 83
  masenderanus Bunge 83
  massagetovii B.Fedtsch. 125
  maverranagri Popov 130
  maximus Willd. 52
  maymanensis Podl. 72
  medicagineus Boiss. 114
  medorum Bornm. 40
  medullaris Boiss. 84
  megalacmus Freyn & Sint. 52
  megaloceras Eig 57
  megalocystis Bunge 111
  megalomerus sensu Podl. 104
  megalotropis Bunge 52
  melaleucus Bunge 52
  melanocalyx Boiss. & Buhse 130
  melanochiton Deml 48
  melanodon Boiss. 107
  melanogramma Bornm. 111
  melanostachys Bunge 60
  melanostictus Freyn 92
  melanosticus Freyn 92
  memnonius A.A.R.Maassoumi & Podl. 107
  mercklinii Boiss. & Buhse 100
  meridionalis sensu auctt. 52
  meridionalis Bunge 52
  merkianus Aitch. & Baker 111
  merxmuelleri Podl. 119
  meschedensis Bunge 34
  meschhedensis Bunge 34
  meseleios Boiss. 34
  mesoleios Boiss. & Hohen. 34
  mesopotamicus Boiss. 93
  meyeri Boiss. 34
  michauxianus Boiss. 34
  michelsoni B.Fedtsch. 57
  micracme Boiss. 111
  micrancistrus Boiss. & Hausskn. 100
  microcalycinus Sirj. & Rech.f. 130
  microcephalus Willd. 34
  microcephalus subsp. pycnocladus (Boiss. & Hausskn.)Eig 34
  microdontus Baker 81
  microphysa Boiss. 113
  microphysa var. durudensis (Sirj. & Rech.f.)Parsa 113
  microphysa var. paucijugus Sirj. & Rech.f. 113
  microthamnus Boiss. & Hausskn. 122
  migpo Kamelin 114
  mikrophyton Sirj. & Rech.f. 127
  minutifoliolatus Wendelbo 78
  minutissimus Freyn & Bornm. 90
  minuto-foliolatus Wendelbo 78
  minutulus A.A.R.Maassoumi 90
  minutus Boiss. 119
  miralamensis Podl. 72
  mirus Sirj. & Rech.f. 114
  miseriflorus Sirj. & Rech.f. 60
  mishouensis Turrill 41
  mobayenicus Parsa 68
  modestus Boiss. & Hohen. 72
  mokurensis Sirj. & Rech.f. 79
  molestus Rech.f. 48
  mollis M.Bieb. 108
  mollis var. iranicus (Bunge)Boiss. 108
  monanthemus Boiss. 72
  monozyx Bornm. 100
  morgani Freyn 35
  mossulensis Bunge 115
  mossulensis var. complicatus Eig 115
  mossulensis var. multiflorus Eig 115
  movlavicus Parsa 71
  mozaffarianii A.A.R.Maassoumi 72
  mucronifolius Boiss. 104
  mucronifolius subsp. robustus Sirj. & Rech.f. 104
  mukusiensis Rech.f. 91
  multijugus DC. 72
  multijugus Grossh. 117
  multispinus Freyn & Bornm. 35
  mundulus Podl. 123
  murinus Boiss. 111
  muschianus Kotschy & Boiss. 35
  musilii Velen. 120
  myriacanthus Boiss. 35
  myrianthus G.Beck 41
  myriocladus Sirj. & Rech.f. 49
  myriocystis Bornm. 122
  myriocystis var. albiflorus Parsa 122
  nachitschevanicus Rzazade 58
  nadirius Parsa 88
  nadjafabadensis Parsa 52
  naftabensis Sirj. & Rech.f. 93
  nahavandicus A.A.R.Maassoumi 108
  nawabianus Aitch. & Baker 118
  nedjefabadensis Parsa 52
  negevensis Zoh. & Fertig 88
  neglectus Freyn 34
  neilreichianus Freyn & Sint. 129
  neo-mobayenii A.A.R.Maassoumi 72
  neo-podlechii A.A.R.Maassoumi 72
  neoalbovillosus Kitam. 57
  neostipitatus Kitam. 125

Astragalus (*cont.*)
neoverticillatus Kitam. 72
nephtonensis Freyn 72
nervistipulus Boiss. & Hausskn. 92
neubauerianus Sirj. & Rech.f. 72
neurocentron Franchet 104
neurophyllus Franchet 104
nicharensis var. longepedunculatus (Sirj., Rech.f. & Aellen)Parsa 78
nigrescens Popov 55
nigricans Barneby 55
nigritus Sirj. & Rech.f. 84
nigrivestitus Podl. & Deml 48
nigrolineatus Sirj. & Rech.f. 130
ninae Gontsch. 56
nishapurensis Sirj. & Rech.f. 111
nitens Boiss. & Heldr. 130
nitens Boiss. & Buhse 131
nitidulus Hand.-Mazz. 96
nivalis Karelin & Kir. 96
noemiae Eig 32
novus Grossh. 128
noziensis Sirj. & Rech.f. 111
nudicarpus Rech.f. 121
nudus Gontsch. 70
nummularius sensu auctt. 74
nurabadensis A.A.R.Maassoumi & Podl. 72
nurensis Boiss. & Buhse 95
nuristanicus Sirj. & Rech.f. 76
nuristanicus Kitam. 80
nuristanicus var. elasoonensis Sirj. & Rech.f. 76
obcordatus Boiss. 52
obtusifolius DC. 50,52
ochreatus Bunge 82
ochrobius Bunge 35
ochrochlorus Boiss. & Hohen. 44
ochtoranensis Freyn 35
octopus C.Towns. 35
odoratus Lam.
oleifolius sensu Rawi 33
oleifolius var. kurdistanicus Sirj. 33
olginii Bunge 115
oligophyllus Boiss. 55
olivieri Bunge 93
oloricus Manden. 52
olufsenii Freyn 95
oncotrichus Bunge 61
onobrychioides M.Bieb. 114
onobrychis L. 115
oocephalus Boiss. 53
oocephalus subsp. stachyophorus Huber-Mor. & Chamberlain 53
oocephalus var. stachyophorus (Huber-Mor. & Chamberlain)Ponert 53
ophiocarpus Bunge 116
ophiocarpus var. plurijugus Sirj. 116
ophiocarpus var. vermicularis (Bornm.) Sirj. 116
oplites Parker 50

orbiculatus Ledeb. 73
ornithopodioides Lam. 117
oroproselius Rech.f. 100
orthanthoides Boriss. 96
orthanthus Freyn 96
orthocarpoides Sirj. & Rech.f. 58
orthocarpus Boiss. 58
orthorhynchus Bornm. 118
otiporensis Boiss. 123
ourmitanensis Franchet 51
ovczinnikovii Boriss. 123
ovigerus Boiss. 44
ovinus Boiss. 73
ovoideus Sirj. & Rech.f. 104
oxyglottis M.Bieb. 124
oxyglottoides Rech.f. & Gauba 123
oxyrrhynchus Fischer & Meyer 89
pachyacanthus Bunge 35
pachyrachis Sirj. & Rech.f. 122
pachystachys Bunge 35
paghmanensis Sirj. & Rech.f. 49
paktiensis Podl. 123
palaestinus ssp. heteranthesmus Eig 120
palmyrensis Post 120
pamiricus B.Fedtsch. 64
pamphylicus Boiss. 120
panjaoensis Sirj. & Rech.f. 100
papillosus Podl. 73
paradoxus Bunge 105
paralipomenus Bunge 41
paralurges Bunge 93
paraplesius Bunge 45
parimanicus Parsa 84
parrowianus Boiss. & Hausskn. 36
parvistipulus Rech.f. 79
parvulus Bornm. 73
parwanicus Podl. & Deml 48
patentipilosus Kitam. 73
patrius A.A.R.Maassoumi 108
patulepilosus Sirj. & Rech.f. 130
paulsenii Freyn 116
pauperiflorus Bornm. 108
pecten-erinis Kress-Deml 48
pecten-hystricis Kress-Deml 49
pectinatus Boiss. 51
peduncularis Royle 128
pellitus Bunge 73
pellitus var. orbicularis Parsa 73
peltatus Podl. & Deml 119
pendanthus Boiss. 86
penetratus A.A.R.Maassoumi 90
penicillatus Podl. 73
pentanthus Boiss. 86
pentapetaloides Bunge 67
pentapetaloides var. blepharophyllus Bunge 67
perdurans Podl. 73
peregrinus Vahl 120
peristereus Boiss. & Hausskn. 36

Astragalus (cont.)
  perpexus A.A.R.Maassoumi 95
  perplexans Podl. 73
  perpusillus Bertol. 119
  persepolitanus Boiss. 118
  persicus (DC.)Fischer & Meyer 93
  phalacrophyton Deml 49
  phaulanthus Turrill 59
  phlomoides Boiss. 103
  phyllokentrus Hausskn. & Bornm. 104
  phyllostachys Boiss. & Hausskn. 41
  pichleri G.Beck 41
  pictus Boiss. 59
  piestolobus Bunge 73
  piletocladus Freyn & Sint. 41
  pindreensis (Baker)Ali 76
  pinetorum Boiss. 73
  pinetorum subsp. alamutensis A.A.R.Maassoumi & Podl. 74
  pinetorum subsp. declinatus Podl. 74
  pinetorum var. drusorum Eig 63
  pinetorum subsp. pinetorum 74
  piptocephalus Boiss. & Hausskn. 41
  piranshahricus A.A.R.Maassoumi & Podl. 74
  pirimukurunicus Eig 34
  pish-chakensis A.A.R.Maassoumi 95
  pithyusarum Bornm. 97
  plagiophacos A.A.R.Maassoumi & Podl. 120
  platyraphis Fischer 64
  platysematus Bunge 100
  plebeius Boiss. 108
  podlechii Deml 49
  podocarpus C.Meyer 108
  podosphaerus Boiss. & Hausskn. 55
  polemius Boiss. 49
  poliotrichus Bornm. 127
  polyacanthus sensu Baker 49
  polyanthus Bunge 36
  polybotrys Boiss. 74
  polyphyllus Bunge 62
  ponticus Pallas 53
  porphyrobaphis Fischer 112
  porphyrocystis Bornm. 113
  porphyrodon C.Towns. 93
  porphyrophysa Bornm. & Gauba 122
  procerus Boiss. & Hausskn. 100
  prolixus Bunge 91
  protractus Boriss. 108
  prusianus Boiss. 36
  pseudaureus Sirj. & Rech.f. 36
  pseudo-beckii Sirj. & Rech.f. 108
  pseudoangustifolius Sirj. & Rech.f. 44
  pseudoasterothrix Hand.-Mazz. 68
  pseudobrachystachys Sirj. & Rech.f. 74
  pseudobrachytropis Gontsch. 117
  pseudobuchtormensis Sirj. & Rech.f. 71

pseudocancellatus Grossh. 114
pseudocyclophyllus Rech.f. 100
pseudofragiferus S.Tietz 113
pseudofragrans C.Towns. 95
pseudogompholobium Podl. 74
pseudohofmeisteri Sirj. & Rech.f. 123
pseudoibicinus A.A.R.Maassoumi & Podl. 74
pseudoibicinus subsp. kowlikoshensis 74
pseudoibicinus subsp. pseudoibicinus A.A.R.Maassoumi & Podl. 74
pseudoindurascens Sirj. & Rech.f. 74
pseudojohannis A.A.R.Maassoumi & Podl. 75
pseudokurrumensis Sirj. & Rech.f. 75
pseudolanuginosus Gontsch. 67
pseudomongholicus Sirj. & Rech.f. 80
pseudomossulensis Nab. 115
pseudomultijugus Podl. 75
pseudoparrowianus Sirj. & Rech.f. 41
pseudopellitus Podl. 75
pseudopendulinus Sirj. & Rech.f. 130
pseudopsilacanthus Ali 49
pseudopsilacanthus subsp. polyneurus Deml 49
pseudopsilacanthus subsp. pseudopsilacanthus 49
pseudosquarrosus Sirj. & Rech.f. 55
pseudosulfuratus Podl. 57
pseudoszovitsii Sirj. & Rech.f. 111
pseudotomentellus Podl. 75
pseudoutriger Grossh. 75
pseudovinus A.A.R.Maassoumi & Podl. 75
pseudozagrosicus A.A.R.Maassoumi & Podl. 75
psilacanthus sensu auctt. 47
psilacanthus Boiss. 49
psilacanthus subsp. psilacanthus 49
psilacmos Bunge 28
psilocentros Fischer 49
psilodontius Boiss. 36
psiloglottis DC. 124
psilopterus Bunge 48
psilostylus Bunge 36
ptilocephalus Baker 36
ptychophyllus Boiss. 113
ptychophyllus var. longepedunculatus Sirj. & Rech.f. 113
ptychophyllus var. xerxis Sirj. & Rech.f. 113
pubifolius V.V.Nikitin 73
pulchellus Boiss. 84

Astragalus (cont.)
puligumrensis Rech.f. 130
puligumuriensis Rech.f. 130
pulvinatus Bunge 36
punctatus Bunge 100
punjabicus Sirj. 87
purpurascens Bunge 75
pushtashanicus C.Towns. 41
pycnocephalus Fischer 37
pycnocephalus var. nabelekii Sirj. 37
pycnocladus Boiss. & Hausskn. 34
pycnophyllus Steven 34
pyrrhotrichus Boiss. 75
quadrisulcatus Bunge 88
quercetorum Rech.f. 129
quinquefoliatus Bunge 100
quinquefoliolatus Bunge 100
quinquejugus Sirj. & Rech.f. 55
quisqualis Bunge 112
racemulosus Boiss. & Hausskn. 61
racemulosus var. leptorhachis Rech.f. 61
raddei Basilevsk. 88
radkanensis Bunge 41
radschirdensis Sirj. & Bornm. 37
rahensis Sirj. & Rech.f. 41
ramiaensis Sirj. & Rech.f. 62
ramianensis Sirj. & Rech.f. 62
ramicaudex Chamberlain 74
raphiodontus Boiss. 49
rarus Sirj. & Rech.f. 71
rassoulii Podl. 75
raswendicus Hausskn. & Bornm. 111
rauwolfii Pallas 122
rauwolfii Vahl 122
rawianus C.Towns. 75
rawlinsianus Aitch. & Baker 108
rayatensis Eig 32
razicus Parsa 85
recognitus Fischer 93
recollectus Rech.f. 74
refractus Boiss. & Buhse 97
refractus C.A. Meyer 101
regelii Trautv. 51
regestus A.A.R.Maassoumi 90
registanicus Rech.f. 85
remotiflorus Boiss. 111
remotifolius Boiss. & Hausskn. 126
remotijugus Boiss. & Hohen. 76
remotijugus var. pumilus Parsa 67
renzianus Podl. 76
reshadensis Podl. 79
retamocarpus Boiss. 58
retamocarpus subsp. albiflorus Freyn & Sint. 58
reticulato-venosus A.A.R.Maassoumi & Podl. 76
reticulatus M.Bieb. 89
reuterianus Boiss. 113
rhabdophorus Bornm. 76
rhizanthus Benth. 76

rhizanthus subsp. rhizanthus 76
rhizocephalus Baker 76
rhizocephalus subsp. hindukushensis (Wendelbo)Podl. 76
rhizocephalus subsp. rhizocephalus 76
rhodochrous Boiss. & Hausskn. 37
rhodosemius Boiss. & Hausskn. 42
rhodosenus Boiss. & Hausskn. 42
rimarum Bornm. 95
riouxii Rech.f. 101
rivashensis A.A.R.Maassoumi 108
robustus Bunge 101
rollovii Grossh. 101
rosae C.Agerer-Kirchhoff 58
rosellus Sirj. & Rech.f. 42
rostratus C.Meyer 101
rotundifolius Royle 84
rubriflorus Bunge 93
rubrocalycinus A.A.R.Maassoumi & Podl. 76
rubrolineatus Sirj. & Rech.f. 111
rubromarginatus Czerniak. 76
rubromarginatus subsp. rubromarginatus 77
rubrostriatus Bunge 93
rudbaricus Bunge 98
rudehendicus Sirj., Rech.f. & Aellen 52
rufescens Freyn 77
rugosus Boiss. 73
rugosus var. humilis Parsa 62
rugosus var. pilosus Bornm. 73
ruscifolius Boiss. 130
russelii Banks & Sol. 122
russellii Banks & Sol. 122
russellii var. hirsutus Bornm. 122
rusticus A.A.R.Maassoumi 108
rytilobus Bunge 118
saadius Parsa 81
sabzakensis C.Agerer-Kirchhoff 59
saccatus Boiss. 88
sachanewii Sirj. 109
saetiger R.Becht. 53
saganlugensis Trautv. 125
sahendi Buhse 46
salangensis Podl. 77
samamensis Boiss. & Buhse 74
samarkandinus Freyn 71
sanandajianus S.Tietz 113
sanctus var. hamadensis Velen. 128
sangcharakensis Podl. 77
sangimashensis Rech.f. 55
sangonensis Sirj. & Rech.f. 125
sapozhnikovii Schischkin 109
sarae Eig 42
sarcocolla Dymock 86
sarobiensis Rech.f. 131
sarypulensis B.Fedtsch. 131
savellanicus Podl. 77
saxatilis Freyn & Bornm. 96

Astragalus (cont.)
   saxifractor Rech.f. & Gilli 102
   schachdarinus Lipsky 82
   schahrudensis Bunge 53
   schahrudicus Bunge 53
   scharudensis Bunge 53
   schemachensis Karj. 77
   scheremetevianus B.Fedtsch. 123
   schimperi Boiss. 118
   schimperi subsp. magnus Eig 118
   schirasicus Fischer 112
   schirazicus Fischer 112
   schirkuhicus Bornm. 101
   schistocalyx Bunge 44
   schistocalyx var. bizgimontanus Sirj. & Rech.f. 44
   schistocalyx subsp. schistocalyx 44
   schistocalyx subsp. sclerocladus (Bunge)Deml 44
   schistosus Boiss. & Hohen. 117
   schmalhausenii Bunge 124
   schmidii Podl. 77
   schugnanicus B.Fedtsch. 60
   schurabicus Bunge 121
   sciureus Boiss. & Hohen. 93
   scorpiurus Bunge 87
   scorpius Boiss. 122
   sefinensis Bornm. 29
   segazicus Parsa 42
   seidabadensis Bunge 93
   seidlitzii Bunge 74
   selemiensis Mout. 63
   semilunatus Podl. 77
   semipellitus Bunge 37
   semnanensis Bornm. & Rech.f. 104
   senganensis Bunge 34
   senilis Bornm. 84
   sericostachys Stocks 53
   serpentinicus Sirj. & Rech.f. 37
   sesameus Sibth. & Sm. 118
   sesamoides Boiss. 118
   sessiliceps Bornm. 51
   setulosus Gontsch. 125
   shahpouricus Parsa 109
   sharestanicus Podl. & Deml 49
   sharifii Sirj. & Rech.f. 42
   shatuensis Podl. 77
   shebarensis Podl. 85
   shelkovnikovii Grossh. 126
   shirkuhicus Bornm. 101
   siahderrensis Sirj. & Rech.f. 59
   sieberi DC. 82
   sieversianus Pallas 59
   sikaramensis (Sirj. & Rech.f.)Ali 144
   silachorensis Bornm. 62
   siliquosus Boiss.1846 73
   siliquosus Boiss. 126
   siliquosus subsp. siliquosus 126
   siliquosus subsp. stramineus (Boiss.)E.Kozik & Podl. 126
   sinaicus Boiss. 118
   singarensis Boiss. & Hausskn. 84
   sirdjanicus Parsa 81
   sirensis Turrill 93
   sitiens Bunge 130
   sivandi Parsa 113
   sivasicus Bunge 33
   soficus Bunge 37
   sojakii Podl. 77
   sorkhabadensis Sirj. & Rech.f. 142
   sp.(sect.Rhacophorus) S.Collenette 30
   sp.2408 S.Collenette 86
   spachianus Boiss. 109
   sparsus Decne. 82
   speciosus Boiss. & Hohen. 53
   sphaeranthus Boiss. 125
   sphaerocephalus Steven 52
   spinellifer Rech.f. & Gilli 48
   spinellus Boiss. & Hausskn. 45
   spinescens Bunge 81
   spinescens subsp. aitchisonii Eig 81
   spinosus (Forsskal)Muschler 122
   spinosus var. hamrinensis Eig 122
   spinosus var. kneuckeri (Freyn)Tackh. & Boulos 122
   splendidissimus Sirj. & Rech.f. 37
   spongocarpus Meff. 58
   spruneri Boiss. 101
   sprunerianus Bunge 101
   squarrosus Bunge 55
   ssahandi Fischer 46
   ssahendi Fischer 46
   ssyrtchensis Bunge 44
   staintonianus Ali 645
   stalinskyi Sirj. 56
   stapfii Sirj. 42
   steineranus Podl. 130
   stella Gouan 118
   stenanthus Freyn 67
   stenolepis Fischer 42
   stenonychioides Freyn & Bornm. 37
   stenopterus Sirj. & Rech.f. 50
   stenostachys G.Beck 106
   stenostegius Boiss. & Hausskn. 45
   stephanianus Aitch. & Baker 57
   stephenianus Aitch. & Baker 57
   steppicola Sirj., Rech.f. & Aellen 42
   stepporum Podl. 53
   stevenianus DC. 117
   stipitatus Bunge 125
   stipitatus subsp. angustifructus 125
   stipitatus subsp. shatuensis Podl. 125
   stipitatus subsp. stipitatus 125
   stocksii Bunge 79
   stocksii var. elongatus Bunge 79
   stocksii var. honigbergeri Sirj. 79
   stojani Nab. 53
   stramineus Boiss. & Kotschy 126
   straussii Bornm. 93
   striatellus M.Bieb. 59

Astragalus (cont.)
strictifolius Boiss. 37
strictipes Bornm. 115
strigosostipulatus Rech.f. & Koie emend. Deml 50
strobiliferus Benth. 37
subalpinus Boiss. & Buhse 99
subalpinus Freyn 101
subangustidens V.V.Nikitin 65
subconduplicatus Ali 71
suberosus Banks & Sol. 120
suberosus subsp. ancyleus (Boiss.) V.Matthews 120
suberosus subsp. mersinensis (Sirj.) V.Matthews 120
suberosus subsp. suberosus 120
subinduratus var. pseudoindurascens (Sirj. & Rech.f.)Parsa 74
subinduratus var. pseudoinduratus (Sirj. & Rech.f.)Parsa 74
submaculatus Boriss. 100
submitis Boiss. & Hohen. 111
subrobustus Boriss. 101
subrosulariformis Sirj. & Rech.f. 77
subscaposus Boriss. 60
subsecundus Boiss. & Hohen. 103
subulatus M.Bieb. 131
subuliformis DC. 131
subumbellatus Klotzsch 87
suffalcatus Bunge 131
sulfuratus Rech.f. & Gilli 59
sulfureus Bunge 126
sultanabadensis Sirj. & Rech.f. 62
sultanensis Bunge 120
sultani Ali 68
sulukliensis Freyn & Sint. 108
superbus Bunge 51
superbus var. ovalifoliolatus Sirj. & Rech.f. 51
superfluus Rech.f. & Koie 100
supervisus (O.Kuntze)E.Sheldon 10
supinus C.A.Meyer 46
supraglaber Kitam. 60
supralanatus Freyn & Sint. 65
susianus Boiss. 60,61
susianus subsp. sericeus S.Tietz 61
susianus subsp. susianus 61
sykesiae N.Simpson 100
sympileicarpus Rech.f. 86
syrtschensis Bunge 44
szovitsii Fischer & C.Meyer 111
tabrisianus Boiss. & Buhse 94
taftanicus Parsa 104
takharensis Podl. 77
takhtadjanii Grossh. 109
takhtadzjanii Grossh. 109
talagonicus Boiss. 42
talarensis Sirj. & Rech.f. 38
talbotianus Aitch. & Baker 55
talimansurensis Sirj. & Rech.f. 104
taluquanensis Podl. 58

talyschensis Bunge 74
tarumensis Sirj. & Rech.f. 56
taschkendicus Bunge 101
tashqurghanicus Rech.f. & Podl. 85
tauricolus Boiss. 109
tavernieri Boiss. 72
tawilicus C.Towns. 103
tecti-mundi Freyn 79
teheranicus Boiss. 116
tempskyanus Freyn & Bornm. 101
tenax Bunge 94
tenellus Bunge 102
tenuijugis Boiss. 88
tenuirugis Boiss. 88
tenuirugis var. brevipedunculatus Parsa 88
tenuiscapus Freyn & Bornm. 109
tephreschensis Rech.f. et al. 94
tephrosioides Boiss. 59
tephrosoides Boiss. 59
terekliensis Gontsch. 78
termeanus A.A.R.Maassoumi & Podl. 109
terrestris Kitam. 50
tetanocarpus Bornm. & Gauba 99
tetragonocarpus Boiss. 126
thaumasios Podl. 126
thessalus Boiss. 101
thionanthus Bornm. 102
thlaspi Lipsky 127
thyrsiflorus Sirj. & Rech.f. 94
tianschanicus var. pamiricus B.Fedtsch. 64
tibetanus Bunge 95
tigridis Boiss. 101
timuranus Franchet 51
tolgorensis Sirj. & Rech.f. 131
tomentellus Podl. 78
toppinianus Ali 78
torrentum Bunge 62
tortuosus DC. 111
totchalensis Bornm. 38
totschalensis Bornm. 38
touranicus Freitag & Podlech 78
trachoniticus Post 131
trachyacanthus Fischer 42
tribuloides Del. 119
tribuloides var. kirghizicus (Stschegl.) Sirj. 119
tricholobus DC. 127
trigonelloides Boiss. 117
triocholobus DC. 127
triquetrer Bornm. & Gauba 131
triradiatus Bunge 119
trottianus Parsa 99
truncato-alatus Sirj. & Rech.f. 102
tscharikarensis Sirj. & Rech.f. 59
tuberculosus DC. 120
tuberculosus subsp. mersinensis Sirj. 120
tumidus M.Bieb. 67

Astragalus (cont.)
  tumidus Willd. 122
  turbat-haidariensis Sirj. & Rech.f. 71
  turbinatus Bunge 53
  turcomanicus var. elegans Bornm. &
    Gauba 54
  turczaninovii Karelin & Kir. 126
  turkestanus Bunge 59
  turrillianus Parsa 42
  turrillii Eig 27
  typhaeformis A.A.R.Maassoumi 109
  uliginosus M.Bieb. 87
  ulodjensis Sirj. & Rech.f. 125
  ulothrix G.Beck 102
  unifoliolatus Sirj. & Rech.f. 124
  uranolimneus Boiss. 94
  urbanus Podl. &
    A.A.R.Maassoumi 78
  urgunensis Podl. 56
  urgutinus Lipsky 103
  urmiensis Bunge 78
  ustiurtensis Grossh. 114
  vanensis Sosn. 127
  vanillae Boiss. 82
  variegatus Freyn & Bornm. 77
  variegatus Franchet 131
  variifolius Freyn & Sint. 129
  variistipula Turrill 117
  vegetus Bunge 116
  velenovskyi Nab. 94
  velenowskyi Nab. 94
  venulosus Boiss. 123
  venustus A.A.R.Maassoumi &
    Podl. 79
  vereskensis A.A.R.Maassoumi &
    Podl. 78
  vermicularis Bornm. 116
  versipilus Rech.f. & Koie 131
  verticillaris Bunge 72
  verus Olivier 43
  vespilus Rech.f. & Koie 131
  vessalae A.A.R.Maassoumi &
    Podl. 109
  vexans Rech.f. & Koie 78
  vexillaris Boiss. 95
  vicarius Lipsky 124
  vicia Sirj. & Rech.f. 84
  viciaefolius M.Bieb. 96
  viciaefolius G.Beck 105
  viciaefolius var. haematinus (Sirj. &
    Rech.f.)Parsa 95
  viciifolius DC. 96
  vidaeus Parsa 128
  vimineus Pallas 129
  virgaeformis Sirj. & Rech.f. 82
  virgaurea Kitam. 58
  viridiformis Sirj. 131
  viridis Bunge 131
  vogelii (Webb)Bornm. 90
  vogelii subsp. fatimensis Maire 91

  vogelii subsp. prolixus
    (Bunge)Maire 91
  vogelii subsp. vogelii 91
  volkii Rech.f. 78
  vulcanicus Bornm. 78
  wagneri Bunge 88
  warackensis Eig 58
  wartoensis Boiss. & Kotschy 28
  webbianus Benth. 78
  weirianus Aitch. & Baker 80
  wendelboi Deml 50
  wiesneri Bornm. 110
  wilhelminae Kress-Deml 119
  wrangelii Sirj. 43
  xanthinus Freyn & Bornm. 46
  xanthogossypinus Hand.-Mazz. 32
  xanthomeloides Korovin & M.Popov
    104
  xanthoxiphidiopsis Rech.f. 131
  xanthoxiphidium Freyn & Sint. 129
  xiphidiopsis Bornm. 102
  xylobasis sensu Rawi 114
  xylobasis Freyn 116
  xylocladus Rech.f. & Gilli 86
  xylorrhizus Freyn & Sint. 116
  yawnuensis Podl. 57
  yazdekhast Parsa 112
  yosiianus Kitam. 158
  zagrosicus Boiss. & Hausskn. 78
  zanskarensis Bunge 50
  zanskarensis subsp. zanskarensis 50
  zarghumensis Rech.f. 57
  zerdanus Boiss. 90
  ziaratensis Podl. 79
  zoharyi Eig 38
  zohrabi Bunge 94
  zubairensis Eig 82
Azukia radiata (L.)Ohwi 183

Bauhinia acuminata L. 7
  ellenbeckii Harms 7
  inermis Forsskal 7
  purpurea L. 7
  racemosa Lam. 7
  thonningii Schum. 7
  tomentosa L. 7
  variegata L. 7

Cadia purpurea (Picciv.)Aiton 186
Caesalpinia bonduc (L.)Roxb. 1
  decapetala (Roth)Alston 1
  erianthera Chiov. 1
  gilliesii (Hook.)D.Dietr. 2
  mexicana A.Gray 2
  pulcherrima (L.)Sw. 2
  sepiaria Roxb. 1
Cajanus cajan (L.)Millsp. 178
  indicus Sprengel 178
Calophaca cuneata (Benth.)Komarov 133

Index

Canavalia africana Dunn 179
  virosa sensu auctt. 179
Caragana acaulis Baker 133
  afghanica Kitam. 132
  aurantiaca Koehne 132
  brachyantha Rech.f. 121
  cuneata (Benth.)Ali 133
  decorticans Hemsley 132
  gerrardiana Benth. 133
  grandiflora (M.Bieb.)DC. 133
  grandiflora Bunge 132
  maimanensis Rech.f. 133
  nuristanica Rech.f. & Edelb. 133
  prainii C.Schneider 132
  pygmaea sensu auctt. 133
  ulicina Stocks 133
  versicolor Benth. 133
Cassia absus L. 4
  acutifolia Del. 5
  adenensis Benth. 5
  alexandrina (Miller)Thell. 5
  angustifolia Vahl 5
  artemisioides DC. 5
  aschrek Forsskal 6
  auriculata L. 5
  corymbosa Lam. 5
  didymobotrya Fresen. 5
  fallacina Chiov. 4
  fistula L. 3
  glauca Lam. 6
  hochstetteri Ghesq. 4
  holosericea Fresen. 5
  italica (Miller)Sprengel 6
  mimosoides L. 4
  nigricans Vahl 4
  obovata Colladon 6
  obtusifolia L. 6
  occidentalis L. 6
  oocarpa Baker 5
  pubescens R.Br. 5
  senna L. 5
  sophera L. 6
  surattensis Burm.f. 6
  tora L. 6
Ceratonia oreothauma Hillc., G.P.Lewis & Verdc. 4
  oreothauma subsp. oreothauma 4
  siliqua L. 4
Cercis griffithii Boiss. 8
  siliquastrum L. 8
  siliquastrum subsp. hebecarpa (Bornm.)Yalt. 8
  siliquastrum subsp. siliquastrum 8
Chamaecrista absus (L.)Irwin & Barneby 4
  dimidiata (Roxb.)Lock 4
  fallacina (Chiov.)Lock 4
  mimosoides (L.)Greene 4
  nigricans (Vahl)Greene 4
Chesneya acaulis (Baker)Popov 133
  afghanica Rech.f. & Koie 133
  astragalina Jaub. & Spach 133
  crassipes Boriss 133
  cuneata (Benth.)Ali 133
  gaubaeana Bornm. 134
  gaubeana sensu Parsa 134
  kopetdaghensis Boriss 134
  kotschyi Boiss. 134
  macranica Rech.f. & Esfand. 134
  makranica Rech.f. & Esfand. 134
  microphylla Vahl 134
  oliverii Jaub. & Spach 134
  olivierii Jaub. & Spach 134
  parviflora Jaub. & Spach 134
  rytidosperma Jaub. & Spach 134
  tadzhikistana Boriss 134
  ternata (Korsh.)Popov 134
  vaginalis sensu Rawi 134
  vaginalis Jaub. & Spach 134
  velutina Jaub. & Spach 134
  vesiculifera C.Towns. 134
  volkii Rech.f. 134
Chondrocarpus dilankuri (Lipsky) Nevski 98
Cicer acanthophyllum Boriss. 18
  anatolicum Alef. 19
  arietinum L. 19
  bijugum Rech.f. 19
  caucasicum Bornm. 19
  chorassanicum (Bunge)Popov 19
  cuneatum A.Rich. 19
  echinospermum P.Davis 19
  ervoides (Sieber)Fenzl 19
  fedtschenkoi Lincz. 19
  flexuosum sensu Koie 20
  garanicum Boriss. 18
  glutinosum Alef. 19
  incisum (Willd.)K.Maly 19
  incisum var. libanoticum (Boiss.) Bornm. 19
  jacquemontii Jaub. & Spach 20
  kermanense Bornm. 19
  kopetdaghense Lincz. 21
  longearistatum Rech.f. 21
  macracanthum Popov 19
  microphyllum Benth. 20
  minutum Boiss. & Hohen. 19
  multijugum Maesen 20
  nuristanicum Kitam. 20
  oxyodon Boiss. & Hohen. 20
  pinnatifidum Jaub. & Spach 20
  pungens Boiss. 20
  rechingeri Podl. 20
  sintenisii Bornm. 20
  songaricum sensu Jaub. & Spach 19
  songaricum var. pamiricum Paulsen 19
  songaricum var. schugnanicum Popov 19
  songaricum var. spinosum Aitch. 19
  spinosum Popov 20
  spiroceras Jaub. & Spach 20
  stapfianum Rech.f. 20

Cicer (*cont.*)
  straussii Bornm. 21
  subaphyllum Boiss. 20
  tragacanthoides Jaub. & Spach 21
  trifoliatum sensu Parsa 19
  trifoliolatum Bornm. 19
  yamashitae Kitam. 21
Clitoria ternatea L. 179
Colutea afghanica Browicz 135
  buhsei (Boiss.)Shap. 135
  buhsei var. densiflora Browicz 135
  cilicica Boiss. & Bal. 135
  gifana Parsa 135
  gracilis Freyn & Sint. 135
  halepica Lam. 135
  haleppica Lam. 135
  istria Miller 135
  karakoramensis Kitam. 135
  komarovii Takht. 135
  mesantha Ali 135
  nepalensis Sims 135
  paulsenii Freyn 135
  paulsenii var. menantha (Ali) Browicz 135
  paulsenii subsp. mesantha (Ali) Ali 135
  paulsenii subsp. paulsenii 135
  persica Boiss. 135,136
  persica var. buhsei Boiss. 135
  persica var. buhsei sensu Rech.f. 135
  persica var. gracilis (Freyn)Parsa 135
  porphyrogramma Rech.f. 136
  rostrata Gilli p.p. 135
  triphylla Boiss. 137
  uniflora G.Beck 136
Coronilla cappadocica Willd. 175
  coronata L. 170
  montana Scop. 170
  orientalis Miller 175
  parviflora Willd. 175
  scorpioides (L.)Koch 170
  securidaca L. 175
  varia L. 175
Crotalaria aculeata De Wild. 21
  aegyptiaca Benth. 21
  astragalina A.Rich. 22
  azaisii Sacl. 23
  burhia Benth. 21
  deflersii Schweinf. 21
  emarginella Vatke 22
  hirta Willd. 22
  impressa Walp. 22
  incana L. 22
  incana subsp. incana 22
  incana subsp. purpurascens (Lam.)Milne-Redh. 22
  juncea L. 22
  laxa Franchet 22
  leptocarpa Balf.f. 22
  leptocarpa subsp. leptocarpa 22
  lupinoides Hochst. 23

  medicaginea Lam. 22
  microphylla Vahl 22
  mysorensis sensu Pelt. 22
  natalitia Meissner 23
  oocarpa Baker 23
  oocarpa subsp. microcarpa Milne-Redh. 23
  persica (Burm.f.)Merr. 23
  platycalyx Baker 23
  plowdenii Baker 23
  pteropoda Balf.f. 23
  pycnostachya Benth. 23
  quartiniana A.Rich. 23
  rathjensiana Schwartz 22
  retusa L. 23
  saltiana Andrews 23
  schweinfurthii Deflers 23
  senegalensis (Pers.)DC. 24
  sennii Cuf. 22
  sp. S.Collenette 24
  spinosa Benth. 24
  squamigera Deflers 24
  thebaica (Del.)DC. 24
  wissmannii Schwartz 21
Cullen corylifolia (L.)Medikus 184
  drupacea (Bunge)Stirton 184
  jaubertiana (Fenzl)Stirton 184
  plicata (Del.)Stirton 184
Cyamopsis senegalensis Guillemin & Perrottet 165
  tetragonoloba (L.)Taubert 166
  tetragonolobus (L.)Taubert 166
Cytisus calycinus M.Bieb. 149

Dalbergia sissoo Roxb. 24
Delonix elata (L.)Gamble 2
  regia (Hook.)Raf. 2
Desmodium elegans DC. 25
  gangeticum (L.)DC. 25
  ospriostreblum Chiov. 26
  repandum (Vahl)DC. 26
  tiliaefolium G.Don 25
  tiliifolium G.Don 25
  triflorum (L.)DC. 26
  triquetrum DC. 26
Dichrostachys cinerea (L.)Wight & Arn. 15
Didymopelta turkestanica (Regel & Schmalh.)Regel & Schmalh. 84
Dipelta turkestanica Regel & Schmalh. 84
Dolichos axillaris E.Meyer 180
  biflorus sensu auctt. 180
  lablab L. 180
  sericeus E.Meyer 179
  sericeus subsp.formosus (A.Rich.)Verdc. 179
  trilobus L. 179
Dorycnium calycinum Stocks 114
  haussknechtii Boiss. 170
  intermedium Ledeb. 170

Dorycnium (cont.)
    pentaphyllum Scop. 170
    pentaphyllum subsp. haussknechtii
        (Boiss.)Gams 170

Ebenus ferruginea Jaub. & Spach 151
    horrida Jaub. & Spach 151
    lagopus (Jaub. & Spach)Boiss. 150
    stellata Boiss. 151
    tragacanthoides Jaub. & Spach 151
Eriosema longipedunculatum (A.Rich.)
    Baker 179
Ervum kotschyanum Boiss. 217
    lens L. 212
    orientale Boiss. 212
Eversmannia hedysaroides Bunge 152
    subspinosa (Fischer)B.Fedtsch. 151

Faba vulgaris Moench 215
Factorovskya aschersoniana (Urban)Eig 189
Faidherbia albida (Del.)A.Chev. 14
Flemingia grahamiana Wight &
    Arn. 179

Galega officinalis L. 136
    persica (Boiss.)E.Small 136
Genista raetam Forsskal 150
    tinctoria L. 150
Gleditsia caspica Desf. 2
    horrida subsp. caspica (Desf.)Paclt 2
    horrida var. caspica (Desf.)C.Schneider 2
    triacanthos L. 2
Glycine javanica L. 181
    max (L.)Merr. 180
    soja Siebold & Zucc. 180
    ussuriensis Regel & Maack 180
    wightii (Wight & Arn.)Verdc. 181
    wightii subsp. longicauda (Schweinf.)
        Verdc. 181
    wightii subsp. wightii 181
Glycyrrhiza aspera Pallas 136
    asperrima L.f. 136
    bucharica Regel 136
    echinata L. 136
    erythrocarpa (Vassilcz.)Abdull. 16
    glabra L. 137
    glabra var. violacea (Boiss.)Boiss. 137
    glandulifera Waldst. & Kit. 137
    triphylla sensu Parsa 136
    uralensis Fischer 137
Goebelia alopecuroides (L.)Boiss. 186
    alopecuroides var. alopecuroides 186
    alopecuroides var. tomentosa Boiss. 186
    pachycarpa (C.Meyer)Boiss. 187
Gueldenstaedtia cuneata Benth. 133

Haematoxylon campechianum L. 3
Haematoxylum campechianum L. 137
Halimodendron argenteum (Lam.)DC. 137
    halodendron (Pallas)Voss 151
Hedysarum astragaloides Benth. 151
    atropatanum Boiss. 151
    bellevii (Prain)Bornm. 151
    brahuicum Boiss. 151
    bucharicum B.Fedtsch. 151
    callithrix Boiss. 151
    cappadocicum Boiss. 151
    cephalotes Franchet 153
    criniferum Boiss. 152
    damghanicum Rech.f. 152
    ecbatanum Beck 152
    elbursense Bornm. & Gauba 152
    elymaiticum Boiss. & Hausskn. 152
    falconeri Baker 152
    fallacinum Rech.f. & Aellen 152
    farinosum Parsa 152
    formosum Basiner 152
    glaucescens Ledeb. 153
    grandiflorum subsp. cappadocicum
        Boiss. 151
    halophilum Bornm. & Gauba 152
    hyrcanum Bornm. & Gauba 152
    ibericum M.Bieb. 153
    kopetdaghi Boriss 153
    kotschyi Boiss. 153
    lehmannianum Bunge 153
    mahrense Rech.f. 153
    maitlandianum Aitch. & Baker 153
    micropterum Boiss. 153
    minjanense Rech.f. 153
    pannosum sensu auctt. 154
    papillosum Boiss. 153
    plumosum Boiss. & Hausskn. 153
    praticolum Rech.f. 153
    purpureo-pilosum Kitam. 151
    renzii Rech.f. 153
    sauzakense Rech.f. 154
    sericatum Kitam. 154
    sericeum M.Bieb. 154
    singarense Boiss. & Hausskn. 154
    syriacum Boiss. 154
    varium Willd. 154
    varium subsp. syriacum (Boiss.)
        C.Towns. 154
    volkii Rech.f. 154
    wakhanicum Podl. & Anders 154
    wrightianum Aitch. & Baker 154
Helminthocarpum abyssinicum A.Rich. 176
Hippocrepis bicontorta Lois. 170
    biflora Sprengel 171
    bisiliqua Forsskal 171
    bornmulleri Hausskn. 171
    ciliata Willd. 171
    constricta Kunze 171
    cornigera Boiss. 170
    cyclocarpa Murb. 171
    multisiliquosa L. 171
    unisiliquosa L. 171
    unisiliquosa subsp. biflora
        (Sprengel)O.Bolos & Vigo 171
    *unisiliquosa subsp. bisiliqua (Forsskal)
        Bornm.* 171

Hymenocarpos circinnatus (L.)Savi 171
  circinnatus subsp. circinnatus 171
  circinnatus subsp. nummularius (DC.)
    Chrtek & B.Slavik 172
  nummularius (DC.)Boiss. 172

Indigofera amorphoides Jaub. & Spach 166
  anabaptista Steudel 168
  arabica Jaub. & Spach 166
  argentea Burm.f. 166
  argentea sensu auctt. 167
  arrecta A.Rich. 166
  articulata Gouan 166
  articulata sensu auctt. 166
  atriceps Hook.f. 166
  atriceps subsp. kaessneri (Baker f.)
    J.B.Gillett 166
  atropurpurea Hornem. 166
  brevicalyx Baker f. 166
  burmannii Boiss. 166
  caerulea Roxb. 167
  coerulea Roxb. 167
  colutea (Burm.f.)Merr. 167
  contorta J.B.Gillett 167
  cordifolia Roth 167
  costata Guillemin & Perrottet 167
  costata subsp. gonioides (Baker)
    J.B.Gillett 167
  deflersii Baker f. 167
  disjuncta J.B.Gillett 167
  gerardiana Baker 167
  *gerardiana var. heterantha (Brandis)
    Baker* 167
  heterantha Brandis 167
  himalayensis Ali 167
  hochstetteri Baker 168
  insularis Chiov. 168
  intricata Boiss. 168
  linifolia (L.f)Retz. 168
  nephrocarpa Balf.f. 168
  oblongifolia Forsskal 168
  parviflora Wight & Arn. 168
  phillipsiae Baker f. 168
  pseudointricata J.B.Gillett 168
  schimperi Jaub. & Spach 168
  semitrijuga Forsskal 168
  sessiliflora DC. 169
  sp.3234 S.Collenette 169
  sp.aff.volkensii S.Collenette 169
  spicata Forsskal 169
  spiniflora Boiss. 169
  spinosa Forsskal 169
  tinctoria L. 169
  tribuloides Boiss. 169
  trita L.f. 169
  tritoides Baker 169
  viscosa Lam. 167

Kerstania nuristanica Rech.f. 174
Keyserlingia griffithii (Stocks)Boiss. 187

Lablab niger Medikus 180
  purpureus (L.)Sweet 180
Lathyrus altaicus var. humilis Ledeb. 209
  amoenus Fenzl 209
  amoenus var. lineatus Dinsm. 209
  annuus L. 208
  apaca L. 208
  aphaca L. 208
  bijugus Boiss. & Noe 208
  bithynicus L. 213
  boissieri Sirj. 209
  cassius Boiss. 209
  chloranthus Boiss. 209
  cicera L. 209
  cicera var. lineatus Post 209
  cyaneus (Steven)K.Koch 209
  erectus Lagasca 209
  erectus var. stenophyllus Boiss. 209
  gorgoni Parl. 209
  hirsutus L. 209
  hispidulus Boiss. 209
  humilis (Ser.)Sprengel 209
  humilis Guillemin & Perrottet 209
  inconspicuus L. 209
  incurvus (Roth)Willd. 209
  latifolius L. 210
  laxiflora sensu Rech.f. 210
  laxiflorus (Desf.)Kuntze 210
  laxiflorus subsp. laxiflorus 210
  marmoratus Boiss. & Blanche 210
  miniatus Steven 211
  mulkak Lipsky 210
  multijugus (Ledeb.)Czefr. 210
  nervosus (Boiss.)Boiss. 209
  neurolobus Boiss. & Heldr. 210
  nissolia L. 210
  nivalis Hand.-Mazz. 210
  ochrus (L.)DC. 210
  odoratus L. 210
  pannonicus (Jacq.)Garcke 210
  pannonicus subsp. multijugus
    (Ledeb.)Bassler 210
  *pannonicus var. paucijugus (Ledeb.)
    Schischkin* 210
  pratensis L. 211
  pseudo-cicera Pampan. 211
  pseudocicera Pampan. 211
  roseus Steven 211
  rotundifolius Willd. 211
  rotundifolius subsp. miniatus
    (Steven)P.Davis 211
  sativus L. 211
  sativus var. stenophyllus Boiss. 211
  saxatilis (Vent.)Vis. 211
  sphaericus Retz. 211
  *subvillosus (Ledeb.)Aitch. &
    Hemsley* 218
  szowitsii Boiss. 211
  tuberosus L. 211
  vernus (L.)Bernh. 212
  vernus subsp. vernus 212

Lathyrus (cont.)
   vinealis Boiss. & Noe 212
Lens culinare Medikus 212
   culinaris Medikus 212
   cyanea (Boiss. & Hohen.)Alef. 212
   cyaneum (Boiss. & Hohen.)Alef. 212
   esculenta Moench 212
   kotschyanus (Boiss.)Nab. 217
   montbretii (Fischer & C.Meyer)P.Davis & Plitm. 217
   orientalis (Boiss.)Hand.-Mazz. 212
   pygmaea Grossh. 212
Lespedeza aitchisonii Ricker 26
   cuneata (Du Mont.)G.Don 26
   juncea (L.)Pers. 26
   nuristanica Rech.f. 26
Leucaena glauca (L.)Benth. 15
   leucocephala (Lam.)de Wit 15
Lotononis dichotoma (Del.)Boiss. 24
   leobordea Benth. 24
   platycarpa (Viv.)Pichi-Serm. 24
Lotus aegeus (Griseb.)Boiss. 172
   aleppicus Boiss. 173
   angustissimus L. 172
   arabicus L. 172
   arabicus var. glabrescens Schweinf. 173
   arabicus var. trigonelloides (Webb)Webb 173
   brachycarpus var. lalambensis (Schweinf.)Brand 173
   corniculatus L. 172
   corniculatus subsp. afghanicus Chrtkova-Zert. 172
   corniculatus var. alpinus DC. 172
   corniculatus subsp. corniculatus 172
   corniculatus subsp. frondosus Freyn 172
   corniculatus subsp. fruticosus Chrtkova-Zert. 173
   corniculatus subsp. tenuis (Kit.) Briq. 174
   frondosus (Freyn)Kuprian. 172
   garcinii DC. 173
   gebelia Vent. 173
   gebelia var. tomentosus Boiss. 174
   glinoides sensu Baker 172
   glinoides Del. 173
   halophilus Boiss. & Spruner 173
   krylovii Schischkin & Serg. 173
   lalambensis Penzig 173
   lanuginosus Vent. 173
   laricus Rech.f. 173
   makranicus Rech.f. & Esfand. 175
   michauxianus Ser. 174
   mossamedensis Baker 172
   palustris Willd. 174
   pumilus Parsa 173
   pusillus Viv. 173
   quinatus (Forsskal)J.B.Gillett 174
   roseus Forsskal 172
   schimperi Boiss. 174
   sharifii Rech.f. & Esfand. 173
   sp.aff.arabicus S.Collenette 174
   sulphureus Boiss. 172
   tenuifolius (L.)Reichb. 174
   tenuis Waldst. & Kit. 174
   trigonelloides Webb & Berth. 174
   villosus Forsskal 173
Lygos raetam (Forsskal)Heyw. 150

Macrotyloma axillare (E.Meyer)Verdc. 180
Medicago arabica (L.)Hudson 188
   aschersoniana Urban 189
   astroites (Fischer & C.Meyer) Trautv. 188
   biflora (Griseb.) E.Small 188
   borealis Grossh. 189
   brachycarpa M.Bieb. 188
   ciliaris (L.)All. 189
   coerulea Ledeb. 192
   constricta Durieu 188
   coronata (L.)Bartal. 188
   coronata var. brevipedunculata Eig 188
   coronata var. multiflora Eig 188
   crassipes (Boiss.)E.Small 188
   denticulata var. apiculata (Willd.)Posp. 191
   difalcata Sinsk. 189
   edgeworthii Sirj. 188
   erecta Kotov 189
   falcata L. 189
   fischeriana (Ser.)Trautv. 189
   gerardi Willd. 191
   globosa sensu auctt. 188
   hispida Gaertner 188
   hispida var. apiculata (Willd.)E.Burnat 191
   hispida var. denticulata (Willd.)E.Burnat 191
   hypogaea E.Small 189
   intertexta (L.)Miller 189
   laciniata (L.)Miller 189
   ladak Vassilcz. 192
   lanigera Winkler & O.Fedtsch. 189
   laxispira Heyn 89
   littoralis Lois. 189
   lupulina L. 190
   maculata Willd. 188
   medicaginoides (Retz.)E.Small 190
   minima (L.)Bartal. 190
   monantha (C.Meyer)Trautv. 190
   monspeliaca (L.)Trautv. 190
   noeana Boiss. 190
   orbicularis (L.)Bartal. 190
   orthoceras (Karelin & Kir.) Trautv. 190
   persica (Boiss.)E.Small 191
   phrygia (Boiss. & Bal.)E.Small 191

Medicago (cont.)
  polyceratia (L.)Trautv. 191
  polychroa Grossh. 191
  polymorpha L. 191
  polymorpha var. coronata L. 188
  *polymorpha subsp. denticulata
    (Del.)Boiss.* 191
  *polymorpha var. apiculata
    A.al-Rawi* 191
  pubescens Aylmer 188
  quasifalcata Sinsk. 189
  radiata L. 191
  retrorsa (Boiss.)E.Small 191
  rigidula (L.)All. 191
  rotata Boiss. 192
  sativa L. 192
  *sativa subsp. coerulea
    (Ledeb.)Schmalh.* 192
  sativa subsp. microcarpa Urban 192
  sativa subsp. sativa 192
  serratifolia A.al-Rawi 191
  tenderlensis Opperman 189
  tribuloides Desr. 192
  truncatula Gaertner 192
  tuberculata Willd. 192
  turbinata (L.)All. 192
  × varia Martyn 192
Melilotus alba Medikus 193
  altissima Thuill. 193
  elegans Ser. 193
  indica (L.)All. 193
  messanensis (L.)All. 193
  officinalis (L.)Pallas 193
  sicula (Vitman)B.D.Jackson 193
  suaveolens Ledeb. 193
*Melissitus adscendens (Nevski)
    Ikonn.* 203
Meristotropis erythrocarpa Vassilcz. 136
  xanthioides Vassilcz. 136
Mimosa himalayana Gamble 15
  pudica L. 15

Neonotonia wightii (Wight &
    Arn.)J.A.Lackey 181
  wightii subsp. wightii 181

Onobrychis acaulis Bornm. 155
  aequidentata (Smith)Urv. 155
  afghanica Sirj. & Rech.f. 155
  afghanica subsp. afghanica 155
  afghanica subsp. brachycalyx
    Rech.f. 155
  afghanica subsp. codringtonii (Sirj.
    & Rech.f.)Rech.f. 155
  altissima Grossh. 155
  alyassinicus Parsa 155
  amoena Popov & Vved. 155
  amoena subsp. amoena 155
  amoena subsp. meshhedensis Sirj. &
    Rech.f. 155
  andalanica Bornm. 156

arnacantha Boiss. 156
atropatana Boiss. 156
aucheri Boiss. 156
aucheri subsp. aucheri 156
aucheri subsp. psammophila
    (Bornm.)Rech.f. 156
aucheri subsp. teheranica
    (Bornm.)Rech.f. 156
aurantiaca Boiss. 158
aurantiaca var. velutina Post 158
baldzuanica Sirj. 162
belangeri Boiss. & Buhse 161
bicolor Bornm. 158
bicornis Vassilcz. 156
buhseana Boiss. 156
bungei Boiss. 156
cadmea Boiss. 156
caloptera Aitch. & Baker 162
cana (Boiss.)Hand.-Mazz. 157
caput-galli (L.)Lam. 157
carduchorum C.Towns. 157
charikarensis Podl. 158
chorassanica Boiss. 157
cornuta (L.)Desv. 157
cornuta subsp. cornuta 157
cornuta subsp. leptacantha
    Rech.f. 157
crista-galli (L.)Lam. 157
*crista-galli subsp. trilocarpa Quezel &
    Santa* 157
dasycephala Baker 157
dealbata Stocks 157
depauperata Boiss. 157
echidna Lipsky 158
elegans Franchet 162
elymaitiaca Boiss. & Hausskn. 158
eubrychidea Boiss. 158
fallax sensu auctt. 157
fallax Freyn & Sint. 158
freitagii Rech.f. 158
gaillardoti Boiss. 162
galegifolia Boiss. 158
*galegifolia var. luristanica Sirj.
    & Rech.f.* 158
gaubae Bornm. 158
grandis Lipsky 158
gypsicola Rech.f. 158
haussknechtii Boiss. 158
heliocarpa Boiss. 159
heterophylla C.Meyer 159
hohenackeriana C.Meyer 159
hypargyrea Boiss. 159
iberica Grossh. 159
iranica Bornm. 159
iranshahrii Rech.f. 159
kachetica sensu auctt. 159
kermanensis (Sirj. & Rech.f.)
    Rech.f. 159
koeieana Sirj. & Rech.f. 155
kotschyana Fenzl 159

Onobrychis (*cont.*)
   lahidjanicus Parsa 159
   lanata Boiss. 162
   laxiflora Baker 160
   laxiflora subsp. kabulica Rech.f. 160
   laxiflora subsp. laxiflora 160
   laxiflora subsp. macrodonta Rech.f. 160
   laxiflora subsp. shahrestanica Rech.f. 160
   laxiflora subsp. taftanica Rech.f. 160
   linearis Pau & Vicioso 161
   lipskyi Korovin 164
   longipes Bunge 164
   lunata Boiss. 160
   luristanica Rech.f. 160
   macrorrhiza Rech.f. 160
   major (Boiss.)Hand.-Mazz. 160
   major var. angustifolia Sirj. & Rech.f. 160
   marginata Beck 163
   mazanderanica Rech.f. 161
   megalobotrys Aitch. & Baker 157
   megataphros Boiss. 161
   melanotricha Boiss. 161
   melanotricha var. kermanensis Sirj. & Rech.f. 159
   merxmuelleri Podl. & Rech.f. 161
   michauxii sensu Sirj.,p.p. 159
   michauxii DC. 161
   micrantha Schrenk 161
   microptera Baker 161
   nummularia Boiss. 161
   olivieri Boiss. 161
   oxyptera Boiss. 161
   oxypteroides Sirj. & Rech.f. 160
   persica Sirj. & Rech.f. 161
   picta Bornm. 163
   pindicola sensu Rawi 157
   pinnata (Bertol.)Hand.-Mazz. 162
   plantago Bornm. 161
   poikilantha Rech.f. 162
   porphyrogramma Rech.f. 158
   psammophila Bornm. 156
   psoraleifolia Boiss. 161
   ptolemaica (Del.)DC. 162
   ptolemaica subsp. macroptera C.Towns. 162
   ptychophylla Sirj. & Rech.f. 162
   pulchella Schrenk 162
   radiata (Desf.)M.Bieb. 162
   rechingerorum Wendelbo 162
   samanganica Rech.f. 162
   saravschanica B.Fedtsch. 162
   sativa Lam. 164
   sativa var. subinermis Boiss. 164
   sauzakensis Sirj. & Rech.f. 162
   schahuensis Bornm. 163
   schugnanica B.Fedtsch. 160
   scrobiculata Boiss. 163
   semnanensis Sirj. & Rech.f. 158
   seravschanica B.Fedtsch. 162
   shahpurensis Rech.f. 163
   sintenisii Bornm. 163
   sirdjanicus Parsa 163
   sirinae Nab. 155
   sirinae var. behboudii Sirj. & Rech.f. 155
   sirjaevi Nab. 162
   sojakii Rech.f. 163
   spinescens Bornm. 156
   spinosissima Baker 163
   splendida Rech.f. & Podl. 163
   squarrosa Viv. 157
   subacaulis Boiss. 163
   subnitens Bornm. 163
   susiana Nab. 163
   sykesiae N.Simpson 157
   szovitsii Boiss. 164
   talagonica Rech.f. 164
   tavernieraefolia Boiss. 164
   teheranica Bornm. 156
   transcaspica V.V.Nikitin 164
   transcaucasica Grossh. 164
   unicornis Pau 155
   verae Sirj. 164
   verae var. rechingeri Sirj. 164
   viciaefolia Scop. 164
   viciaefolia var. persica Sirj. 155
   viciifolia Scop. 164
   wettsteinii Nab. 164
Ononis afghanica Sirj. & Rech.f. 194
   antiquorum L. 195
   arvensis L. 195
   arvensis subsp. arvensis 195
   biflora Desf. 194
   campestris Koch & Ziz 195
   chorassanica Bunge 194
   columnae All. 195
   hircina Jacq. 195
   hirta Poiret 194
   leiosperma Boiss. 195
   mitissima L. 194
   natrix L. 194
   natrix subsp. stenophylla (Boiss.)Sirj. 194
   nuristanica Podl. 194
   pubescens L. 194
   pusilla L. 195
   reclinata L. 195
   reclinata var. minor Moris 195
   reclinata var. mollis (Savi)Heldr. 195
   repens L. 195
   repens subsp. antiquorum (L.) Greuter 195
   repens subsp. arvensis (L.)Greuter 195
   repens subsp. leiosperma (Boiss.) Greuter 195
   repens subsp. spinosa Greuter 195
   serrata Forsskal 195
   sicula Guss. 195

Ononis (*cont.*)
  sicula subsp. *microcarpa* Milne-Redh. 195
  spinosa Benth. 195
  spinosa subsp. *afghanica* (Sirj. & Rech.f.)Kitam. 194
  spinosa subsp. *antiquorum* Franchet 195
  spinosa subsp. *leiosperma* Willd. 195
  viscosa L. 195
  viscosa subsp. *breviflora* (DC.) Nyman 196
  viscosa subsp. *sicula* (Guss.) Huber-Mor. 195
Ophiocarpus *paulsenii* (Freyn)Ikonn. 116
Oreophysa *microphylla* (Jaub. & Spach) Browicz 137
  triphylla Boiss. 137
Ormocarpum bibracteatum sensu auctt. 18
  dhofarense Hillc. & J.B.Gillett 18
  yemenense J.B.Gillett 18
Ornithopus *compressus* L. 174
Orobus albus var. multijugus (Ledeb.) Ledeb. 210
  cyaneus Steven 109
  lacteus var. multijugus Ledeb. 210
  triflorus Stapf 218
Oxytropis *admiranda* Rech.f. 137
  aellenii Vassilcz. 137
  afghanica Rech.f. & Koie 138
  alavae Rech.f. 138
  andersii Vassilcz. 138
  assadliensis Vassilcz. 138
  asterocarpa Vassilcz. 138
  astragaloides Boriss. 138
  aucheri Boiss. 138
  baburi Vassilcz. 138
  bella O.Fedtsch. 138
  bicornis Vassilcz. 138
  binaludensis Vassilcz. 138
  boguschi B.Fedtsch. 139
  cabulica (Boiss.)Boiss. 139
  callophylla Vassilcz. 139
  caraganetorum Vassilcz. 139
  carduchorum Hedge 139
  chiliophylla Benth. 139
  chionophylla Schrenk 139
  chitralensis Ali 139
  chrysocarpa Boiss. 139
  cinerea Vassilcz. 139
  crassiuscula Boriss. 139
  czapan-daghi B.Fedtsch. 140
  danorum Rech.f. 140
  darvasica B.Fedtsch. 139
  dashtinavarensis Vassilcz. 140
  farsi Vassilcz. 140
  fohlenensis Vassilcz. 140
  fuscescens Vassilcz. 140
  glacialis Benth. 140
  gorbunovii Boriss. 140
  gracillima Vassilcz. 140
  griffithii Boiss. 140
  gubanovii Vassilcz. 140
  guntensis B.Fedtsch. 141
  gypsophila sensu Parsa 141
  hedgei Vassilcz. 141
  heratensis Bunge 141
  hindukushensis Vassilcz. 141
  hirsutiuscula Freyn 141
  hypsophila Bunge 141
  immersa (Baker)B.Fedtsch. 141
  incanescens Freyn 141
  iranica Vassilcz. 141
  karjaginii Grossh. 141
  kazidanica Vassilcz. 141
  kermanica Freyn & Bornm. 141
  khinjahi Vassilcz. 142
  kopetdaghensis Gontch. 142
  kotschyana Boiss. & Hohen. 142
  kuchanensis Vassilcz. 142
  kukkonenii Vassilcz. 142
  kunarensis Vassilcz. 142
  lapponica (Wahlenb.)Gay 142
  laxiracemosa Vassilcz. 142
  liliputa Vassilcz. 142
  linczevskii Gontch. 142
  lupinoides Grossh. 142
  luteo-coerulea (Baker)Ali 143
  lydiae Vassilcz. 143
  marco-poloi Vassilcz. 143
  margacea Vassilcz. 143
  masanderanensis Vassilcz. 143
  microsphaera Bunge 143
  minjanensis Rech.f. 143
  neo-rechingeriana Vassilcz. 143
  nuristanica Vassilcz. 143
  oreophila Vassilcz. 143
  oroboides Vassilcz. 143
  pagobia Bunge 144,147
  pakistanica Vassilcz. 144
  pamirica Danguy 141
  panjshirica Podl. & Deml 144
  parvanensis Vassilcz. 144
  persica Boiss. 144
  platonychia Bunge 144
  podlechii Vassilcz. 144
  poincinsii Franchet 144
  poncinsii Franchet 144
  proxima Boriss. 144
  pseudohirsutiuscula Vassilcz. 144
  puberula Boriss. 144
  pusilloides Vassilcz. 145
  rechingeri Vassilcz. 145
  rhodontha Vassilcz. 145
  riparia Litv. 145
  rudbariensis Vassilcz. 145
  salangensis Podl. & Deml 145
  salicetorum Vassilcz. 145
  saperlebulensis Vassilcz. 145
  sata-kandaonensis Vassilcz. 145
  savellanica Boiss. 145

Oxytropis (cont.)
    shirkuhi Vassilcz. 146
    siah-sangi Vassilcz. 146
    sikaramensis (Sirj. & Rech.f.)Ali 146
    sojakii Vassilcz. 146
    spec.I Podl. & Anders 138
    spec.II Podl. & Anders 143
    spec.III Podl. & Anders 138
    spec.V Podl. & Anders 147
    straussii Bornm. 146
    suavis Boriss. 146
    surculosa Rech.f. 146
    surmandehi Vassilcz. 146
    szovitsii Boiss. & Buhse 146
    tachtensis Franchet 146
    takhti-soleimanii Vassilcz. 146
    tatarica Baker 146
    thaumasiomorpha Rech.f. 147
    tianschanica Bunge 147
    trichosphaera Freyn 138,147
    tunnellii Vassilcz. 147
    vadimii Vassilcz. 147
    vakhdzhiri Vassilcz. 147
    vavilovii Vassilcz. 147
    viae-amicitiae Vassilcz. 147
    volkii Rech.f. 147
    vositensis Vassilcz. 147
    wendelboi Vassilcz. 147
    yazdi Vassilcz. 148
    zangolehensis Vassilcz. 148

Parkinsonia acleata L. 3
    aculeata L. 3
Phaca salsula Pallas 148
Phaseolus aconitifolius Jacq. 183
    aureus Roxb. 183
    lunatus L. 181
    mungo sensu auctt. 183
    mungo L. 183
    radiatus L. 183
    vulgaris L. 181
Piliostigma thonningii (Schum.) Milne-Redh. 7
Pisum arvense sensu Guest 212
    elatius M.Bieb. 212
    formosum (Steven)Alef. 212
    humile Boiss. & Noe 212
    sativum L. 212
    sativum subsp. elatius (M.Bieb.)Asch. & Graebner 212
    sativum subsp. humile (Holmboe) Greuter 212
    sativum subsp. sativum 213
Pithecellobium dulce (Roxb.)Benth. 14
Pithecolobium littorale Rec. 14
Podolotus hosackioides Benth. 174
Poinciana pulcherrima L. 2
Prosopis africana (Guillemin & Perrottet) Taubert 15
    chilensis (Molina)Stuntz emend. Burkart 16

    cineraria (L.)Druce 16
    farcta (Banks & Sol.)J.F.Macbr. 16
    glandulosa Torrey 16
    juliflora (Sw.)DC. 16
    juliflora var. glandulosa (Torrey) Cockerell 16
    koelziana Burkart 16
    oblonga Benth. 15
    sp. Chaudhary 16
    spicigera L. 16
Pseudolotus makranicus (Rech.f. & Esfand.) Rech.f. 175
Psoralea corylifolia L. 184
    drupacea Bunge 184
    jaubertiana Fenzl 184
    plicata Del. 184
Pterolobium stellatum (Forsskal)Brenan 3
Ptycholobium plicatum (Oliver)Harms 176
    plicatum subsp. arabicum Brummitt 176

Retama raetam (Forsskal)Webb 150
Rhynchosia buramensis Hutch. & Bruce 181
    elegans A.Rich. 181
    flava (Forsskal)Thulin 181
    malacophylla (Sprengel)Bojer 181
    memnonia (Del.)DC. 182
    minima (L.)DC. 182
    pulverulenta Stocks 182
    schimperi Boiss. 182
    sp.nov. S.Collenette 182
    sublobata (Schum.)Meikle 182
    totta (Thunb.)DC. 182
    usambarensis Taubert 182
Robinia pseudacacia L. 185
Rudua aurea (Roxb.)F.Maek. 183

Scorpiurus muricatus L. 175
    subvillosus L. 175
    sulcatus L. 175
Securigera coronilla DC. 175
    orientalis (Miller)Lassen 175
    parviflora (Desv.)Lassen 175
    securidaca (L.)Degen & Doerfler 175
Senna acutifolia (Del.)Batka 5
    alexandrina Miller 5
    artemisioides (DC.)Randall 5
    auriculata (L.)Roxb. 5
    corymbosa (Lam.)Irwin & Barneby 5
    didymobotrya (Fresen.)Irwin & Barneby 5
    dimidiata Roxb. 4
    holosericea (Fresen.)Greuter 5
    hookeriana Batka 5
    italica Miller 6
    italica subsp. italica 6
    obtusifolia (L.)Irwin & Barneby 6

Senna (cont.)
  occidentalis (L.)Link 6
  sophera (L.)Roxb. 6
  surattensis (Burm.f.)Irwin & Barneby 6
  tora (L.)Roxb. 6
Sesbania aculeata (Willd.)Pers. 185
  bispinosa (Jacq.)W.Wight 185
  cannabina (Retz.)Poiret 185
  concolor J.B.Gillett 185
  grandiflora (L.)Poiret 185
  leptocarpa DC. 185
  pachycarpa DC. 185
  pachycarpa subsp. pachycarpa 185
  sericea (Willd.)Link 185
  sesban (L.)Merr. 185
  sesban subsp. sesban 186
  sinuo-carinata Ali 185
Sewerzowia turkestanica Regel & Schmalh. 124
Smirnowia iranica Vassilcz. 186
  turkestana Bunge 186
Sophora alopecuroides L. 186
  alopecuroides subsp. alopecuroides 186
  alopecuroides subsp. tomentosa (Boiss.)Bornm. 186
  alopecuroides var. tomentosa (Benth.) Brenan 186
  gibbosa (DC.)Yakovlev 187
  griffithii Stocks 187
  japonica L. 187
  jauberti Spach 187
  korolkowii Koehne 187
  lehmanni (Bunge)Yakovlev 187
  mollis (Royle)Baker 187
  mollis subsp. griffithii (Stocks)Ali 187
  pachycarpa C.Meyer 187
Spartium junceum L. 150
Sphaerophysa microphylla Vahl 137
  salsula (Pallas)DC. 148
Stylosanthes fruticosa (Retz.)Alston 18
Swainsonia salsula (Pallas)Taubert 148

Tamarindus indica L. 1
  indicus sensu Miller 15
Taverniera aegyptiaca Boiss. 164
  albida Thulin 164
  brevialata Thulin 165
  cuneifolia (Roth)Arn. 165
  echinata V.Mozaffarian 165
  glabra Boiss. 165
  glauca Edgew. 165
  lappacea (Forsskal)DC. 165
  multinoda Thulin 165
  nummularia DC. 165
  persica Boiss. & Hausskn. 165
  spartea (Burm.f.)DC. 165
  stefaninii Chiov. 165
Tephrosia apollinea (Del.)Link 176

  apollinea subsp. persica (Boiss.)Bornm. 177
  desertorum Scheele 176
  dura Baker 176
  elata Defl. 176
  elata subsp. elata 177
  geminiflora Baker 177
  haussknechtii Bornm. 177
  heterophylla Vatke 177
  humilis Guillemin & Perrottet 177
  nubica (Boiss.)Baker 177
  nubica subsp. arabica (Boiss.) J.B.Gillett 177
  nubica subsp. nubica 177
  pentaphylla (Roxb.)G.Don 177
  persica Boiss. 177
  pumila (Lam.)Pers. 177
  purpurea (L.)Pers. 177
  purpurea subsp. leptostachya (DC.) Brummitt 177
  quartiniana Cuf. 178
  schweinfurthii Defl. 178
  senticosa sensu auctt. 177
  strigosa (Dalz.)Santapau & Maheshw. 178
  subtriflora Baker 178
  tomentosa Forsskal 178
  uniflora Pers. 178
  uniflora subsp. petrosa J.B.Gillett & Ali 178
  vicioides A.Rich. 178
  villosa (L.)Pers. 178
Teramnus labialis (L.f.)Sprengel 182
  labialis subsp. arabicus Verdc. 182
  repens (Taubert)Baker f. 182
Thlaspidium thlaspi (Lipsky)Rassul. 127
Tipuana tipu (Benth.)Kuntze 25
Trifolium agrarium L. 196
  alexandrinum L. 196
  alpestre L. 196
  amani Dinsm. 200
  amani var. glabrescens Thieb. 200
  ambiguum M.Bieb. 196
  ambiguum var. majus Hossain 196
  angustifolium L. 196
  arvense L. 196
  aureum Pollich 196
  badium Schreber 197
  badium subsp. rivulare A.Rich. 197
  badium subsp. rytidosemium Benth. 197
  boissieri Soyer-Will. & Godron 197
  bonnevillei Mout. 202
  bullatum Boiss. & Hausskn. 197
  campestre Schreber 197
  canescens Willd. 197
  caucasicum Tausch 197
  cherleri L. 197
  clusii Godron & Gren. 197
  compactum Post 201
  curvisepalum Tackh. 203

Trifolium (cont.)
  dasyurum C.Presl 198
  dubium Sibth. 198
  echinatum M.Bieb. 198
  elegans Ser. 198
  filiforme L. 198
  fontanum Bobrov 200
  formosum Urv. 197
  fragiferum L. 198
  fragiferum var. pulchellum Lange 198
  glaucescens Hausskn. 197
  glomeratum L. 198
  grandiflorum Schreber 198
  guestii Blakelock 200
  haussknechtii Boiss. 198
  hirtum All. 198
  humboldtianum A.Brown & Asch. 199
  hybridum L. 198
  lappaceum L. 198
  lappaceum var. rhodense (Pampan.) Rech.f. 198
  leucanthum M.Bieb. 199
  leucanthum var. declinatum Boiss. 199
  lucanicum 201
  maritimum sensu Rawi 196
  mazanderanicum Rech.f. 199
  medium L. 199
  medium subsp. banaticum (Heuffel) Hendrych 199
  medium subsp. majus Baker 199
  meneghinianum Clem. 199
  micranthum Viv. 199
  montanum L. 199
  montanum subsp. humboldtianum (A.Brown & Asch.)Hossain 199
  nigrescens Viv. 199
  nigrescens subsp. nigrescens 199
  nigrescens subsp. petrisavii (Clem.) Holmboe 199
  ochroleucum Hudson 199
  pachypodium O.Schwarz 200
  pamphylicum sensu Rawi 197
  pannonicum sensu auctt. 197
  parviflorum Ehrh. 201
  pauciflorum Urv. 200
  petrisavii Clem. 199
  phleoides Willd. 200
  physodes M.Bieb. 200
  physodes var. psilocalyx Boiss. 200
  pilulare Boiss. 200
  pilulare var. longipedunculatum M.Evenari 200
  pratense L. 200
  procumbens sensu auctt. 197
  pumilum Hossain 197
  purpureum Lois. 200
  radicosum Boiss. & Hohen. 200
  rechingeri Vassilcz. 200
  repens L. 200
  resupinatum L. 201
  resupinatum var. microcephalum Zoh. 201
  retusum L. 201
  rivulare Boiss. & Bal. 197
  ruprechtii Tomasch. & Fed. 197
  rytidosemium Boiss. & Hohen. 197
  rytidosemium var. rivulare (Boiss. & Hohen.)Zoh. 197
  rytidosemium var. rytidosemium 197
  scabrum L. 201
  scutatum Boiss. 201
  sebastianii Savi 201
  semipilosum Fresen. 201
  smyrnaeum Boiss. 202
  spadiceum L. 201
  speciosum Willd. 198
  spumosum L. 201
  squamosum L. 201
  stellatum L. 202
  striatum L. 202
  subterraneum L. 202
  subterraneum subsp. subterraneum 202
  suffocatum L. 202
  supinum Savi 198
  sylvaticum Lois. 202
  talyschense Chalilov 203
  tomentosum L. 202
  tomentosum subsp. bullatum (Boiss. & Hausskn.)Oppenh. 197
  trichocephalum M.Bieb. 202
  trichocephalum var. lonchophyllum Hossain 202
  trichocephalum var. macrophyllum Hossain 202
  tumens M.Bieb. 203
  tumens var. rechingeri Zoh. & Heller 200
  vavilovii Eig 203
Trigonella adscendens (Nevski)Afan. & Gontch. 203
  afghanica Vassilcz. 203
  anguina Del. 203
  aphanoneura Rech.f. 203
  arcuata C.Meyer 190
  aschersoniana Urban 189
  astroites Fischer & C.Meyer 188
  aurantiaca Boiss. 191
  azurea C.Meyer 204
  bactriana Vassilcz. 203
  badachschanica Afan. 203
  brachycarpa (M.Bieb.) Moris 188
  brahuica Boiss. 190
  brevidentata Blakelock 208
  cachemiriana Cambess. 204
  calliceras Fischer 203
  cancellata Pers. 190
  capitata Boiss. 204
  cedretorum Vassilcz. 204
  coelesyriaca Boiss. 204
  coerulescens (M.Bieb.)Hal. 204

Trigonella (cont.)
  corniculata L. 204
  crassipes Boiss. 188
  cylindracea Desv. 204
  cylindrica Desv. 204
  disperma Vassilcz. 204
  edelbergii (Sirj. & Rech.f.)Rech.f. 204
  elliptica Boiss. 204
  emodi sensu Aitch. 203
  emodi Benth. 204
  filipes Boiss. 204
  fischeriana Ser. 189
  foenum-graecum L. 205
  foenum-graecum var. haussknechtii Sirj. 205
  freitagii Vassilcz. 205
  geminiflora Bunge 190
  gharuensis Rech.f. 205
  gracilis Benth. 205
  grandiflora Bunge 205
  griffithii Boiss. 204
  griffithii var. micrantha Gilli 208
  hamosa L. 205
  hamosa subsp. hamosa 190
  hamosa subsp. uncata (Boiss. & Noe) C.Towns. 205
  heratensis Rech.f. 205
  incisa Benth. 190
  ionantha Rech.f. 205
  kafirniganica Vassilcz. 205
  koeiei Sirj. & Rech.f. 205
  komarovii Lipsky 116
  kotschyi Boiss. 206
  laciniata L. 206
  latealata (Bornm.)Vassilcz. 206
  latialata (Bornm.)Vassilcz. 206
  laxiflora Aitch. & Baker 206
  laxissima Vassilcz. 206
  lipskyi Sirj. 206
  lipskyi var. edelbergii Sirj. & Rech.f. 204
  lunata Boiss. 188
  macroglochin Durieu 206
  marco-poloi Vassilcz. 206
  mareschiana Hand.-Mazz. 191
  mareschina Hand.-Mazz. 191
  monantha C.Meyer 190
  monantha subsp. geminiflora (Bunge) Rech.f. 190
  monantha subsp. incisa (Benth.)Ali 190
  monantha subsp. noeana (Boiss.) Huber-Mor. 190
  monspeliaca L. 190
  noeana Boiss. 190
  orthoceras Karelin & Kir. 190
  orthoceras var. anatolica (Boiss. & Bal.) Boiss. 190
  orthoceras var. baylissii Blakelock 190
  pamirica Boriss 206
  persica Boiss. 191
  phrygia Boiss. & Bal. 191
  podlechii Vassilcz. 206
  polyceratia L. 191
  pubescens Aylmer 188
  pycnotricha Rech.f. 206
  radiata L. 191
  rechingeri Vassilcz. 207
  retrorsa Boiss. 191
  salangensis Vassilcz. 207
  spicata Sibth. & J.Smith 207
  spruneriana Boiss. 207
  stellata Forsskal 207
  stenocarpa Rech.f. 207
  strangulata Boiss. 207
  striata L.f. 190
  subenervis Rech.f. 207
  teheranica Bornm. 207
  tenuis M.Bieb. 190
  theranica sensu Parsa 207
  torulosa Griseb. 207
  turkmena Popov 207
  uncata Boiss. & Noe 208
  uncinata Banks & Sol. 208
  verae Sirj. 208
  xeromorpha Rech.f. 208

Vatovaea biloba Chiov. 183
  pseudolablab (Harms)J.B.Gillett 183
Vavilovia formosa (Steven)Al.Fed. 213
Vermifrux abyssinica (A.Rich.)J.B.Gillett 176
Vicia afghanica Chrtkova-Zert. 213
  aintabensis Boiss. 213
  aintabensis sensu auctt. 216
  akhmaganica Kazarjan 214
  amphicarpa Lam. 217
  anatolica Turrill 213
  angustifolia L. 218
  armena Boiss. 214
  articulata Hornem. 213
  assyriaca Boiss. 213
  aucheri Boiss. 213
  aucheri Jaub. & Spach 214
  aurantia (M.Bieb.)Boiss. 214
  bithynica (L.)L. 213
  blakelockiana C.Towns. 216
  boissieri Freyn 214
  bombycina Post 217
  calcarata sensu auctt. 216
  canescens Labill. 214
  canescens subsp. gregaria (Boiss. & Heldr.)P.Davis 214
  canescens subsp. latistipulata P.Davis 214
  canescens subsp. variegata (Willd.) P.Davis 214
  cappadocica Boiss. & Bal. 214
  ciceroidea Boiss. 214
  ciceroidea var. multijuga Boiss. 217
  cinerea M.Bieb. 216
  cracca L. 214
  cracca subsp. stenophylla Velen. 214

Vicia (cont.)
  cracca subsp. tenuifolia (Roth) Gaudin 214
  crocea (Desf.)B.Fedtsch. 214
  cuspidata Boiss. 215
  cypria Unger & Kotschy 215
  dasycarpa sensu Blakelock 219
  ervilia (L.)Willd. 215
  faba L. 215
  galeata Boiss. 215
  gracilis Lois. 216
  grandiflora Scop. 215
  gregaria Boiss. & Heldr. 214
  griffithii Baker 216
  hajastana Grossh. 213
  hirsuta (L.)Gray 215
  hirta DC. 216
  hybrida L. 215
  hyrcanica Fischer & C.Meyer 215
  iranica Boiss. 215
  koeieana Rech.f. 216
  kotschyana Boiss. 216
  lathyroides L. 216
  laxiflora Brot. 216
  lutea L. 216
  lutea subsp. vestita (Boiss.)Rouy 216
  michauxii Sprengel 216
  mollis Boiss. 216
  monantha (L.)Desf. 213
  monantha Retz. 216
  monantha subsp. cinerea (M.Bieb.)Maire 216
  monantha var. cinerea (M.Bieb.)Boiss. 216
  monantha subsp. monantha 216
  montbretii Fischer & C.Meyer 217
  multijuga (Boiss.)Rech.f. 217
  narbonensis L. 217
  narbonensis var. serratifolia (Jacq.)Ser. 218
  narbonensis subsp. serratifolia (Jacq.) Arcang. 218
  noeana Boiss. 217
  palaestina Boiss. 217
  pannonica Crantz 217
  paucijuga (Trautv.)B.Fedtsch. 214
  peregrina L. 217
  persica Boiss. 214
  persica var. stenophylla Boiss. 214
  pseudocassubica Rech.f. 217
  rechingeri Chrtkova-Zert. 214
  rigidula Royle 217
  sativa L. 217
  sativa subsp. amphicarpa (Boiss.) Asch. & Graebner 217
  sativa var. amphicarpa (Boiss.)Alef. 217
  sativa subsp. nigra (L.)Ehrh. 218
  sativa subsp. sativa 218
  sericocarpa Fenzl 218
  serratifolia Jacq. 218
  setidens Bornm. 218
  singarensis Boiss. & Hausskn. 218
  sojakii Chrtkova-Zert. 218
  subvillosa (Ledeb.)Boiss. 218
  tenuifolia Roth 214
  tenuissima (M.Bieb.)Schinz & Thell. 216
  tetrasperma (L.)Schreber 218
  truncatula M.Bieb. 218
  varia sensu Zoh. 219
  varia var. eriocarpa Hausskn. 219
  variabilis Freyn & Sint. 214
  variegata Willd. 214
  variegata var. aucheri (Jaub. & Spach) Bornm. 214
  variegata subsp. bornmulleri Radzhi 214
  variegata var. stenophylla (Boiss.) C.Towns. 214
  venulosa Boiss. & Hohen. 218
  villosa Roth 219
  villosa subsp. eriocarpa (Hausskn.) P.Ball 219
  villosa subsp. villosa 219
Vigna aconitifolia (Jacq.)Marechal 183
  ambacensis Baker 183
  cylindrica (L.)Skeels 184
  heterophylla A.Rich. 183
  luteola (Jacq.)Benth. 183
  membranacea A.Rich. 183
  mungo (L.)Hepper 183
  nilotica (Del.)Hook.f. 183
  pseudolablab Harms 183
  radiata (L.)Wilczek 183
  sinensis (L.)Hassk. 184
  unguiculata (L.)Walp. 184
  vexillata (L.)A.Rich. 184

www.ingramcontent.com/pod-product-compliance
Ingram Content Group UK Ltd.
Pitfield, Milton Keynes, MK11 3LW, UK
UKHW021317180426
11947UKWH00015B/1291